RESEARCH MONOGRAPHS ON HUMAN POPULATION BIOLOGY

General editor: G. AINSWORTH HARRISON

THE CHANGING SAMOANS
Behavior and Health in Transition

Edited by

Paul T. Baker
The Pennsylvania State University

Joel M. Hanna
The University of Hawaii

Thelma S. Baker
The Pennsylvania State University

New York • Oxford • OXFORD UNIVERSITY PRESS • 1986

Oxford University Press

Oxford   New York   Toronto
Delhi   Bombay   Calcutta   Madras   Karachi
Petaling Jaya   Singapore   Hong Kong   Tokyo
Nairobi   Dar es Salaam   Cape Town
Melbourne   Auckland

and associated companies in

Beirut   Berlin   Ibadan   Nicosia

Library of Congress Cataloging-in-Publication Data
The Changing Samoans.
Bibliography: p.
Includes index.
1. Samoans—Anthropometry.   2. Samoans—Health and
hygiene.   I. Baker, Paul T.   II. Hanna, Joel M.
III. Baker, Thelma S.
GN57.S257C45   1986   306'.46   86-3044
ISBN 0-19-504093-7 (alk. paper)

1 2 3 4 5 6 7 8 9

Printed in the United States of America
on acid-free paper

Preface

When the explorers, whalers, and traders of Europe and the Americas first contacted the Polynesians in their dispersed domain of islands, they were, in general, graciously received. The Polynesians not only welcomed their trade goods
but also their religions, their settlements, and their genes. As was usual in such
contacts, the diseases that were introduced and the wars that occurred led to a
rapid decline in the native population on many of the islands. As a result of
these processes both the biological and cultural characteristics of the island
populations were rapidly altered. By 1900 the Samoans and some groups on
smaller islands were the only populations that had not been radically changed.
For a variety of reasons the colonialism imposed on the Samoans in 1900 also
had a less deleterious effect on their biology and culture than on many other
populations in comparable situations.

One colonial attribute that the Samoans did share with other groups was the
foreigners' great interest in all aspects of their natural environment, behavior,
and biology. Robert Louis Stevenson and his wife Fanny provided the reading
public of the late 1800s with some of the earliest descriptions of Samoa and
Samoans. By 1902 a colonial physician, Augustin Kraemer, had published a
highly detailed account of the biology and behavior of Samoans. In addition,
official records and reports, which began to flow from colonial administrators,
provided useful information about Samoans, but it was probably Margaret
Mead's publications on Samoa that stimulated a large number of foreign investigators to examine Samoan culture and behavior.

Today Samoans are themselves examining their biology and behavior with
excellent research on health being conducted by Samoan health scientists and
acute perceptions of behavior and values being recorded and interpreted by
writers such as Albert Wendt.

Despite the great number of investigators who have observed, questioned,
and measured them, Samoans have retained their traditional courtesy and generosity toward strangers. We could never have conducted the research

reported in this book without their generous cooperation. And in the Samoan tradition of reciprocal exchange, we hope that the information provided in this book will be helpful to them in the solution of their particular problems.

Funding for the multiple projects integrated in this research program was obtained from many sources with the primary support for personnel and analysis costs provided by the several universities. NIGMS Grant No. 5-TO1-GMO-1748 and NIMH Grant No. 5T32 MH 15741 provided major support for the program. NIA Grant No. 5-T32-A600048-07, NIH Grant 5-R23-AM 31785-02, and NIH Grant No. HD17133 provided partial funding. NSF Grant Nos. BNS-8204572 and BNS81-11466 helped support research in Western Samoa as did Wenner-Gren Foundation student grants 4368-4237 and 4173. We especially wish to acknowledge the Wenner-Gren Foundation for partially funding the meeting which eventually led to this book. Finally, we would like to note the substantial funding provided through the Hill Fund and the Evan Pugh Funds of The Pennsylvania State University.

At the start of the research program at Penn State, Usiusitai Toso Thurman, a native of Western Samoa, provided a great deal of help as an excellent translator for the large volume of questionnaires generated; she also became a patient and sophisticated teacher of the Samoan language.

In Hawaii Chiefs Vai Ala'ilima, Lilomaiava Galea'i, Tauili'ili Fau'olo, Fonoti Matautia, Agalaua'a Alailefaleula, Pau Paul Eseroma, and Malafoua Tagoai provided the critical organizational skills which made the study possible there. In San Francisco the Office of Samoan Affairs was helpful, and we would like to thank in particular Pat Luce, the director. In the diplomatic area, Chief Leusoluu Leatitufu acted as ambassador, obtaining much needed cooperation from various participants.

In American Samoa the Bureau of Samoan Affairs and the *pulenu'u* of the villages made the surveys possible, while Arieta Mulitauaopele, Toafu Viage, and Diana Pilitati personally assisted in the data collection. The administrators of the L.B.J. Tropical Medical Center were extremely generous in providing us access to data collected within their administrative units. In particular we would like to thank Sumner Cheeseman, James Turner, Mark Beck, Li Aiga, Tofiga Liaiga, and Nofo Siliga. Charles R. McCuddin, Director of American Samoan Health Planning and Development Agency, was a strong supporter of our efforts. Other officials who aided in selected studies included Pemerika L. Touiliili, Faresa Paopas, and Sipaia Fatuesi. Private business employers— Mua'au Aulava, Thomas Seiuli, James Carriere, and James Brittle—also provided their support.

In Western Samoa administrators of the University of the South Pacific, the Health Services, and the Office of the Foreign Secretary were all very cooperative. We would especially like to thank Albert Wendt, Makerita Vaai, Keith Ridings, George Schuster, Vio Annandale, M. Nu'ualofa Tu'u'au, Brenda Sio, Allen Wendt, Felix Wendt, and Greg Urwin. In the villages and Apia a number of excellent helpers worked in projects as translators and assistants: Elisapeta Eteuati, Fausagafou Pulepule, Ricky Ioka, Raymond Vailopa, and Tautalaa'aso Fa'aleava. The aid of many chiefs and pastors is gratefully acknowledged, but we would particularly like to thank Pastor Eteuati Tolova'a and his wife Sisilia

Eteuatai, who were of inestimable help to many of us who worked in Sala-mumu. In Apia private businesses also aided us, and we note especially Karen and Neville Barrett, Mr. & Mrs. Robert Batchelor, and Rudolf Keil.

The genetic markers used in the migration project were kindly analyzed by Robert Kirk of Australian National University and Kevin McLaughlin of Christchurch Hospital in New Zealand. The catecholamine analyses of urine samples collected in Western Samoa were provided by David Jenner and G. Ainsworth Harrison of Oxford University, United Kingdom.

As a final comment, it should be noted that in an effort to reduce the cost of this volume the galleys were produced on the main computer and laser printer at the Pennsylvania State University, supported by Oxford University Press. This method of composition does result in some problems in the orthography of Samoan words and in scientific notation, but we hope readers will not find this too distracting. We would like to thank the entire staff of the Center for Computing Assistance in the Liberal Arts at Penn State for their heroic efforts in teaching a computer to do new tricks.

September 1986 P.T.B.
J.M.H.
T.S.B.

Contents

Contributors

Fay Ali'ilima
Box 4228
Apia, Western Samoa

Paul T. Baker
Department of Anthropology
The Pennsylvania State University
University Park, PA 16802

Thelma S. Baker
Department of Anthropology
The Pennsylvania State University
University Park, PA 16802

James R. Bindon
Department of Anthropology
University of Alabama
University, AL 35486

Vanessa J. Brown
Department of Anthropology
University of Hawaii
Honolulu, HI 96822

Douglas E. Crews
Center for the Study of Population
659 Bellamy Building
Florida State University
Tallahassee, FL 32306

Timothy B. Gage
Department of Anthropology
SUNY Albany
Social Sciences 263
Albany, NY 12222

Lawrence P. Greksa
Department of Anthropology
433 Yost Building
Case Western Reserve University
Cleveland, OH 44106

Joel M. Hanna
Biomedical Sciences Building
University of Hawaii
1960 East-West Road
Honolulu, HI 96822

Sarah F. Harbison
Population Issues Research Center
22 Burrowes Building
The Pennsylvania State University
University Park, PA 16823

Julia A. Hecht
4730 17th Street NE
Seattle, WA 98105

Conrad A. Hornick
Cardiovascular Research Institute
  M-1315
The University of California at San
  Francisco
San Francisco, CA 94143

Alan Howard
Department of Anthropology
University of Hawaii
Honolulu, HI 96822

Gary D. James
Cardiovascular Center
New York Hospital-Cornell Medical
  Center
525 East 68th Street
New York, NY 10021

Craig R. Janes
Department of Anthropology
University of Colorado-Denver
Denver, CO 80202

Joann M. Martz
Department of Anthropology
University of Hawaii
Honolulu, HI 96822

Stephen T. McGarvey
Department of Medicine
Brown University
825 Chalkstone Avenue
Providence, RI 02908

Martin Orans
Department of Anthropology
University of California, Riverside
Riverside, CA 92521

Ivan G. Pawson
Department of Epidemiology &
  International Health
University of California at San
  Francisco
San Francisco, CA 94142

David L. Pelletier
Center for Social Research
University of Malawi
P.O. Box 278
Zomba, Malawi

Diana E. Schendel
Department of Anthropology
409 Carpenter Building
The Pennsylvania State University
University Park, PA 16802

Mary Stover
Department of Anthropology
University of Hawaii
Honolulu, HI 96822

Shelley Zansky
Department of Anthropology
State University of New York, Albany
Social Science 263
Albany, NY 12222

THE CHANGING SAMOANS

Chapter 1

Rationale and Research Design

PAUL T. BAKER

To understand the biological and cultural evolution of our species, anthropologists have concentrated, in the past, on the study of isolated human groups with relatively stable social and subsistence systems (Steegman 1983; Lee and DeVore 1976). Human geneticists, epidemiologists, demographers, and scientists from other disciplines concerned with people have also found the study of such groups useful for understanding the causes of health problems and the behavior of people in industrial and urban societies (Trowell and Burkitt 1981; Harrison 1977; Harrison and Boyce 1972). Isolated groups in the sense of ones totally lacking in outside contact are now nonexistent, although a significant percentage of the world's population still satisfies a majority of its own basic needs from the local environment resources. Even this behavior is rapidly changing as groups become part of an interlocked world economy and culture. For most of the populations the change in environment is likely to be relatively gradual, but for some, through migration or the sudden input of external wealth, the transformation in their social and physical environment has been almost immediate (Scudder 1982; Kiste 1974; Colson 1971).

An understanding of what happens to the biology and behavior of populations that undergo this rapid change is desirable from several points of view. First, such knowledge can aid in a clear understanding of some of the causes of the biological and behavioral characteristics of the human population that now lives in what is often termed the modern world. Second, it may be of value to the individual population being studied for coping with the health and other practical problems encountered in the transition.

Studies of specific characteristics of individuals undergoing such changes in their environment have a long history. Anthropologists have studied some aspects of the behavioral change under the concept of *acculturation* (Hawthorn 1944; Embree 1941), while other social scientists have studied the change process under such rubrics as *migration, demographic transition,* and *economic development.* The public health scientists have studied these groups in order to improve health delivery systems, while epidemiologists have examined the transitional populations to explore the causes for such diseases as hypertension and diabetes. While these numerous studies have provided a considerable body of data about what happens to particular biological or behavioral traits in a specific population, attempts to generalize the results to other transitional groups have often proved faulty. Furthermore, attempts to extrapolate particular findings from these studies to explain the causes for the health and behavior problems in the modern societies have been severely criticized (Hornabrook 1977; Vorster 1977).

Although there are many reasons for the difficulties in generalizing from such studies, a major one is the method commonly used for generating causal explanations. For example, the discovery that salt intake and blood pressure were both low in an isolated group, but both high after the group migrated to an urban area has led investigators to believe that there was a causal link between salt intake and hypertension (Dahl 1960). Such migrations, however, probably affected many other biological and behavioral characteristics such as average body weight, physical activity, psychological states, and fat intake, to name a few (Page 1979). Lacking appropriate data on these additional variables, any hypothesized causal link between salt intake and blood pressure based on this kind of association is unlikely to be verified (Henry and Cassell 1969; Dawber et al. 1967).

During the early 1970s a few investigators decided that in order to utilize the research potential offered by the population changes that were occurring in the world, it would be necessary to develop long-term studies that involved a broad spectrum of measurements. Prior's (1970; Prior et al. 1977) study concentrating on the health changes of Tokelauan migrants to New Zealand is a prime example. In 1974 Hanna and I decided to attempt a similar study on Samoans, hoping to attract a cooperating group of additional investigators in human biological and social science. While related research activities are continuing, we believe that a large enough information base now exists to attempt the synthesis presented in this book. In this introductory chapter, I present the conceptual base on which the project was developed, the rationale underlying the sequence of studies, and the evolving research strategies.

THEORETICAL BACKGROUND

Because the early evolution of *Homo* occurred in small bands that collected and hunted for their food in the savannas of Africa, it has often been suggested that departures from this life-style should be detrimental to health and well-being (Boyden 1970). While such a hypothesis is logical based on the mechanisms of natural selection on genes, the often striking demographic success of plant and animal species introduced to new environments shows that there are many exceptions to this generalization from evolutionary theory. *Homo*, even in the Pleistocene, certainly proved to be a versatile genus by adapting to a broad range of natural environments. In more recent times, *Homo sapiens* through the sequential development of agriculture, urbanization, and industrialization has caused the development of even more variability in natural and social environments, so that as measured demographically we have adapted successfully to a wide variety of niches (Baker 1984a).

While this suggests great flexibility, it should not be assumed that a given human population can survive all types of environmental change or that there may not be substantial variation in a population's health and well-being in different environments. For example, the Tierra del Fuegan and Tasmanian populations were not, in historical times, able to cope with the natural and social environmental changes that occurred with the arrival of outsiders, and they became genetically and culturally extinct. For many of the other isolated populations the altered natural and social environments that resulted from contact with other groups brought a sharp increase in the death rate, which resulted in lower population numbers (Cook and Borah 1971; McArthur 1967; Hunt et al. 1954).

By most measures some of these groups also had poorer health and a depressed population size for hundreds of years after the change. Even groups with relatively sophisticated technology often suffered from severe adjustment problems when placed in new physical and social environments, as shown by the many examples of high mortality for European colonizing groups during the sixteenth and seventeenth centuries (Monge 1948). On the other hand, once established these colonies sometimes showed high demographic growth rates with better longevity than that found at the same time in the areas of Europe from which they migrated (Swedlund 1984, 1980). I have previously suggested that the adverse effects of environmental change, from whatever source, should be proportional to the amount of change (Baker 1977a). This hypothesis involves definitional problems, since it is difficult to conceive of units of environmental change. Clearly, if a population was, for the first time, exposed to venereal disease, it would be likely to suffer a greater decline in health and survival ability than if it adopted new religions or converted from eating rice to eating wheat. Nevertheless, the probability of encountering serious health threats is likely to be greater when a population must cope with a massively altered physical, biological, and social environment than it is when only a few aspects of its environment are altered. By the same logic a rapidly altered environment poses more serious health threats than one altering slowly. Given sufficient time, human populations may be altered by genetic selection mechanisms in a manner that increases the percentage of individuals who survive or are able to cope biologically with the new environment. A few examples of such adaptation (Cavalli-Sforza and Bodmer 1971) are known, but in general human populations have adjusted or adapted to new environments through mechanisms already possible within their existing genetic structure.

While such processes as physiological acclimatization and short-term learning are obvious adjustment mechanisms, some adjustment processes take longer. Thus some types of physiological and morphological adjustments to a new environment may be possible only during growth (Frisancho 1975); also some aspects of learning require long-term involvement (Scribner and Cole 1973). Given the significance of these age factors in adjustment, a third hypothesis is that in situations of environmental change, the health of children is less likely to be adversely affected than is that of adults. The time aspect of the adjustment process suggests the fourth hypothesis that the adverse effects on health will be most acute immediately after the environmental change and will then decrease over time.

While these four hypotheses seem reasonably supported by what we know about the impact of environmental change prior to 1900, they are not all supported by the studies of change in recent years. Since 1900, and particularly since the 1950s, isolated or traditional subsistence economy populations, whether impinged upon by outside influences or migrating to new areas, have tended to show many signs of improved health. Life expectancy has almost always risen and the impact of infectious disease declined (Preston 1976). Even nutritional status has sometimes but not always improved. These differences from previous experience can be mostly attributed to the fact that such changes usually involve increased contact for the individuals with "modern" society. Unlike contact experiences in earlier epochs, this usually but not always means entering a natural and cultural environment where most serious infectious disease is controlled, food supplies are often adequate, and attitudes often less overtly hostile.

While these indicators suggest that the recent transitions from a traditional life-style to a modern one improve health in some regards, a variety of findings suggest that in a broad sense many aspects of health decline. More specifically, many studies report that the individuals themselves perceive the changes as difficult and psychologically stressful. Often they do not understand the new cultural norms and cannot initially determine appropriate behavior (Inkeles and Smith 1974). Questionnaires indicate increased levels of ill-health-related symptoms (Dutt and Baker 1978), while behavioral measures often show high frequencies of antisocial behavior, including crime and alcoholism (Graves and Graves 1974). Accidental death and injuries also frequently increased (Dutt and Baker 1981).

Morbidity and mortality from infectious disease generally decline, as often does sustained physiological work capacity (Shephard 1978). In most of the groups studied cardiovascular disease risk factors, including smoking (Kagan et al. 1974), hypertension (Scotch et al. 1961), and increased blood cholesterol levels (Keys 1975), have been reported. An increase in the frequency of certain other middle- and old-age disease states, such as type II diabetes (Zimmet 1981, 1979a, 1979b; Bennett et al. 1976), has also been reported for several groups. In addition, there are a number of biological characteristics shared by the population that affects both the degree of health change and the significance of the change. These include growth and maturation rates in children, weight gain in adults (Eveleth and Tanner 1976), dietary changes (Jerome et al. 1980; Durnin 1976) and changes in fertility patterns (Knodel et al. 1984; Nag and Kak 1984).

PROBLEMS AND DEFINITIONS

Theory derived from evolutionary generalizations or empirical observation can provide reasonable expectations about the nature of the behavioral and biological changes one might expect in a human population changing from a traditional to a modern life-style. It is of little value, however, in predicting what will happen in a particular population. It is also doubtful whether any of the specific changes that have been reported are universal. More important such a theory does not provide insights into the proximate causes or the mechanisms involved.

One obvious problem is definitional. It may be argued that such terms as traditional or subsistence agriculturalist have little meaning in describing a population's total environment. For most anthropologists the latter term, when applied to traditional populations, connotes some common attributes such as residence in relatively small kinship groups or villages, subsistence food production, and a minimum of participation in a cash economy or a nation state. It also implies strong continuity in the values and social structures of the group (Spicer 1971).

As the term *Western* society has become invalid, the term *modern* has become more widely used in the social sciences to describe the complex of social characteristics found throughout the world. A society is considered modern if it has a cash economy, a formal education system, is secular in governance and approach to problems, and contains urban units (Levy 1966). By this definition most human populations are today part of modern states. Large populations within the states may, however, be little affected by the state's socioeconomic characteristics. Furthermore, individuals within a community may be affected to differing degrees by these societal attributes. In order to make the contrast a usable research tool many investigators, as did those involved in the Samoan research project, have therefore ranked regions, communities, and individuals on comparative scales from traditional to modern (Sexton and Woods 1977).

While these derived scales are useful for categorizing the degree of group or individual participation in the two types of sociocultural systems, it is doubtful that the definitional traits are proximate causes of any health or biological differences. This does not necessarily reduce their utility, since the component traits may be the easiest or best measure available for assessing the proximate cause. For example, it is doubtful that marriage ever caused pregnancy, but it is the easiest and most reliable measure for exposure to intercourse in most societies. While the definitional traits of modernization can, therefore, only indirectly detect the causes for change in the biology and health of a population, they may contain rather direct causes for behavioral changes. With modernization marital behavior may be modified by modern contraceptive practices, while in traditional societies this would not be a potential behavioral choice. Viewed in this perspective, we can see the causal linkages for understanding why the changes in both behavior and health found during many of the population transitions occurred. A research design can thus be developed that can help us not only understand why new health problems arise for the transitional populations but also possible causes for some of the health problems in modern populations.

To help visualize how this approach may serve as a research strategy Fig. 1.1 attempts to show how modernization may contribute in many populations to the average body weight gain that has been reported for many populations when rapid modernization occurs. The figure is based on the assumption that the diet was nutritionally adequate in the traditional society and that modernization has provided adequate income opportunities.

This simple model (read from left to right) is constructed from a combination of known presumably causal associations and many links that are no better than reasonable assumptions. It is doubtful that in any of the postulated links it would, at present, be possible to estimate the quantitative contributions. The model is also not applicable in its present form to a specific population, since the genetic structure and the original natural and cultural environments of a specific traditional group would also have quantitative effects on the amount of weight gain. Even the presumed contribution of the high body weights to various types of ill health may not occur in a specific group, since it is known that for any individual a number of other attributes contribute to the particular health trait.

The virtues of such a model are twofold. First, it is a strong reminder that the mere association of a health change with one or more of the environmental or behavioral changes probably won't prove a causal relationship. This is because if one of the presumed links did not contribute to changes, a correlation would still exist. As the diagram shows, activity and diet often covary in their response to modernization. Thus, the demonstration that in a modernizing population, a diet change was coincident with an average increase in weight, blood pressure, or type

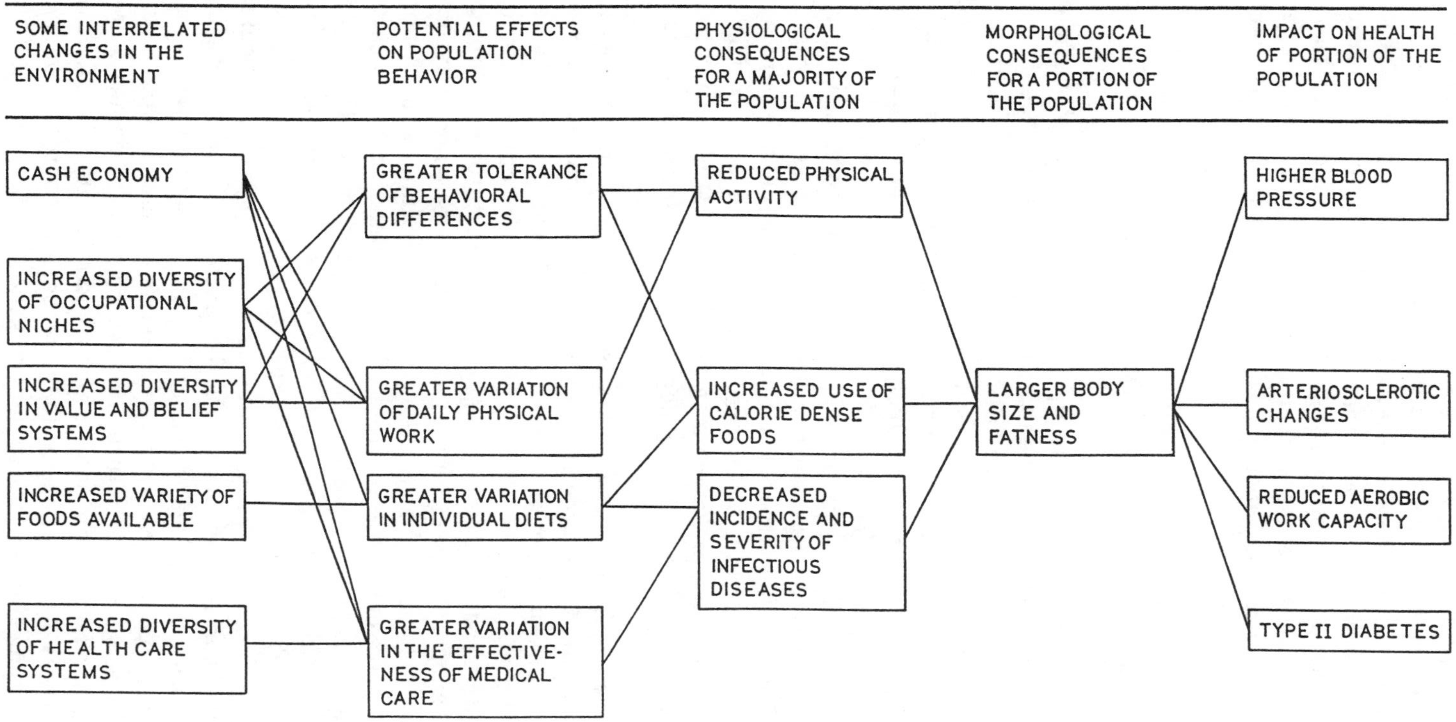

Fig. 1.1 A hypothetical schema of how the socioeconomic changes resulting from modernization may be linked to an increase in average body weight, thus affecting health. The presumed links from left to right are generally based on observed associations rather than known mechanisms.

9

II diabetes would not necessarily prove causal linkage. Indeed, without additional evidence the data would equally support the conclusion that change in activity level was the causative factor.

Second, the model provides a series of guidelines for specific research designs, by suggesting the care that must be taken to control for unstudied variables when testing simple hypotheses. More to the point, the model suggests that the assessment of a wide array of environmental, behavioral, physiological, and morphological variables is the most likely to yield causal insights.

JUSTIFICATION OF SAMOANS AS A STUDY POPULATION

An ideal study of how the change from a traditional life-style to a modern one affects the behavior and health of a population would require following such a group through a complete life cycle in the new environment. Such a comprehensive study has never been conducted, although in the South Pacific area Prior and his coworkers have followed many health and social changes among the Tokelauans for nearly 20 years (Prior et al. 1977). Friedlaender and others have also been able to track a number of Bougainville natives from nearly complete isolation to partial involvement in modern society (Friedlaender et al. in press; Page et al. 1977; Friedlaender 1975). A limited number of health measures were included in this study. Most studies have in fact been restricted to cross-sectional data, and in developing the Samoan studies, Dr. Hanna and I felt that we would have to depend heavily on a cross-sectional approach.

From several research design perspectives the Samoan population seemed appropriate for the type of study we wished to develop. First, the Samoans were an island population so that they had in the past been a reasonably self-contained unit. Second, a considerable body of information already existed on the characteristics of the Samoan natural environment, culture, and demography (see Pereira 1985 and Holmes 1984 for comprehensive bibliographies.) Finally, recent economic changes in American Samoa and massive outmigration from all of the islands had now placed a significant percentage of Samoans in a variety of modern cultural settings. In the following paragraphs I elaborate on some of the factors that were pertinent to the overall research design.

As noted by McArthur (1956), the Samoan population appeared to demographically survive the arrival of the Europeans and others in the nineteenth century with less disruption than most Polynesian island groups. The total population probably never fell below 40 000, and as I have reported elsewhere, admixture with other groups remained quite low compared to other Polynesian populations (Baker 1984b). The traditional marriage practices do not appear to have been ones that would have produced geographical or social barriers to gene flow (Ember 1971; Mead

1930). Movement between the islands of the archipelago appears to have been frequent both pre- and post-European contact. Samoan subsistence behavior and social organization also appear to have been similar between villages, even though the available resources and the relative social status of villages were reported to vary (see Chapter 3). Thus the record suggests that the villages on the various Samoan islands probably had nearly identical gene pools and similar behavior patterns as of 1899.

The imposition of colonial control at that time began to produce differences between the islands, but the available reports and descriptions suggest that life-style in the majority of villages remained quite similar until at least the period of World War II (see Chapter 3). A major population change that did begin in this period was the sudden increase in numbers. As described in greater detail in Chapter 4, the population in American Samoa, after a long period of stability, started growing in about 1900, while significant growth did not begin in Western Samoa until after 1920. Although some minor outmigration from the islands probably occurred continuously after European contact, the rate began to increase in 1950 and remains high today. The post-World War II period also marks the beginning of an increasingly sharp differentiation in economic structure between Western Samoa and American Samoa such that in recent years American Samoans are reported to have one of the higher per capita incomes in the world, while Western Samoans have one of the lowest (American Samoa Government 1980; Western Samoa Dept. of Statistics 1978).

Given these processes, the distribution of Samoans by the 1980 midpoint of the studies reported in this book was approximately as shown in Fig. 1.2. The size of the various migrant and nonmigrant groups meant that from a basically similar gene pool and reasonably similar life-style, sufficiently large numbers of Samoans (for demographic analysis) lived in situations that varied from relatively traditional life-styles to a variety of natural and social environments in modern societies. While these characteristics suggested that a variety of meaningful comparisons could be made between the Samoan subpopulations, we were unsure of the accuracy of the demographic data available on these groups. As of the beginning of the study in 1975, we knew that census and vital statistics were available for Western Samoa and American Samoa. Demographic data on Samoans in Hawaii and the mainland United States were a more difficult problem, since Samoans had not been separately enumerated in the U.S. Census up to that point and we were not certain of the accuracy of the ethnic identification on vital statistics records. Fortunately for project purposes, Samoans were separately identified in the 1980 U.S. Census, although as described in Chapters 4 and 5 the Western Samoan data were less accurate than hoped for.

Within the political units a substantial variation in life-style also existed, thus making comparisons possible. The major contrast in Western Samoa was between the rural villages and the urban complex

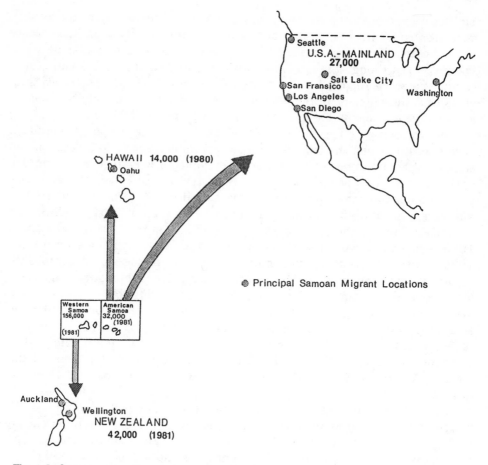

Fig. 1.2 Samoan migration routes and approximate subpopulation numbers in 1980.

around Apia. In American Samoa the villages on the Manu'a islands retained a more traditional life-style than that found in the Pago Pago harbor area. On Oahu in Hawaii, the Samoan population tended to be distributed in clusters in rural towns and urban housing complexes.

RESEARCH OBJECTIVES AND DESIGN

The initial research objectives and strategies of the Samoan Studies Project were developed in 1975 as part of the international Man and Biosphere effort and were described in a MAB Technical Note entitled *Human population problems in the biosphere: Some research strategies and designs* (Baker 1977b). For this initial formulation, a very broad set of objectives and methods was proposed. The specific aims were designed to examine how Samoan utilization of the natural and social environment related to their current socioeconomic, behavioral, and biological characteristics. The migrants and two separately governed political units in the Samoan islands were visualized as subpopulations we could study to help us understand the consequences of modernization and migration for biological fitness, social adjustment, and psychological well-being. The initial research strategy was to compile and analyze the available pertinent information on traditional life-style and behavior, on the contemporary situation in American Samoa, and on Samoans in Hawaii, supplemented by population surveys. Specialized studies were envisaged, as the initial results indicated specific facets of change that warranted more detailed approaches. A very broad battery of measurements was planned.

The first research phase: 1975 through 1978

Given the extensive body of published research on the ethnographic, social, economic, and historical characteristics of traditional Samoans in their islands (Pereira 1985; Holmes 1984), we felt that the initial field research should concentrate on the migrants and their population of origin. We therefore focused our efforts from 1975 through 1978 on Samoans living in Hawaii and American Samoa. Again given the wealth of social scientific literature and the interests of the professionals associated with the project, we decided to concentrate the initial field research on the demographic, biological, and health characteristics of the populations.

The surveys undertaken included a set of anthropometric measures plus skinfold and blood pressure measurements of all age and sex groups. Questionnaires, interview schedules, and observational checklists in English and Samoan were developed to elicit information on the fertility, migration, education, genealogies, and occupations of individuals. Household socioeconomic characteristics were determined by the same methods. The Cornell Medical Index translated into Samoan was administered to adults in an effort to determine health problems. The surveys in Hawaii were conducted in three rural communities and one urban housing area where Samoans had concentrated. In American Samoa the several surveys included people from most of the villages on Tutuila and the Manu'a islands. Additional information was included in some of the surveys so that sample sizes for particular kinds of data varied significantly. In addition to those surveys carried out by The Pennsylvania State University and University of Hawaii staff and graduate students, many individuals, as noted in the Preface, kindly gave us access to a variety of survey data collected by the U.S. Public Health Service and other departments at the LBJ Tropical Medical Center in American Samoa.

In order to further describe the biology and health of these Samoan groups, a number of specialized studies was undertaken. Because of the complexity of the method and the specificity of the objective, these studies were restricted to small sample sizes. These specialized projects included work capacity, nutrition, blood lipids, traditional subsistence behavior, attitudes toward obesity, and genetic structure. Because of the health risks of some measurements and cultural attitudes affecting level of cooperation, the samples were often not representative of the total population but only of select segments.

Assessment

The preliminary analysis of the data collected during the first year revealed that the Samoans in Hawaii appeared to show many of the characteristics that might be expected in a rapidly modernizing population, but the changes were in some aspects unusual. The body weight and skinfold measurements were uniquely high, while maximal oxygen consumption capacities per unit weight were extremely low. Blood pressures in young adults were quite high, yet blood cholesterols were not. Demographically the migration patterns were also somewhat unusual since, unlike most migrations, they appeared to be an age and sex cross section of the American Samoan population. In Hawaii the migrants also had as high or higher fertility levels than those found in the areas from which they migrated. These findings led subsequent project efforts during this first phase to focus on determining the causes for the unusual characteristics of the Samoan migrants. The results of these studies are discussed in detail in the following chapters, but it is pertinent to the rationale for the next phases of the project to note some of the second- and third-year results. First, it was found that the Samoans from the Pago Pago harbor area appeared to be very similar in biological and health characteristics to the migrants. While the Samoans from such remote areas as the Manu'a islands were somewhat lighter in weight and lower in blood pressure than the other groups, they were in all biological regards and some behaviors much different from the descriptions of traditional Samoan populations (Kraemer 1902). It also appeared to me that on some topics our interview and questionnaires often yielded unexpected responses. For example, the most frequent reason given by individuals for migration was to improve their education, yet few of them continued schooling. On the Cornell Medical Index, most individuals reported very few psychological stress symptoms and indeed the group reporting the highest number of symptoms was among those residing in the more traditional village settings.

One of the important questions that arose at this phase of the research program was whether the Samoan migrants were biologically similar or different from those who did not migrate. The problem was examined by Parsons (1982, 1979) with the technical analysis of the genotypes provided by K. McLauglin of Christchurch Hospital, New Zealand, and R. Kirk of The Australian National University (Parsons 1982). The results of this study are not examined elsewhere in this book, so it is important to summarize the findings, since the degree of genetic similarity between the migrants and nonmigrants is significant for interpreting the possible causes for similarities and differences in behavior and health between these groups.

For her analyses Parsons used two bodies of data. The first was anthropometric and blood pressure data collected on 1083 adult Samoans in Hawaii, American Samoa, and Western Samoa. The second was blood type survey information on approximately 108 Samoans in Hawaii and 162 Samoans in American Samoa. In order to investigate the differences between the groups, Parsons divided the anthropometric survey data into environmentally labile and environmentally stable traits and then applied both univariate and multivariate statistics to the three bodies of data. From the results she obtained, the most pertinent for the present chapter is probably the information derived from the discriminant function analyses of the environmentally stable traits and the blood-type data, which included seven red cell antigen systems, five serum protein systems, and twenty-one red cell enzyme systems.

Parsons reported (1982) that when she divided the sample into early migrants, recent migrants, and nonmigrants, a discriminant function based on stable traits would correctly classify 54.3 percent of the men and 51.95 percent of the women. While this was not a very high percentage, it clearly indicated differences between the groups. She suggests that this difference, while partially related to body fat difference, may also be the result of the migrant having relatively smaller limb size to trunk size.

In the blood genetics, the groups were only divided into migrants and nonmigrants, and the discriminant function correctly classified 66.67 percent of the individuals. It was particularly accurate in classifying 78.7 percent of the nonmigrants. The genotypes found for the total sample fit in well with the ranges published for Pacific populations (Kirk 1976). However, marker genes for European, Melanesian, and possibly some East Asian admixture were encountered particularly in the migrants.

Parsons concluded: "Since the blood genetic data unequivocally indicate genetic admixture in migrants, it is reasonable to hypothesize that this same admixture is at least partially responsible for the distinctive migrant phenotype. Samoans with European or Melanesian influenced morphology may be less well integrated into mainstream Samoan culture or may have greater exposure to modernizing or foreign influences, and thus may be more likely to migrate" (Parsons 1982:268). Even so, she finally notes: "However, since migration between American Samoa and Hawaii is so massive, fluid, and accessible, only slight differences between migrants and non-migrants should be expected" (1982:269).

The second research phase: 1979 through 1983

The results obtained in the first phase of the research convinced us that in our next efforts we needed to pursue several new directions. First, it was obvious that if we wanted to fully describe the changes that had occurred in the behavior and health of Samoans, we would have to obtain information on the Samoans in Western Samoa who were living in the most traditional fashion and to analyze any longitudinal data available to obtain historical perspective. Second, the similarity of the behavioral and health characteristics of the Samoans in Hawaii to those who were in the harbor area of American Samoa led us to believe that a study of the Samoans in California was very important in order to determine if migration into the differing physical and social environment found on the mainland would produce a different set of biological and health responses. We believed that a study of Samoan migrants in a California urban setting would be the most useful.

Finally, we realized that the questionnaires and protocols we were using to examine psychological stress and behavioral changes were not eliciting adequate data. We therefore decided that new methods were needed for examining the changing attitudes of the Samoans and the degree to which they were experiencing psychological stress as a consequence of life-style changes.

To fulfill these perceived needs, the focus of the Penn State group was shifted to Western Samoa. I. G. Pawson joined the research program and initiated a study of the Samoans in the San Francisco Bay Area, while the research group in Hawaii developed a new set of methods for studying stress in American Samoa. Fortunately, during this second phase, A. Howard of the University of Hawaii became actively interested in the problem of stress measurement among the Samoans. M. Orans of the University of California, Riverside, who had a long-standing interest in studies in Western Samoa provided considerable data and aid on questionnaire studies for the Penn State research team.

Funding shortfalls prevented both the Western Samoan and the San Francisco surveys from being as large as was desirable, but as reflected in later chapters, the surveys were adequate to provide the necessary ends of the continuum for adult body size and blood pressure. Because of the limited funds available, most of the research in Western Samoa then shifted from a survey approach to problem-focused studies. These concentrated on how the changes in behavioral variables of diet, activity levels, and occupation related to biological characteristics such as morphology, work capacity, stress-hormone excretion rate and cardio-vascular disease risk factors. This team also examined how these factors were interrelated to selected social, educational, and attitudinal variables.

To obtain detailed information on the variables, sample sizes in the studies were necessarily small. Furthermore, to control the number of variables that could affect the relationships, the samples were chosen from restricted age and sex groups, while maximal contrasts in such variables as occupation and traditional as against modern life-styles were sought out. As a consequence of the research designs, the survey information available is restricted to two villages, and as the data in the results chapters indicate, the people in the villages were significantly different from each other in such traits as average blood pressure and body morphology.

The problem-oriented studies were almost competely restricted to young adult men, and while the results provide a significant perspective of how the factors involved in modernization may influence behavior and health in Western Samoa, it is impossible to judge whether similar relationships are valid for women. How the environment during childhood may have affected the adult was not explored.

The research strategy used by the University of Hawaii group for studying stress in American Samoa is described in Chapters 8 and 9. That design did permit extrapolation to a broader base of the population, but as with the Western Samoan studies, the sample sizes were necessarily small and the detection of subtle interrelationships difficult.

THE SYNTHESIS

By 1984 it was judged that the quantity of data and variety of specialized studies had reached a stage where an overall assessment of the findings might provide insights that were not possible by examining the component parts separately. Although the studies undertaken by the three research groups had been loosely coordinated, many of the specialized studies had been the product of individual pre- and postdoctoral student efforts. As a consequence, the methods used in the study of one group of Samoans often varied from that used in another. Furthermore, many sets of data had not been published, since they had limited significance outside of the broad context of the overall research design. To initiate the synthesis a set of working papers summarizing the results available on the behavioral and biological characteristics studied was prepared by project participants. Working from the prepared papers, most of the project researchers met in a conference to discuss the major issues and the structure necessary for a synthesis of results.

The major issue was the utility of the concepts of traditional and modern for structuring the comparisons between and within the subpopulations studied. While the definitions and admonitions given earlier in this chapter expressed the general view of the participants, some felt that slightly variant definitions were needed for their analysis, while others were concerned that we do not in the results imply that modern society per se is a cause for behavioral and biological change. Some of these variations in perspective will, of course, be obvious in the subsequent chapters.

When we examined the comparative results of the studies, it was apparent that the variability in research techniques, in questionnaire structures, and analytical techniques all presented difficulties for making comparisons among the subpopulations. Some of these difficulties could have been overcome by the reanalysis of parts of the data. Since the data were not in a single data bank, such a reanalysis would be costly. We therefore instructed the authors responsible for the book chapters to make the best possible comparisons based on the extant results.

Finally, we agreed that the aggregate research had not achieved all of the objectives sought in the original design. Instead, we felt that the major findings of the project fell into three integrated categories. First, were the studies that examined how, over time, the Samoans' use of their environment changed along with the changes in their social settings and demographic characteristics (Chapters 2-5). Second, there were the studies of how the Samoans responded socially and psychologically to those changes, with particular emphasis on the extent of psychological and physiological stress created by the changes (Chapters 6-9). Third, there was an array of studies that examined the biological changes of the Samoan subpopulations with an emphasis on health-related characteristics (Chapters 10-15). We have therefore structured the book to follow these three themes, and have added two final chapters that present integrated perspectives on the results as viewed by social and human-biological researchers.

As a final note concerning the management of this 10-year effort, I would like to personally comment on certain aspects of such a coordinated effort. As this chapter reflects, no long-term project can be fully justified or designed at its inception. Even if this were possible, it would probably not be desirable, since science is always a process of progression from one level of inquiry to another of greater specificity. Nevertheless, if most of the larger and more significant problems we wish to solve are to be explored, integrated studies requiring a broad array of skills and perspectives are necessary.

The number of projects, such as the present one that attempt to integrate aspects of the social and human-biological sciences have been rare. The reasons for this are complex, but an obvious one is related to the way scientific research is funded. For such projects the initial research is necessarily exploratory and cannot be framed in terms of precise problems and aims. As the project proceeds substantial costs for coordination and data banking occur, and the final process of producing a synthesis also involves high costs in both reanalysis and personnel coordination. Given the current short-term and specific-problem orientation of scientific funding sources in most countries, long-term projects will inevitably be uncommon. Whether the effort and costs are worthwhile must be judged by readers of the syntheses of research programs such as this one.

Chapter 2

Environment and Exploitation

TIMOTHY B. GAGE

The physical and biotic environments of a tropical island are primarily a result of the interaction of four factors: the distance from other landmasses, particularly other large landmasses; the size of the island; the age and composition of the rocks forming the island; and the island's relationship to the trade winds. This chapter briefly describes the environment of the Samoan archipelago in terms of geography, geology, climate, soils, natural biota, domesticated flora and fauna, and human exploitation patterns. The geographical and seasonal gradients of the physical environment are discussed first. These provide the context within which the variations in biota and human exploitation patterns can be evaluated.

ENVIRONMENT

Geography

The Samoan archipelago consists of nine volcanic islands located between 13° and 15° south latitude and 169° and 173° west longitude (Fig. 2.1). Its nearest neighbors are Tonga, 560 km to the South-Southwest and Fiji, 960 km to the West-Southwest. The nearest large landmasses are New Zealand and Australia 2330 km and 4420 km respectively to the Southwest.

The islands have a combined area of only 3017 sq. km. Savai'i, the largest of the islands, has the highest elevation, 1800 m. The remaining islands vary between 600 m and 920 m, except for the small islands, Apolima, Manono, and Aun'u, which are less than 100 m high. Only the largest islands, Savai'i, Upolu, and Tutuila have large areas of level land.

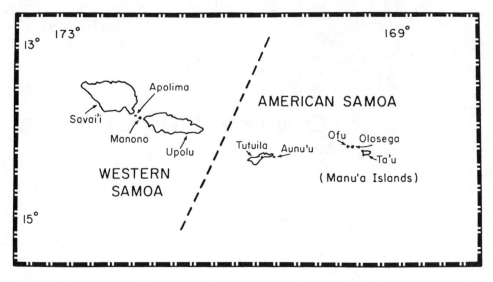

Fig. 2.1 The Samoan archipelago.

Even here, the coastal plain seldom extends inland more than 1.5 km, reaching a maximum of 6 km in the Northwest corner of Upolu (Kear and Wood 1962). In general, the islands rise abruptly with only a narrow level shelf along the coast. Most Samoan villages are located on these level areas, close to the ocean (Watters 1958b).

Geology

The Samoan islands are a result of relatively recent volcanic activity. The earliest formation, the Fagaloa volcanics, are of Pliocene to mid-Pleistocene age. Since this initial event there have been four or five additional flows. Listed from oldest to youngest these are: (1) the Salani series, (2) the Mulifanua series and Lefaga series (of approximately equal age), (3) the Puapua series, and (4) the Aopo series (1760 to the present) (Kear and Wood 1962).

The formations differ in degree of erosion, hydrology, and soil development, depending on their different ages. Fagaloa volcanics, being the oldest, are the most highly eroded. The cones and original surfaces are no longer apparent and the landforms tend to consist of steep weathered slopes above basaltic plateaus. The Salani flows are more gently sloping formations, with identifiable craters and cones. Deep gorges and water courses (*alia*) are common. Judging from the differences in degree of erosion, there appears to have been a considerable hiatus between these first two geological events. The Mulifanua and Lefaga formations are similar to Salani volcanics. The cones and original surfaces are more clearly defined, however, and the alia are not as deep and rarely contain flowing water. Only the oldest formations, Fagaloa

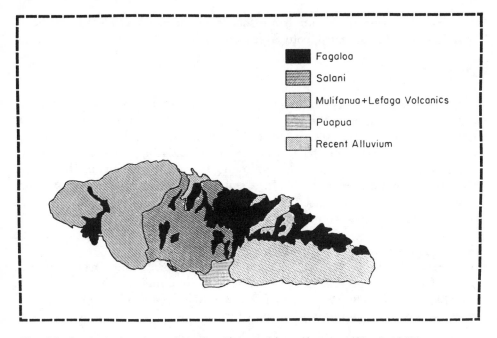

Fig. 2.2 Geological regions of Upolu. (Adapted from Kear and Wood, 1962)

and Salina, support permanent streams. The Puapua volcanics are recent prehistoric flows. They are of low gradient with only a few shallow water courses and are distinguished by their relatively even surfaces. The Aopo series, which occurred since European contact, are also gently sloping but are completely unweathered and support little, if any, vegetation (Kear and Wood 1962).

All of the islands consist of several different flows. On some, older flows are dominant, while on others the younger flows are most common. Upolu and Tutuila are the oldest with Tutuila containing predominately Fagaloa formations, except the southwestern plain, which is younger. In Upolu, however, Fagaloa, Salani, Mulifanua, and Lefaga formations predominate. The distribution of these flows is shown in Fig. 2.2. The Manu'a group of islands appear to be of intermediate age, where Mulifanua and Lefaga formations are common. A historic, submarine eruption occurred between Olosega and Ta'u in 1866. On Savai'i, Mulifanua, Lefaga, Puapua, and Aopo volcanics predominate, and three historic eruptions have occurred, Mauga Afi (1760), Mauga Mu (1902), and Matavanu (1905-1911) (Kear and Wood 1962; Thompson 1921.

The coasts of the Samoan islands are fringed by coral reefs, except where Puapua or Aopo volcanics reach the sea. Puapua and Aopo lavas are too young to have well-developed coral deposits, but two types of formation are supported on the older lavas. Fagaloa coasts are commonly steep and rugged, with fringing reefs less than 100 m wide. Salani, Mulifanua, and Lefaga formations are gently sloping and consequently, reefs have developed further from shore, as much as 2.5 km in parts of Savai'i and northeastern Upolu (Kear and Wood 1962).

Climate

The Samoan climate is hot and wet. The mean annual temperature at sea level is 27° C and the mean monthly variation is less than plus or minus 1° C from the mean annual temperature, although the winter months are consistently cooler than summer months. The mean diurnal range is 5.5° C during the summer months and 6.3° C during the winter (Currey 1962; Coulter 1941). More significant variations in temperature are due to elevation. Assuming an adiabatic lapse rate of 6° per km elevation, the highest points in American Samoa and Upolu have a mean annual temperature of 21° C, while the highest point on Savai'i has a mean annual temperature of about 15° C.

Samoa is located in a region with intermittent trade winds. Southeasterly trades occur from April to September, while winds are variable from October to March (Fig. 2.3) (Currey 1962; Coulter 1941). Wind speeds are commonly 4 to 12 miles per hour from January to

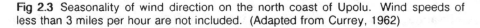

Fig 2.3 Seasonality of wind direction on the north coast of Upolu. Wind speeds of less than 3 miles per hour are not included. (Adapted from Currey, 1962)

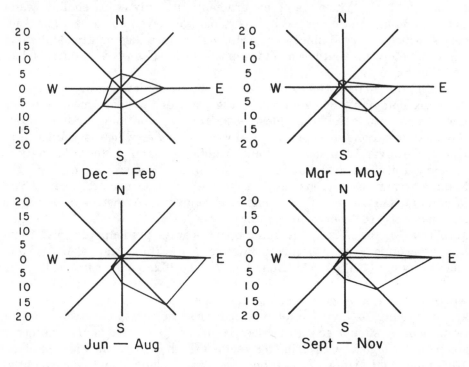

Table 2.1 Precipation at Six Stations in Samoa

Location	Yearly average	Monthly average	Monthly SD
Tutuila (Pago Pago)[a]	497.8	41.5	12.0
N. Coast Upolu (Mulinu'u)[b]	292.2	24.4	10.8
W. Coast Upolu (Mulifanua)[b]	227.3	18.9	7.1
S. Coast Upolu (Mulivai)[b]	373.6	31.1	4.9
E. Coast Upolu (Aleipata)[b]	418.6	34.9	9.2
Inland Upolu (Afiamalu)[b]	608.5	50.7	12.0

[a]Coulter 1941.
[b]Currey 1962.

March, but drop to 0 to 4 miles per hour from June through August. Tropical storms occur from December through May, but the storms are localized so that a storm may be very disruptive in one area while not severely affecting another (Coulter 1941). Between 1900 and 1936, seven storms affected parts of American Samoa.

Precipitation tends to decrease from East to West within the archipelago. Mean annual precipitation on Tutuila is significantly greater than on Upolu, only 100 km to the West (Table 2.1). Savai'i, only a few km further West, receives substantially less precipitation than Upolu (Currey 1962). Both Savai'i and Upolu have marked geographical variations in rainfall, which tend to be lightest on the north and west coasts, heavier on the east and south coasts, and heaviest at higher elevations (Table 2.1). The smaller islands do not throw a marked rain shadow, and precipitation at sea level is more uniformly distributed. Precipitation still increases with elevation, however.

Rainfall has a seasonal trend only in rain-shadowed areas. On the north and west coasts of Upolu, November to April are substantially wetter than May to October, while on the south and east coasts average monthly rainfall varies from month to month but no seasonal trend is apparent (Fig. 2.4). Most of the geographical and seasonal variation in rainfall within Samoa can be attributed to the southeasterly trade winds.

Soils

Samoan soils are classified into two distinct groups, basaltic soils occurring on the island massif and calcareous sands adjacent to the littoral reef. The basaltic soils are all derived from volcanic flows of similar composition. They vary in degree of development, however, a result of their relative ages. Four series are distinguished all corresponding to the geological formations previously discussed. Fagaloa soils, the oldest and deepest, are heavily eroded and tend to be plastic, sticky clays. The Salina soils are more moderately eroded clays 45 to 90

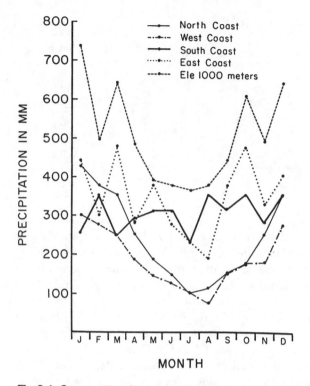

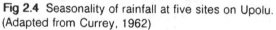

Fig 2.4 Seasonality of rainfall at five sites on Upolu.
(Adapted from Currey, 1962)

cm deep. Lefaga and Mulifanua soils tend to be slightly eroded bouldery
clay loams 15 to 45 cm deep. The Puapua soils are lithosols and too
shallow to support extensive plant growth. The historic lava flows (Aopo)
do not support soils at all (Wright 1962).

The calcareous sands are a result of the erosion of reef formations.
They are level areas, 1 to 3 m above sea level and as much as 200 m
wide (Kear and Wood 1962). They tend to be fertile due to the interaction
of the littoral environment and the calcium carbonate, but are
nutritionally unbalanced (Wright 1962; Setchell 1924). Only a few of the
Samoan cultigens are adapted to these soils.

Temperature and rainfall are the most important factors affecting the
fertility of Samoan soils. Temperature is positively correlated with the
rate of decay of humus. The coastal temperatures are sufficient to ensure
rapid decomposition of humus. At higher elevations humus accumulates,
increasing the acidity of the soil and decreasing its agricultural utility.
Precipitation is positively correlated with the rate of leaching that also
increases the acidity and decreases the utility of the soil. As a result the
fertility of Samoan soils tends to be low at high altitudes, moderate at low
altitudes, and high in rain-shadowed areas (Wright 1962). These gener-

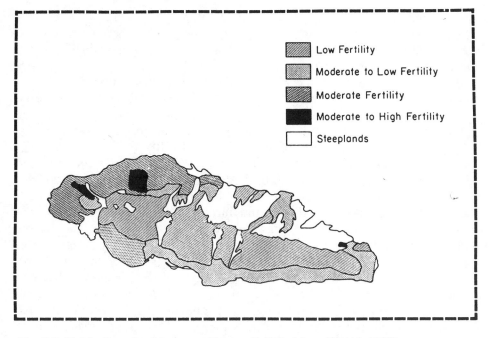

Fig. 2.5 Distribution of soil types on Upolu. (Adapted from Wright, 1962)

alities are further influenced by secondary factors, however. Leaching tends to increase with the age of a soil and decrease as the topography becomes more rugged. In Samoa these factors are confounded, for example, young soils are often gently sloping while the older soils are often steeply dissected. As a consequence, the most fertile soils tend to be the younger soils (derived from Lefaga and Mulifanua volcanics), which have sufficient depth, but because of their recent formation have not been severely leached; the oldest soils (derived from Fagaloa volcanics), which are highly eroded and leached, have fertile pockets of alluvium: while the soils of intermediate age are often least suitable for agricultural production (Wright 1962). The agricultural utility of Upolu's soils, a combination of all these factors, is shown in Fig. 2.5.

Biota

The flora and fauna of the Samoan islands are characterized by a paucity of species and low stocking densities (Cameron 1962). The small number of species is not surprising due to the great distances between Samoa and any major landmass (MacArthur 1972). Several omissions in the Samoan assemblage are puzzling, however. In particular, these are the absence of *Agathis spp.* and the rarity of two-strand species *Casuarina equisetifolia* and *Xylocarpus mollucensis*, which are common in the neighboring Fiji islands and most other high islands of Oceania. In Samoa, these species are replaced by species that usually occur only at higher elevations on other Pacific islands (Cameron 1962). The low stocking densities reported for Samoa are even more anomalous. Islands with low species numbers and hence, little interspecific competition, tend to have high densities of those species that are present (MacArthur 1972). Both of these anomalies may be due to human intervention (Amerson, Whistler, and Schwaner 1982; Cameron 1962), as most of the species useful to the human population were introduced (Mangenat 1963).

There are no distinct floral assemblages in Samoa that vary with soil type, rainfall regimes, or elevation. Most species occur on coastal and upland areas alike. On the basis of their physiogamy, four vegetation types are recognized: (1) the strand forest, (2) the rainforest, (3) the montane forest, and (4) the mist forest. The structure of the native strand forest is largely conjecture, since it has been destroyed by human habitation. Above 230 to 300 m in elevation, however, the rainforest still exists. The largest trees grow in these areas and the vegetation has the layered pattern typical of tropical forests. The trees in the montane forest (elev. 460 to 900 m) are shorter and smaller in diameter. No layering is discernible. At higher elevations, in the cloud layer, a mist forest develops. It consists of similar species, but they are little more than shrubs. There is an increase in mosses, lichens, epiphytes, and ferns. In general, only the lower two forest types, the strand forest and the rainforest, were frequently utilized or even visited by Samoans (Amerson et al. 1982; Cameron 1962).

The nonmarine faunal assemblages of the Samoan islands are even more limited than the floral assemblages, probably for the same biogeographical reasons. Several species of snakes, lizards, land crabs, and fruit bats are native. Polynesian rats, dogs and feral pigs were introduced prehistorically. Seventy species of birds have been identified (Amerson et al. 1982).

The marine environment, on the other hand, contains a wide variety of species in a highly complex ecosystem. While reefs are among the biologically most productive of natural communities, this is not due to the reef forming animals. Large areas of the reef are tidal flats, paved with calcium carbonate in which living corals and calcareous algae are few or absent. The coral communities are adjacent to these flats and form the seaward edge of the reef. The high rates of primary productivity are due to the mats of algae that grow on the tidal flats. These in turn support an extremely wide variety of fish and shellfish (Johannes 1975). The wide variety of organisms at higher trophic levels is probably a function of the high primary productivity and the planktonic larval stages characteristic of many of these species. The larvae can drift for many months at sea before settling in a particular reef. Consequently, the effects of distance from other similar ecosystems, so apparent in the terrestrial flora and fauna, are not as severe in the marine environment.

HISTORY OF HUMAN OCCUPATION

Archaeological evidence indicates that Fiji, Tonga, and the Samoan islands were settled before 1000 B.C. Pottery fragments found in early sites suggest that these populations were probably Austronesian speaking peoples derived from ancestral Lapita communities of Melanesia, 1000 km to the West. The relative isolation of the western Polynesian islands and the environmental differences between these islands and Melanesia suggests that the immediate origins of Polynesian culture took place *in situ*, before spreading to the remaining Polynesian islands. The evidence for horticulture during the period consists of: (1) the location of settlements near arable land, (2) population sizes comparable to present-day communities, (3) domestication of the pig, and (4) the presence of artifacts suggesting agriculture (Kirch 1979). Similarly, the presence of fishhooks suggests a dependence upon ocean resources. Although the details are unclear, it is likely that the Samoan population has maintained both the agricultural and the marine components of their subsistence system since Samoa was first colonized (Kirch 1979; Bellwood 1978).

The Samoan islands were not contacted by Europeans until 1722 when Roggeveens visited the islands. He was followed by Bougainville in 1768 and La Pérouse in 1787. The first habitual contact with Europeans was established in 1836 with the founding of mission stations on Tutuila and Manu'a. Between 1836 and 1900, the indigenous population was converted to Christianity, while the United States, Germany, and Great Britain all attempted to gain hegemony over the islands. In 1899, the islands were subdivided by agreement among the three world powers: Savai'i, Upolu, Apolima, and Manono became a German colony (Western Samoa, encompassing much of the archipelago's natural resources); while Tutuila, Aunu'u and Manu'a became an American territory (American Samoa). Since this time, the two polities have experienced different degrees of European influence.

The German administration's colonial aspirations were primarily economic. They supported the private commercial plantations owned by Europeans, which had developed since the 1850s. At first these operations produced copra, but they soon diversified into cotton, rubber, coffee, cocoa, banana, and cattle. Labor to work the plantations was partially imported, mainly from Melanesia and China (Lewthwaite 1962). After World War I, the administration of Western Samoa passed to New Zealand. The new administration introduced social reforms as well as new economic programs. Health care and education were actively promoted, while the commercial plantations were alienated from their European owners and became Trust Estates, administered by the government. The Chinese and Melanesian indentured laborers were mostly repatriated. In addition, the administration attempted to encourage Samoan participation in commercial agriculture. Governmental marketing programs were set up for copra, cocoa, and banana and the traditional family-based subsistence system began to take on a cash component. Western Samoa became an independent nation in 1962, with development continuing along the lines laid down by the previous administration. Although attempts to diversify the economy primarily through tourism and light industry have been made, agriculture still represented 75 percent of the national product in the early 1970s (Western Samoa Dept. of Statistics 1976; Fairbairn 1973).

In contrast to the German interests in Western Samoa, the American considerations were military, in particular, the deep water harbor at Pago Pago, Tutuila. The Department of the Navy administered the islands from 1900 to 1951, with minimal interference in local affairs. Nevertheless, the navy did provide a governmentally organized copra marketing program that operated until World War II and used significant amounts of wage labor at the naval base (Coulter 1941). Administration of the islands was taken over by the Department of the Interior in 1951 when the naval base was dismantled. Economic and technological aid began in the early 1960s and continues to increase. The rapid advances that have been made in transportation, community services, health care, education, and in the private sector of the economy are discussed in subsequent chapters. Several private enterprises, two fish canneries and a tourist hotel, have been established. Nevertheless, the government is still the major employer. More details on the socioeconomic conditions in both Western and American Samoa are provided in Chapter 3.

The different administrative regimes and the unequal division of natural resources between Western and American Samoa have created large differences in the current patterns of exploitation. In Western Samoa, commercial cropping is conducted both by governmentally owned plantations and by an extension of the traditional family-based subsistence system. In American Samoa, commercial cropping is almost nonexistent, with only the traditional subsistence system prevailing where agriculture is still practiced, while cash income is generated through wage labor. The exploitation patterns described in the following sections deal with traditional family subsistence systems, since these still provide most of the locally derived foodstuffs in both polities.

EXPLOITATION

Natural flora and fauna

The natural flora provide emergency foods, timber for houses, and wood for fuel. At one time a variety of plants were utilized whenever the domesticated crops normally consumed were destroyed by tropical storms or other disasters. These included giant taro (ta'amu, *Alocasia macrorrhiza*, which is also cultivated), wild yams (ufi, *Dioscorea bulbifera*), papaya (esi, *Corica papaya*), tahitian chestnut (ifi, *Inocarpus edulis*), the heart of the tree fern (olioli, *Cyathea plagiostegia*), arrowroot (masoa, *Tacca pennatifida*), and cassava (tapioka, *Manihot esculenta*). Today these plants are seldom consumed even in an emergency. Their function has been largely replaced by governmental relief programs. A few Samoan trees provide good timber. In the past, Ifi-lele *(Intsis bijuga)*, tomanu *(Calphyllum samoerrse)*, poumuli *(Securinega samoana)*, gasu *(Palaguiem stehlinii)*, and asi *(Syzygium inphylloides)* were highly sought after for the construction of houses. Today, milled lumber, whether locally produced or imported, is more often used. Wood from many species, however, is still extensively used as fuel for cooking.

The nonmarine fauna of Samoa are not extensively exploited. Fowling for pigeons and doves was an occasional chiefly sport, and even fruit bats, rats, and wild dogs are rarely hunted or eaten today (Watters 1958a; 1958c). In the past, feral pigs were hunted with specially trained dogs, but dogs are no longer trained and the pigs are only hunted when they disturb agricultural plantations. Land crabs, considered to be a delicacy, are still actively sought, and are trapped using a partially opened coconut shell. In general, however, the nonmarine fauna are not an important resource, and hunting is regarded more as a diversion than as an important economic activity.

In contrast, the marine environment provides a wide variety of foods representing an important dietary component. Diverse fauna taken include mollusks, echinoderms, crustaceans, and fish. The species most commonly caught or collected are listed in Table 2.2 along with the most common methods of exploitation. Foraging from the reef is less important now than it was in past years. Canned fish and other imported meats and milk products have replaced traditional fishing as the major protein source, particularly in American Samoa. In American Samoa, marine animals currently provide only 4.4 to 5.7 percent of the calories consumed, depending upon the area. Nevertheless, in the 1970s they remained an important source of high-quality protein, that is, 18 to 28 percent of the total protein (Bindon 1984b; Hill 1977).

Little information is available concerning the amount of time and energy Samoans regularly expend hunting and fishing, but both sexes are involved, women gathering invertebrates on the reef and men catching vertebrates with line, spear, or net (Hill 1977). In three different surveys (conducted in 1950, 1961, and 1966) Western Samoan men averaged 6, 3, and 3 hours per week, respectively (Lockwood 1971). These figures suggest a decline in subsistence fishing, but in any event, the expenditures are small. No time estimates are available for women or for energy expenditures by either sex.

Domestic flora and fauna

Only a few of the Samoan cultivars listed in Table 2.3 are commonly produced and consumed. Coconut, taro, giant taro, bananas, yams, and

Table 2.2 Marine Species Commonly Caught in Samoa

Genus	English vernacular	Samoan	Method of capture[a]
Mollusks			
Gafrarium	clam	*pipi*	GD, D?
Tridacna	gaint clam	*faisua*	GD, D?
Turbo	turban	*alili*	GD, D?
Polypus	octopus	*fe'e*	GD, GN, D
Patellidae	limpet	*matapisu*	GD, D?
Echinoderms			
Echinometra	short-spined urchin	*tuitui*	GD
Holathurai	sea cucumber	*loli*	GD
Holathuria	sea cucumber	*sea*	GD
Diadema	long-spined urchin	*wana*	GD
Echinothrix	long-spined urchin	*wana*	GD
Crustaceans			
Panulirus	spiney lobster	*ulatai*	GN, D
Scyllarides	slipper lobster	*ula*	GN, D
Carpilius	Samoan crab	*pa'a*	GN, D
Scylla	Samoan crab	*pa'a*	GN, D
Fish			
Myripristis	squirrel fish	*malau*	GN, LF, NG
Holocentrus	squirrel fish	*malau*	GN, LF, NG
Epinephelus	sea bass	*gatala*	GN, LF, D, NG
Lutjanus	snapper	*ta'iva*	GN, LF
Lutjanus	snapper	*savane*	GN
Monotaxis	snapper	*mu*	GN, NG
Mulloidichthys	goat fish	*afulu*	GN, NG
Upeneus	goat fish	*vete*	GN, NG
Pseudepeneus	goat fish	*matalau*	NG
Acanthurus	surgeon fish	*alogo*	GN, D
Acanthurus	surgeon fish	*manini*	D, TN
Acanthurus	surgeon fish	*paloni*	D
Acanthurus	surgeon fish	*siusina*	D
Ctenochaetus	surgeon fish	*pone*	D, NG
Zebrasoma	surgeon fish	*maono*	D
Naso	surgeon fish	*ume*	D, NG
Callyodon	surgeon fish	*fuga*	NG
Gymnothorax	moray eel	*pusi*	GN, D
Lethrinus	emperor fish	*'awa'awa*	LF
Caranx	jack	*lupo*	LF
Caranx	jack	*malauli*	LF
Caranx	jack	*ulua*	LF
Trachurops	mackerel	*atule*	LF, TN
Selar	mackerel	*atule*	LF, TN
Decapterus	mackerel	*opelu*	LF, TN
Scarus	parrot fish	*fuga usi*	D
Scarus	parrot fish	*lae'a*	D
Scarus	parrot fish	*ufa*	D
Hyporhamphus	halfbeak	*ise*	D, TN
Harenaula	sardine	*pelupelu*	TN
Mugil	mullet	*fuafua*	TN
Crenimugil	mullet	*agae*	TN
Thalassoma	wrass	*sugale*	NG, D
Hemipteronotus	wrass	*sugale*	NG, D
Cheilinus	wrasses	*lalafi*	NG, D

Source: Adapted from Hill 1977.
[a] GD = gleaning during the day; GN = gleaning during the night; D = diving; LF = line fishing; NG = gill netting; TN = throw netting.

breadfruit are the significant dietary sources. The remainder are minor foods or are produced for ritual and/or medicinal purposes, roof thatch, and textiles. Coconut, taro, breadfruit, and bananas make up the bulk of the subsistence agricultural production (Gage 1984; Lockwood 1971).

Taro, bananas, giant taro, and yams are cultivated under a slash-and-burn system, whereas breadfruit and coconuts are produced by arboriculture. Slash-and-burn agriculture consists of clearing a plot, planting and weeding it for a period of time, and then allowing it to lie fallow and reforest until the process is repeated. This pattern is, of course, variable. Some plots are cropped continuously, while others are used for only a short period. The number of plantings and length of the fallow period vary, depending upon the distance to and from the plantation, the crops planted, the fertility of the soil, and the land/population ratio. The expected number of plantings on a plantation just cleared is almost two for taro and approximately six for bananas (Gage 1982). Estimates of expected fallow periods range from 5 to 18 years (Gage 1982; Lockwood 1971; Farrell and Ward 1961). The fallow period serves several functions: (1) it allows the soil to recover its natural fertility without using artificial fertilizers, and (2) it keeps crop pests (both competitors and parasites) in check. Cultivation of breadfruit and coconut is less labor-intensive since the trees produce over a period of 50 to 70 years. They are allowed to reseed themselves, although occasionally a breadfruit tree or coconut palm will be planted in order to introduce a favored variety to a new area (Watters 1958a; Coulter 1941). The only task that occurs frequently is harvesting.

The spatial organization of a village's plantations and the geographical variation in cropping regimes may be due to differences in the physiological requirements of the major cultigens (Barrau 1965). Taro requires a minimum monthly rainfall of 17 cm, moderate temperatures (21° C to 27° C), and prefers heavy wet clays with high moisture holding capacities (Onwueme 1978). Bananas, on the other hand, tolerate less rain (10.2 cm per month), slightly higher temperatures (27° C and up), and well-drained soils, although the calcareous soils are not optimal for them (Williams 1975). Little is known about the trees, however. Breadfruits seem to prefer low elevations and may be adversely affected by drought (Ochse et al. 1961; Watters 1958a). Only breadfruit and coconut are adapted to the nutritionally unbalanced calcareous sands.

Each crop is grown in a geographical location that matches its physiological requirements. The average elevations and distances from the village at which taro, breadfruit, and bananas are produced appear in Table 2.4. Taro is grown at high elevations, where rainfall tends to be heavier and temperatures lower than at sea level. These conditions are generally optimal for taro production. Bananas are usually produced at lower elevations just inland of the calcareous coastal sands and in close proximity to the village. Here rainfall is generally sufficient and the

Table 2.3 Traditional Samoan Cultigens

Botanical nomenclature	English vernacular	Samoan	Food type[a]	Cultivation[b]	Use[c]
Monocotyledons					
Araceae					
Colocasia esculenta	taro	*talo*	T	S	SF
Alocasia macrorrhiza	giant taro (elephant ear)	*ta'amu*	T	S	EF
Diacoraceae					
Dioscorea alata	greater yam	*ufi*	T	S	EF
Broussonetia papyrifera	paper mulberry	*u'a*	-	H	TX
Musaceae					
Musa (Eumusa Simmonds)	banana	*fa'i*	F	S	SF
Musa (Australinusa)	banana	*soa'a*	F	S	OF
Palmae					
Cocosnucifera	coconut	*niu*	O	A	SF
Pandanus sp.	pandanus	*fala*	-	H	TX
Gramineae					
Saccharum officinarum	sugar cane	*tolo fau lou*	-	S	BM, OF
Nicotiana sp.	tobacco	*tapa'a*	N	S	R
Dycotyledons					
Moraceae					
Artocarpus altilis	breadfruit	*ulu*	F	A	SF
Piperaceae					
Piper methysticum	kava	*ava*	N	S	R
Ceiba pentandra	kapok	*vavae*	-	A	TX

a T = farinaceous tuber F = fruit; N = narcotic; O = nut;
b S = slash-burn cultivation A = arboculture; H = horticulture;
c SF = staple food OF = occasional food; EF = emergency food; TX = textile; BM = building material; R = ritual use.

Table 2.4 Locations of Plantations Rural American Samoa, 1978

| | Distance [a] | | Elevation [b] | |
	Mean	SD	Mean	SD
Taro	2.34	0.94	103	107
Banana	1.17	1.28	43	39
Breadfruit	0.02	0.00[c]	36	16

Source: Adapted from Gage 1982.

[a] From household in kilometers.

[b] From sea level in meters.

[c] All breadfruit are associated with the household residences which were arbitrarily assigned a distance.

temperature regimes and soil characteristics are more suitable for bananas than at higher elevations. Breadfruits are grown at low elevations, usually within the village itself and on the calcareous soils of the strand. Coconuts are produced in most locations, where they serve as boundary markers and as a convenient source of fluids while working in the plantations, as well as in continuous stands along the coast. Except for some commercial crops, particularly cocoa, crops are not interplanted (Gage 1982).

Despite the placement of these cultivars in generally favorable ecological locations, geographical and seasonal variabilities still remain. The rain shadows on the leeward sides of Savai'i and Upolu are sufficient to seriously reduce taro production during the dry season. On the north coast of Upolu there are four months during which the average precipitation is below the minimum requirement of 17 cm (Fig. 2.4), while on the west coast five months fall below this requirement. Although the rainfall regimes at higher elevations would not be as unfavorable, the practice in these areas is to produce taro during the wetter months and switch to yams or giant taro during the drier months (Farrell and Ward 1962; Watters 1958a). Villages on the south and east coasts of Upolu and Savai'i receive sufficient rainfall during all months and produce taro continuously, as do the islands of Tutuila and the Manu'a chain, which are not large enough to produce a rain shadow.

Seasonality of banana production is affected by rainfall in conjunction with soil depth. Although monthly average precipitation on the west coast of Upolu, falls below the minimum monthly requirement of 10.2 cm for one month (Fig. 2.4), there is usually sufficient water storage in the soil to maintain production. Only in northwestern Savai'i, where the soils are young and shallow, does banana production fall during years of average rainfall (Ward 1959).

A final source of seasonality, which varies by location and variety, is breadfruit. The months of the year during which ripe breadfruits have been reported in the Manu'a islands are illustrated in Fig. 2.6. In these islands there are three breadfruit-bearing cycles per year, each consisting

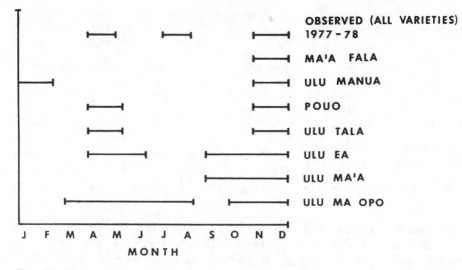

Fig. 2.6 The fruiting season of seven varieties of breadfruit and the seasonal availability observed in Manu'a, November 1977 to August 1978. (Adapted from Gage, 1982)

of about four months, two months without breadfruit and two months with breadfruit (Gage 1982). Most areas of Samoa, particularly Western Samoa, however, are reported to experience two fruiting seasons a year, each consisting of four months without breadfruit and two months with breadfruit (Farrell and Ward 1962; Watters 1958a).

The seasonality of breadfruit alters the consumption ratios of taro, bananas, and breadfruits. When breadfruit is not available, taro and bananas represent 16 percent and 10 percent of the total average diet, respectively, in a rural American Samoan village, but when breadfruit is available, taro and banana contribute 0 percent and 5 percent of the total average diet with breadfruit contributing 66 percent. These shifts cannot be attributed to changes in the productivity of taro or bananas (Gage 1984).

The results of four studies estimating the contribution of taro, breadfruit, and banana to the total Samoan diet are listed in Table 2.5. The average portion of the diet supplied by subsistence crops is about 50 percent in all studies. The proportions vary considerably from sample to sample, however, even within the same study. This variability has been attributed to differences in exposure to modern lifeways among the villages sampled (Bindon 1984b; Wilkins 1965). A reanalysis of their data, using the proportion that breadfruit contributes to the total diet to control for the breadfruit season, however, suggests that at least some of the differences among the studies are due to the seasonal availability of breadfruit. A reanalysis also indicates that total caloric intakes may fluctuate with the breadfruit season (Gage 1984).

The consumption of coconuts is considerably more difficult to estimate

Table 2.5 Proportion of the Diet by Agricultural Staples: Estimates from Four Studies

		Variation	
Study	Mean	Minimum	Maximum
Gage (1982)[a]	0.46	0.26	0.66
Lockwood (1971)	0.58	-	-
Bindon (1981)[b]	0.36	0.24	0.47
Wilkins (1965)[b]	0.47	0.37	0.64

[a]Seasonal differences (rural American Samoa).
[b]Seasonal and geographic differences.

than the other crops, since much of its production goes for pig fodder or commercial purposes. Estimates based on recall data in American Samoa suggest that the contribution to the total diet is about 10 percent (Bindon 1984b).

Three domesticated animals are kept: pigs, chickens, and dogs. They are fed table scraps and are allowed to forage for themselves. The pigs and chickens are also fed grated coconut. In the past, pork was second to fish as the most common source of high quality protein (Watters 1958a). Today, the importance of home-produced pork in the diet is minimal, with its consumption reserved for important occasions. As discussed in Chapter 12 it does not in any recent survey emerge as a significant calorie or protein source. Chickens, however, were reported in American Samoa to be only slightly less important than marine animals (Bindon 1984b). Bindon found that they contributed 3.4 to 5.4 percent of dietary calories and 13.5 to 17.2 percent of dietary protein. Whether these are locally produced or purchased from outside sources is not known but, as reported in Chapter 12, they may not be comparable in Western Samoan diets.

Like the seasonal trends in consumption patterns, there are seasonal trends in the time and energy expended in subsistence production (Gage 1984; Farrell and Ward 1962). Detailed figures on the time and energy expended in subsistence production for a typical rural household on the island of Ta'u are shown in Table 2.6. Overall, the time and energy expended in agricultural tasks decreases 12 percent and 14 percent, respectively, when breadfruit is available. This is contrary to the patterns of consumption; that is, energy intakes are lower when outputs are higher and vice versa.

Geographical variation in the time and energy expended in household agricultural production depends largely on the extent of commercial production. Two studies of villages in Western Samoa where commercial production is conducted found that men and women who performed agricultural chores spent about 16 and 20 hours, respectively, in those activities (Fairbairn 1973; Farrell and Ward 1962). On the other hand, agriculturally active men in an isolated Western Samoan village without

Table 2.6 Monthly Time and Energy Expenditures: Fitiuta 1978

	February		April		Average	
Crop	Time (min)	Energy (Kcal)	Time (min)	Energy (Kcal)	Time (min)	Energy (Kcal)
Taro	2948	21 768	2090	15 206	2519	18 486
Banana	807	5560	769	5350	788	4239
Breadfruit	0	0	365	2739	183	1370
Total	3755	27 328	3224	23 295	3490	25 311

Source: Adapted from Gage 1984.
Typical household = 8.29 individuals.

high levels of commercial production, worked 9.5 hours per week (not including traveling to and from plantations) (Lockwood 1971). Lockwood suggested, however, that these figures were high due to the destruction of a recent hurricane, and considered 5 to 6 hours per week the normal average. Estimates for a rural American Samoan village (with no commercial production) show that agriculturally active men spend perhaps 8.6 hours per week working in agriculture, including walking to and from plantations, and 5.6 hours per week not including travel (Gage 1984). These results all suggest that where commercial agriculture in not widespread, relatively little time (5 to 10 hours per week) is expended in agricultural pursuits. Chapter 13 contains a more detailed analysis of this topic.

Altogether, the 1970s Samoan subsistence system found in Ta'u provided much of the dietary requirement with very little labor expenditure, and supplied about 65 to 70 percent of the total dietary intake. The remaining 30 to 35 percent of the diet was made up of store-bought foods, and of course more time was expended working at wage labor jobs or cash cropping. This pattern undoubtedly varied widely from a very heavy reliance on subsistence production in some Western Samoan villages to total wage labor in many parts of Tutuila. Precontact production and work levels are difficult to estimate, since they must be inferred from incomplete data. Estimates of precontact agricultural production are similar to the hectorages currently maintained in rural American Samoa (Gage 1982; Watters 1958a; Coulter 1941). On the other hand, the contribution of fish to the total diet and time spent fishing may have declined from precontact levels (Hill 1977). Watters believes that prehistorically more time was spent fishing than in agriculture. Given these limitations, Tutuila's current production and work requirements can be used as a model to speculate about precontact work levels. Work requirements must have been considerably higher before the introduction of steel tools, but only the costs of clearing land would have been affected. Even if clearing took three times longer with stone tools, it would only increase the time spent in agricultural tasks from 5 to 10 hours per week to 6 to 12 hours per week. If as much time was spent fishing as in agriculture, the total Samoan expenditures would still be well below the range of work inputs suggested for other Stone Age subsistence systems (28 to 42 hours per week) (Sahlins 1972). In comparison to most other human subsistence systems, the Samoan exploitation system, past or present, is characterized by high rates of production and low levels of work requirement.

OVERVIEW

The Samoan physical environment is surprisingly diverse given its tropical location and relatively small landmass. The geological features and climatological characteristics interact to form a variety of different conditions, each encompassing a relatively small geographical area. These variations are sufficent to influence the natural biotic communities and the human populations' patterns of resource exploitation.

The biotic environment is composed of two contrasting ecosystems, the terrestrial system and the marine system. The terrestrial system is characterized by an impoverished species assemblage and low productivity. Many of the species useful to the human population have been introduced. The marine system, on the other hand, is diverse and very productive. Human modification of its ecosystem has been considerably less than their modification of the terrestrial system.

Location of human settlements during most of Samoa's cultural history has been influenced by the two contrasting ecosystems mentioned above and the availability of fresh water (Watters 1958b). Villages have generally been located adjacent to the coast, minimizing the distance to both important ecosystems. At the time of first contact, villages were smaller and located as close to their fishing reefs and plantations as possible. In recent times, villages have tended to coalesce around churches, increasing the distances to important subsistence resources. Only during periods of intense intergroup hostilities have Samoan villages been relocated further inland to defensible areas.

Exploitation of the local ecosystems is characterized by a primary dependence upon agriculture and a secondary dependence upon marine resources. Agriculture provides the bulk of the locally produced dietary energy. Both the consumption of agricultural products and the agricultural work loads vary seasonally and geographically within Samoa, however. Although variable, work loads in Samoa are low compared to other traditional subsistence systems. Marine fauna and a single domestic animal, the chicken, are secondary sources of energy, and primary sources of high quality protein. Other local food resources are of minimal importance.

It appears likely that prehistoric agricultural production was quite similar to that reported in rural American Samoa today, although the work requirement necessary to achieve this level of production was probably higher in the past. On the one hand, introduction of steel bush knives has reduced the costs of some agricultural tasks; on the other, relocation into larger villages may have increased the energetic cost of transportation costs. The exploitation of marine resources was considerably higher during prehistoric times than it is even in the rural areas today, and the work loads must also have been higher. Despite higher levels of production, however, marine resources were still only a secondary source of energy.

Interestingly, breadfruit receives little mention in most descriptions of Samoan life. Taro, on the other hand, receives a great deal of attention, probably due to its cultural importance. Nevertheless, breadfruit imposes seasonality on the Samoan exploitation system, represents in some areas the largest single dietary component, and is primarily responsible for the minimal work requirement of Samoan exploitation systems. Clearly, the distribution of breadfruit and its seasonality must play a large role in creating the heterogeneity found among Pacific exploitation systems. In particular, subsistence systems with a high dependence upon breadfruit are likely to be more seasonal and less labor-intensive than systems without breadfruit. Perhaps this is one of the factors responsible for the differences between the populations of high islands, where breadfruit may be abundant, and atolls, where breadfruit may not be abundant. Little is known about the quantitative distribution of breadfruit, however, and even less is known of its seasonality. This is an area of Pacific sub-sistence systems that requires additional work.

Chapter 3

Social Settings of Contemporary Samoans

JULIA A. HECHT
MARTIN ORANS
CRAIG R. JANES

Beyond an undeniably shared genetic and cultural heritage, the play of ecology, history, and cultural tradition has ensured variation in Samoan social organization. Anthropologists argue that Samoa is characterized by remarkable linguistic and cultural homogeneity,

> a shared commitment to a large number of political and kinship institutions, a common consciousness among Samoans of their singular identity, origins, language, physical characteristics, and history, and finally a shared set of understandings and categories which serve as common premises for interpreting and orienting behavior . . . [which] does not preclude wide divergences in specific practices and beliefs . . . . (Shore 1983:3, 303, fn. 7)

Shore expresses the view that "studies of change or local variation [often] fail to clarify what may remain constant in spite of such variations, so that it is hard to gauge how deep we are to assume such variation runs" (1982:146). Yet the broad and relatively constant themes in Samoan culture are thoroughly considered in the existing ethnography (Shore 1982; Holmes 1974, 1958; Mead 1969, 1968; Grattan 1948; Kraemer 1902; Turner 1884); here we are surely ready to examine variation.

The objective of this chapter is to examine social variation within and between the sites of the Samoan project (see the Bibliography at the end of this book). The research is oriented primarily toward physical and biological variation, predicating many of its efforts on the apparent variation between Samoans residing in different places: Western Samoa, American Samoa, Hawaii, and the U.S. mainland, on a rough continuum of increasing "modernization."[1]

40

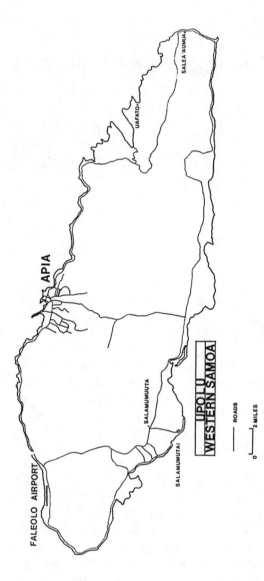

Fig. 3.1 Locations of Samoan project studies on the Samoan islands.

Cultural change in Samoa has been particularly rapid since World War II. Several processes of culture change are associated: increasing universality of education, commercialization and monetization, development of communication and transportation links, increasing concentrations of population, the general availability of formal health services, and the expansion of alternatives to the extended family and kin networks. In the Samoan project studies, "modernization" is used as a rubric for these co-occurring processes without implications of causality.

Baker (1981:7) points out that "studies using between-population designs must also take cognizance of within-population causes for variability." The Samoan project study sites have ranged from remote Western Samoan villages to suburbs of Apia, to Fitiuta in the Manu'a group of American Samoa and the villages clustering around the bay of Pago Pago, to rural and urban Oahu in Hawaii, and finally to the Bay Area of San Francisco (Figs. 3.1 and 3.2). In this chapter, we focus on the sites of various studies in the project in order to consider social variation. In each of the social settings where Samoans live, Samoan culture provides a substantial part of the model for living, constrained in each place by different social contexts, especially by variations in economic and political organization. Different cultural models also come into play,

Fig. 3.2 Locations of Samoan groups studied in Hawaii.

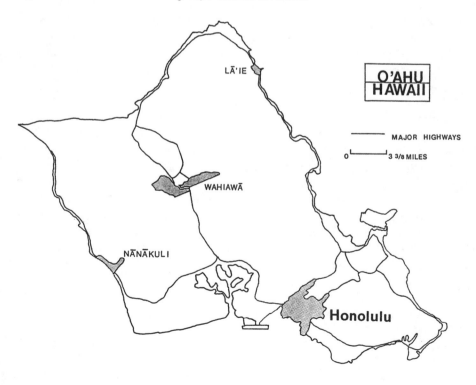

particularly in Hawaii and California. Before considering specific study sites, we review the central features of Samoan culture that are called upon in the many social settings of Samoan culture today.[2]

FA'ASAMOA

Foundations of Samoan culture and society

> Samoa contained no proletariat, none who could not take pride in their family connections, none who in youth could not look forward to the possibility of occupying a responsible position later in life. Samoan society protected self-respect (Davidson 1967:29-30). Although the politics of one village thus differed from those of others—as the result of modern changes or of the character of the old political and social structure—these differences represented no more than variations on the pervasive theme of Samoan custom. Beneath the complex interweaving of means and ends, there was a unified system of values or, at the very least, of proprieties. (Davidson 1967:276)

Samoans have labeled their kinship and political structures, their values of service to 'aiga (kin) and church. They are conscious of sharing these cultural and social forms that they call fa'aSamoa (the Samoan way). Unless otherwise stated, the following paragraphs deal with traditional village and family organization as it still exists in more isolated villages.

Samoan social relationships are grounded in the idioms of kinship. But the culture focuses on the political form and content of relationships, on concepts such as fa'aaloalo (respect), fa'alupega (proper address, order of precedence), and feagaiga (social contract).

The kin and political systems are inextricably interlocked. Nonunilineal descent categories ('aiga potopoto) associated with land and with titles (suafa) are represented in villages through extended family household groups ('au'aiga).[3] The 'aiga potopoto elects a titleholder (matai) who "wields primary authority over the members of his own household, allocating tasks and receiving food and monetary contributions from members, as well as controlling access to local lands under the control of his title" (Shore 1983:62). Matai organize the pooling and redistribution requisite in fulfilling family and community obligations (fa'alavelave). In most villages, land and considerable other productive property, for example, trucks, canoes, copra dryers, are divided among 'au'aiga, and are thus effectively in the control of the matai. Every member has rights in the 'au'aiga property that he keeps by contributing goods and services (tautua) through the matai (Pitt and Macpherson 1974).

Matai titles are ranked respective to other titles within villages and districts. The village confirms the election of 'aiga leaders. Matais then represent their 'aiga in village and district meetings *(fono)* where a formal charter of address (fa'alupega) is used to order precedence in sitting and speaking. There are two basic kinds of matais: *ali'i* (high chiefs, and *tulafale* (orators).

Senior chiefs have main title houses which, associated with grassy lawns *(malae),* constitute the visual focus of the traditional village *(nu'u)* (Shore 1983:61). Untitled men *(taule'ale'a)* are organized into a working group *('aumaga)* in the service of the chiefs. The 'aumaga was often headed by the *manaia,* ideally the son of the paramount matai of the village. The focus of the young women's group *(aualuma)* was entertaining village guests. The *taupou,* a "ceremonial village virgin" and usually the daughter of the ranking matai, was the leader of the young women's group. Often wives of the titled men *(faletua ma tausi)* constitute another recognized group. Many modern villages include vestiges of these groups as well as such postcontact formal organizations as church groups. The modern Women's Committee, dating from German colonial times and serving such governmental functions as aiding the nursing services, has taken over most of the functions of the traditional women's groups.

Christianity was introduced into Samoa in the mid-nineteenth century. While it did not overtly alter traditional kin and village organization, it added new statuses to the Samoan social structure and new forms to its organization. The minister or *faife'au* is an additional voice to be reckoned with in the village. Ministers cannot hold titles, however, and Congregational ministers, moreover, are never appointed to their own villages. Ministers are not supposed to be active politicians or overtly partisan. Nevertheless, the church is politically influential, through the pastors themselves or "indirectly through the matai, who have strong, active affiliations with one of the major denominations and who take heed of the churches' attitudes to important matters" (Pitt and Macpherson 1974:51).

Beyond the village level

> Samoan political structure and activity . . . showed certain common characteristics from the village to the national level. Decisions were reached through discussion in council. The structure of these councils paid exact regard to the status of their various members. These characteristics were not confined to the political sphere but permeated the whole of Samoan society. (Davidson 1967:29-30)

The districts of Upolu and Savai'i were the centers of supravillage political activity in traditional Samoa. At this level, there was constant jockeying for control over high titles, and the political organization was inherently unstable. Tutuila formed part of the eastern district of Upolu, but actually had little to do with political affairs at the district level

(Davidson 1967). Manu'a was the seat of the highest traditional title, the Tu'i Manu'a, and stood separate from the political turmoil of the Savai'i and Upolu districts (Keesing and Keesing 1956; Grattan 1948).

Some contrasts between Western Samoa and American Samoa

Samoa is a large place by Polynesian standards, one that encompassed diversity traditionally as it does today. While land utilization patterns are considered in more detail elsewhere (Chapter 2), it will suffice here to emphasize the small scale of American Samoa relative to Western Samoa. Western Samoa claims some of the largest landmasses in Polynesia. Upolu totals 430 sq. mi and Savai'i another 700 sq. mi Several tiny islands cluster close to the coasts of the major islands. In contrast, the total land area of American Samoa is approximately 76 sq. mi (Fox and Cumberland 1962). Arable land is limited both by the narrow, mountainous character of Tutuila and the small size of the Manu'a group of islands that lie some 80 mi east of Tutuila.

Today, the general impression of American Samoa is that of a place so transformed by American political and commercial values as to no longer be Samoan. If this is true in some sense, the transformations seems to have occurred subsequent to the war years and especially after the transition to Interior Department administration and the vast infusions of funds that began during the Kennedy era. But American Samoa's susceptibility to alien influence may be more the result of the intrusiveness of the American colonial presence. Because of its small size and resource base and because of the traditionally low status of Tutuila, there always may have been more interest in change and new opportunities in American Samoa than in Western Samoa.

Paradoxically, American Samoa includes what was traditionally the highest status place, the Manu'a group, and the lowest status place in the entire archipelago, Tutuila. In today's world where size and accessibility count for more than titular rank, Tutuila, with its ample harbor, is the gathering place and seat of government; noble but tiny Manu'a is an outpost.

> Manu'a long stood proudly off to one side: In Manu'a the pattern of living has long been one of isolated adherence to traditional Samoan life styles. The Tu'i Manu'a and his coterie of powerful orators have been able to rest in the knowledge that tradition has placed them at the top of the Samoan social order and that any compromise with other areas of Samoa could do nothing to improve this position. (Young 1972:102-103)

In contrast, Young says that Tutuilans "have long endured the abuse of the lowly in Samoan circles."

Samoans are perhaps unique among Polynesians in their self-conciousness about their culture. In their awareness of fa'aSamoa, they occasionally denigrate the expression of that culture in its more remote backwaters; Tutuila is the most removed of cultural estuaries. Stair (1897:71) refers to the practice of banishing tyrannical and oppressive ali'i from Savai'i and Upolu to Tutuila where they were humiliated by "running a gauntlet from the beach to the settlement."

Perhaps partly because of size and scale, partly because of the level of political orientation in Samoa as a whole, that is, because of Samoan attitudes toward Tutuila in particular, American Samoa has been very little studied in contrast to Western Samoa. The major exception in the early years was, of course, Mead's work in the Manu'a group, an exception that tends to overshadow the fact that the overwhelming proportion of ethnographic studies have been done in Western Samoa.

Mead chose to work in Ta'u "as a corrective for the degree of culture contact found on the island of Tutuila, where the U.S. Naval Station was located at Pago Pago" (Mead 1969, xv). For essentially this same reason, most researchers since that time have bypassed American Samoa entirely, in favor of conducting their research in Western Samoa. One wonders, however, if Samoans had some part in directing Mead to Ta'u or approving her choice, not only because it was less influenced but also because it was of higher status, more "cultured" than lowly Tutuila.

Traditional culture and society in Samoa were figured by the interplay of status, geography, and history. In more recent years, political development, the influence of the wage economy, educational patterns, and the nature of the concentrations of population and services in the Apia and Pago Pago areas have differentiated Western Samoa and American Samoa.

Generally, development has been more intrusive in American Samoa than in Western Samoa. Writing of the United States' responsibilities to its territory, James Bishop, a U.S. State Department representative, felt that while

> benign neglect and fa'aSamoa's flexibility preserved the integrity of Samoan folkways during the initial phase of western impact . . ., the equalitarian ideal and the seductive influence of American affluence now prompt many [American] Samoans to reject much of fa'aSamoa. . . . As long as the federal subsidy remains generous, those in American Samoa will enjoy a standard of living quite satisfactory from a material standpoint. But even for those in the islands, the psychic costs of cultural conflict are becoming evident in higher incidents [sic] of hypertension, suicide, etc. (Bishop 1977:52)

Political development

While the traditional political and kinship systems of American and Western Samoa are the same in principle, vagaries of scale have led to somewhat different systems "on the ground," even before the advent of western political influence. Keesing pointed out that "the native socio-political system" in American Samoa had been more "adaptable to Western forms":

> The traditional leadership is more defined, hence more easily utilized. Where, for instance, a Tutuila district will have one outstanding chief or orator who is obviously the person to act as district governor, county chief, or native judge, an Upolu district may have ten to every one such position available. . . . Then, too, the small size of the territory, and the completeness of domination by the naval authorities made possible by the peculiar status of the islands have allowed a more thorough, if exceedingly benevolent, control. (Keesing 1934a:300-301)

Davidson's *Samoa Mo Samoa* (1967) *(Samoa for the Samoans)*, an account of the political history of Western Samoa, considers variations in village political organization. During the early 1950s, Western Samoan villages varied considerably in whether they were divided into subvillages *(pitonu'u)* and in the breadth of authority and frequency of meetings of the village *fono* (Davidson 1967). Similar village divisions and intervillage variations occur in American Samoa.

Economic modernization

Both Western and American Samoa are heavily dependent on imports, but their agricultural productivity differs.

By international standards, Western Samoa is a relatively under-developed nation with an economy that continues to be mainly agricultural. The primary unit of production is still the 'au'aiga. Some household plantations, and commercial concerns, at least partially divorced from 'aiga control, sell raw produce and processed products locally, in the Apia market and stores, and in export to New Zealand and American Samoa (Pirie 1970). Harbison (this volume, Chapter 4) points out, however, that the extended family system is weakening as the unit of production and consumption. Villages differ according to their relationship to the market center of Apia. Wage employment is generally available only in the Apia area, and the large public service is centered there. As Harbison indicates, villages may export labor as well as produce. While some Apia dwellers and overseas migrants attenuate their 'aiga obligations, others continue to make at least some contributions. Consid-erable income flows into the country through remittances from Samoans abroad (Shankman 1976).

In contrast, the American Samoan economy has rapidly become less agricultural. Production is not sufficient for local consumption, and much of the produce in the market at Pago Pago is imported from Western Samoa and Tonga. Where American Samoan plantations are being worked, it appears that many Western Samoan and even Tongan youth are providing the labor. As the economy has become less subsistence oriented, it has expanded in the public and commercial sectors.

The navy was the major employer for a short time in the postwar period. Before 1940, less than a "tenth of the available man-power" in American Samoa was involved in naval employment. But during World War II, nearly "every able bodied male became involved in one way or another. Copra-cutting ceased but stevedoring and construction multiplied, cash flowed with unprecedented freedom, and some young men went off to work in the shipyards of Pearl Harbor" (Lewthwaite et al. 1973:134). Considerable labor was also attracted from Western Samoa during this period of expansion. When the navy departed in 1951, there was a lapse in general employment until expansion of government services under the Interior Department administration led to an increase in public service employment and commercial employment was developed.

The government remains the largest employer. By the mid-1970s, the Government of American Samoa employed about 10 percent of the total population (Bishop 1977), with almost half the salaried workers employed in government by 1979 (Schramm et al. 1981).

Commercial employment opportunities were developed with the opening of the first tuna cannery in 1954. Another cannery and a can manufacturing plant opened in 1963, and the canneries quickly became the second largest employer (Pirie 1970). While the canneries and can manufacturing provide employment opportunities for American Samoans, predominantly Asian migrant labor staffs the fishing fleets.

In the late 1970s, the "phantom of industrialization [had] captured the imagination of several governors, but the territory's industrial park [in Tafuna] stands largely vacant, a mute reminder of American Samoa's distance from both raw materials and markets, as well as of the shortage of vocationally-educated workers" (Bishop 1977:51). Tourism has also been touted as a route toward economic progress, but it has had limited success due to the territory's few tourist-oriented attractions and the damp climate (Bishop 1977; Pirie 1970).

As in Western Samoa, remittances are also a considerable source of income in American Samoa. While Holmes has argued for the basic stability of Manu'an culture, when he returned to Ta'u in 1962, he found that 85 percent of Fitiuta's "*matai* were being supported by donations of money from family members in Tutuila, Hawaii or even in the Mainland United States" (Holmes 1972). The average donation was $50 per month per contributor and some matai were receiving as much as $400 per month from remittances. High prestige accrued to recipients of these donations from abroad.

Education and modernization

For Western Samoa, Harbison and Baker found that once basic literacy is achieved, school attendance falls off drastically. There are some regional differences, Savai'i having the lowest attendance rates, rural Upolu intermediate rates, and Apia the highest levels of attendance. This may be attributable to the basically agricultural economy and differential family labor needs in the rural areas as contrasted with the town (Harbison and Baker 1980). They conclude that education in Western Samoa has not been a major modernizing influence, "since it is provided at the local village level by native Western Samoans, with very little media involvement, in the context of a subsistence economy."

In contrast, education in American Samoa has been more centralized, broadcast media were used heavily for a number of years, and more expatriate teachers continue to participate than in Western Samoa, particularly now in teacher training. Between 1944 and 1954, the annual budget of the Department of Education in American Samoa increased 850 percent. The per pupil expenditure during the same period, however, only approximately doubled as the general trend toward more universal education brought more pupils into the educational system.

The major educational thrust in American Samoa came with the introduction of educational television, to the elementary schools in 1964, and to all the secondary schools by 1966. To the increasing universality of education, this "bold experiment" (Schramm et al. 1981; Harbison and Baker 1980), added greater uniformity. Depending on grade level, from 29 to 35 percent of class time was devoted to the television curriculum. While educational programming was cut back drastically during the early 1970s, pupils of the 1960s and teachers who continued to teach in American Samoa were more generally exposed to English usage and Western ideas than any previously. As educational programming was cut back, American Samoan television increased its commercial programming so that today Tutuila has virtually universal access to television, a further homogenizing influence.

Demography and modernization

Changes in population growth and distribution accompany social structural, political, and economic changes. From typical village sizes of 200 to 300 (Pitt and Macpherson 1974) and a fairly even distribution, population concentrations have developed around the Apia and Pago Pago areas, the focus of new economic opportunities, political influence, and educational and health facilities. Villages with fewer resources have lost population and young adults have also moved in greater proportions than those beyond the childbearing years. Demographic changes build upon themselves and may also have other social consequences. For example, population growth meant more human resources for the 'aiga, and possibilities of migration to various destinations opened options for the allocation of those resources. As population expanded in the port areas, new options attracted even more migrants. Relationships to matai were altered as 'aiga members moved beyond the sphere of immediate influence.

THE STUDY SITES

The Samoan project research strategies have differed according to the location of the studies (Baker, this volume, Chapter 1). Descriptions of the study sites in Western Samoa, American Samoa, Hawaii, and California will vary correspondingly. Research in Western Samoa has focused on specific rural villages and the Apia urban area. Some researchers also distinguish an "urbanizing" or "transitional" area between Apia and Faleolo airport (Shore 1983:2). Rather than focusing intensively on a small set of communities, the studies in American Samoa have selected communities at various removes from the harbor area to represent "traditional," "intermediate," and "modern" categories of exposure to modernizing influences. Work in Hawaii has focused on samples from rural settings and from areas of Samoan population concentration in Honolulu. Pawson and Janes' work in California has concentrated on the northern area, specifically Alameda, San Francisco, San Mateo, and Santa Clara counties. Sampling through churches in the San Francisco Bay Area was used for specific studies.

Study sites in Western Samoa

Samoan project studies have been conducted in Apia and in the more remote Upolu villages of Salea'aumua and Salamumu.

Salea'aumua, Upolu
Pelletier (1984) describes Salea'aumua, a village about 35 mi east of Apia. With adequate reef resources, the village has a mixed agricultural subsistence base with no local wage labor opportunities. Taro and copra are the primary sources of monetary income. Some income from non-agricultural sources comes from remittances, wages of Apia commuters, and proceeds from sales of handicrafts.

Until the completion of a new road in 1979, the options of a rough road and boat travel limited contact with Apia. Salea'aumua is now served by four privately owned buses, each of which travels to Apia once daily. The 2-hour travel time and $0.95 (U.S. dollar) fare make a trip in for a day feasible, to do shopping, family and village business, or to sell crops. But the time and cost involved make daily commuting unattractive and preserve the rural character of the village. Those who are overwhelmingly attracted to the town have the options of moving to Apia permanently for wage labor or education or commuting home on weekends.

Approximately half the villagers are Catholics, and the other half London Missionary Society, with one family each adhering to the Church of the Latter Day Saints (Mormon) and the Seventh Day Adventists.

Salamumu, Upolu
The following description of Salamumu is based on Orans (1981); additional ethnographic background on Salamumu will be found in other

chapters of this volume (Harbison, Chapter 4; T. Baker, Chapter 7). Salamumu was founded in 1911 by some of the former residents of Sale'aula, Savai'i, who had moved after a volcanic eruption in 1908 destroyed their village. Salamumu and a reestablished Sale'aula retain important kin and political ties, continuing to share their fa'alupega. In regard to the matai system, while the traditional division of matai into ali'i and tulafale holds sway in Salamumu, the village is dominated by its tulafale (Orans 1981). Davidson (1967) suggests that the tulafale of Savai'i were the focus of the district political structure of Savai'i. Perhaps the domination of Salamumu's tulafale is at least partly attributable to the Savai'i origin of the village.

Salamumu now has two subsections (pitonu'u), the original beach village (tai), and a new inland section (uta). Salamumu tai and Salamumu uta have separate Methodist churches with their own buildings and ministers. While the tai/uta distinction is based on the location of the churches, "those who attend the tai or uta churches may be described as tai residents or uta residents, even though they may live well into the other section" (Orans 1981:131). There are also dual church organizations and Women's Committees.

Salamumu uta is about 26 mi from Apia and roughly and hour and a quarter, usually traveled in one of the two pickup trucks owned by the villagers; Salamumu tai is an additional 3 mi away. Bus service, instituted in 1978, takes a longer route and an extra half-hour travel time. Of the villagers who frequent Apia to sell produce and make purchases, to attend meetings and special celebrations, a few average more than a day in town per week. But the village is still known as a conservative one under the strict control of its matais. The village sells most of its produce in Apia and also buys most of what it needs from there.

Apia, Upolu

Apia and its surrounding urban area lie in the center of the coastal plain on the north coast of Upolu. Apia is the capital city, central market, and the focus of commercial development in Western Samoa. With bus service on Upolu and connecting boat service to Savai'i, Manono, and Apolima, many Western Samoans commute to Apia on a daily or weekly basis.

Samples taken in the Apia area are not from defined communities. In Pelletier's work, the Apia area includes roughly a 2-mi radius from the center of Apia and is not equivalent to the more restrictive urban Apia area as defined in the Western Samoa Census of Population and Housing (Pelletier 1984). Pelletier includes daily commuters within the 2-mi radius, and weekly commuters, some from as far as Savai'i, in his urban sample.

Variability in household and community organization characterizes the Apia area. While most households are headed by matai, others are not.

linked through a matai to village-level social and economic activities (Pelletier 1984). Most households, however, maintain extensive ties with networks of relatives in Western Samoa and abroad (Pelletier 1984; Tiffany 1975). James (1984) and Harbison (this volume, Chapter 4) found that many households are now composed of young people from one rural village who live together in Apia during the week.

Some individuals remit their entire earnings, apart from subsistence needs, to their matai. Others remit as specific demands are made and particular occasions arise. Generally, when villagers live in town except for weekends, they are unlikely to give all their wages to the matai.

Government aids village agriculture and contributes to education and medical services. Apia differs from rural villages, however, in the nature and number of modern institutions. Apia is the site of the government Fono, central government and mission schools, the National Hospital, and a concentration of commerce, transportation, and marketing. Through these institutions, Apia residents and commuters experience daily contacts with nonvillagers, infrequent occurrences for their rural counterparts. Moreover, since a high proportion of Europeans and part-Samoans live in the Apia area, these contacts are with a wider variety of persons than is possible in rural Western Samoa (Hirsh 1958).

The villages of Apia are all served by urban utilities, including water and power. Pelletier (1984) notes, however, that households have few electrical appliances and many limit their use of expensive utilities by using benzine or kerosene lanterns.

Study sites in American Samoa

The harbor area
Political and commercial centralization in the area around Pago Pago harbor compounded the natural divisions of mountain ridges and sea-lanes in American Samoa. Studies have addressed variability in American Samoa by using distance from the harbor area as the independent variable and indices of modernization as dependent variables. The divisions of *modern, intermediate,* and *traditional* are used in order to examine some of the variation in American Samoa, on the provisional supposition that the harbor area is the major source of modern influence and that exposure to modern influence increases with physical proximity to the harbor area. Thus, the Pago Pago harbor area villages are considered modern. The remainder of the Tutuila villages connected to the harbor area by paved road are in the intermediate category. The traditional rubric includes the remote north shore villages of Tutuila, unconnected by paved road to the harbor area, and the Manu'a group. In fact, we shall see that recent developments are attenuating the centrality of the harbor area.

One's main impression of the harbor area is density of population and activity. And much of that activity is directed toward the ends of commercial establishments and the central government. In 1974, over 75 percent of wage employment was in the Central District (Levin and Wright 1978).

But in the back streets of any of the harbor area villages, one does have the sense of being in a village: cooking and washing are evident in the yards and through open doorways. People in informal lavalavas carry mats or baskets of taro. Meetings take place, village-level decisions are made. Other than along the main road and the bars around the Fagatogo malae, the area usually shuts down quietly at night.

Park suggests that urbanization in American Samoa probably started only with the establishment of the naval station and government buildings in Fagatogo village and took off with World War II and the concomitant economic boom in the bay area: "A cash economy was introduced to the communal society, a class of wage-earners was created and a concentration of population rapidly evolved" (1979:50). The population of the eight Pago Pago Bay area villages, Anua, Atu'u, Aua, Faga'alu, Fagatogo, Leloaloa, Pago Pago, and Utulei increased from 17.4 percent of the total population in 1912 to 28.8 percent of the total population in 1974 (American Samoa Development Planning Office 1976).

The outer districts of Tutuila

The remainder of Tutuila is less easy to characterize than the harbor area. The shopping center ambiance of the main road at Nu'uuli in the Western District with its pizza parlor and bowling alley contrasts to the rural churches and coral sands of the Eastern District, punctuated only occasionally with a video game parlor or nightclub. Toward the far ends of the island, particularly in the Western District, there is evidence of more subsistence agriculture and fewer daily commuters to the harbor area. Improved roads have influenced the distribution of population. "Many of those employed in the Bay area commute daily from their villages. For example, in 1963, out of 719 cannery workers, 43 percent resided outside the Bay area" (Park 1979:51). Commuters are particularly concentrated in the Eastern District, described as "suburban" (Harbison, this volume, Chapter 4) and as a "bedroom community" for the Central District (Levin and Wright 1978:151).

Villages just beyond the bay area, such as Nu'uuli and Tafuna in the Western District, were growing rapidly in 1974. Recognizing the multiple roles of the bay area as "administrative centre, commercial centre, port of entry, industrial area, tourist resort and cultural centre," Park predicted that the area would continue to grow rapidly, but that with increasing pressure on land, "further expansion will necessarily be towards the westward plain. The government has already begun to develop an airport, housing facilities and more recently a 35-acre industrial park in that area" (Park 1979:51). Since Park wrote, an increasing number of retail establishments have also moved into shopping centers in the Western District, particularly in Nu'uuli.

A 1970-1980 intercensal increase in the population of the Tafuna/Nu'uuli area is probably due as much to the increasing avail-ablility of housing for non-Samoans and wealthier Samoans in the Western District as it is due to any impact of industrialization in that area. Nu'uuli had a population of 1804 in 1970 and 2585 in 1980, an increase from 6.6 to 8.0 percent of the total population. Likewise, Tafuna's population increased from 278 in 1970 to 1086 in 1980, from 1.0 to 3.4 percent of the total population (with slight changes in bound-aries and definitions). In contrast to these increases, the harbor area had only 27.8 percent of the total population in 1980, a slight decrease from its 1970 share of 28.3 percent (U.S. Bureau of the Census 1982).

The Manu'a group

Based primarily on his fieldwork in the mid-1950s and early 1960s, Holmes (1974) describes Fitiuta as a quiet traditional village occupied largely in subsistence activities and political ceremonies. Since then, emigration has led to considerable depopulation in the Manu'a district (Park 1979), but the character of the community has not been substantially altered. Scheder (1983:2) describes the "most traditional village in American Samoa" as of 1982.

> experiencing at least some aspects of modernization. . . . Electricity, piped water, tinned fish and corned beef, government sponsored jobs, a shift away from traditional house styles, and at least one television set are among the external accoutrements of modernization. The road to the village was recently paved, and people now travel by jeep and pickup truck to the airstrip 7 miles away. A small propeller airplane makes three loosely scheduled runs each week between this island and the main island of Tutuila. (Scheder 1983:2)[4]

Sampling strategies

The development of the distance from harbor area hypothesis

Within American Samoa, particular attention has been paid to the idea that modernization increases with proximity to the Pago Pago harbor area (Martz 1982; James, McGarvey, and Baker 1983; McGarvey and Baker 1979; Ember 1964).

Ember's fieldwork in the mid-1950s appears to have been the first guided by a "distance from harbor area" hypothesis. His research was directed toward correlating commercialization with political change for three locations in American Samoa, Ofu in the Manu'a group, Malaeloa in the Western District of Tutuila, and Pago Pago in the harbor area of Tutuila.

Ember's results show a linear relationship between increasing political change and increasing commercialization with the least political change and commercialization in Ofu, Malaeloa intermediate, and Pago Pago at the most changed and commercialized end of the continuum. "The more a contemporary village has become commercialized, the more it has discarded the traditional political system" (1964:109).

The McGarvey and Baker paper (1979) concludes that there are significant differences between traditional, intermediate and modern individuals or villages in both sociocultural characteristics, for example, languages spoken, formal education completed, and occupational classification and blood pressure, where the modern group had the highest and the traditional group the lowest blood pressures. For most age/sex groups in the intermediate area

> the blood pressures . . . more closely approximate the modern area than the traditional one. However, during late middle age and old age, the values for women in the intermediate area become almost identical to those of women in the traditional group. (McGarvey and Baker 1979:469)

These "results on the population living in the intermediate area of modernization suggest that an even closer association could be demonstrated if an appropriate measure of participation in modern culture could be devised for individuals" (McGarvey and Baker 1979:474).

Martz (1982) raises important questions about individual variability, particularly in relation to harbor-area contact. She holds that employed individuals in outer villages were undersampled in earlier studies because they were absent in the harbor area, and that bus service made Pago Pago relatively accessible for all villagers. These factors may help explain the more complex results in the intermediate area. She suggests that Tutuila is more homogenous than previous studies with smaller sample sizes had shown and she tenders a hypothesis that will be echoed in the following pages:

> Alternatively, the variables considered may not adequately reflect differential exposure to a more modern life-style. Although this interpretation is speculative based on the data presented here, it suggests that area differences found previously could be making relevant individual differences. . . . further research into measuring the degree to which individuals are differentially following more modern ways of life is needed. (1982:32)

Martz suggests that the intermediate classification may not be as cogent as the modern and traditional groupings in American Samoa and urges us to look for more individual variation in the intermediate communities. Linked to the harbor area but with more of their traditional social structures intact, these communities may allow more leeway for variation in life-style (Martz 1982). Individuals may also take differential advantage of these options through the life cycle, choosing to partake more of modern opportunities in their youth, and perhaps settling down later to traditional options as they gain stature in their community through age and experience. Those women who take up traditional options after finishing their education or leaving the labor force may represent another variation on this pattern.

Community vs. individual

Harbison and Baker (1980:2-3) suggest that "the first distinction in the modernization process to be made is that between individual and community levels of modernization." This distinction, particularly crucial for examining differences in exposure to modernizing influences and their effects in intermediate areas, has not been maintained with precision.

McGarvey and Baker (1979) first identify subjects as modern, intermediate, or traditional, based on whether they lived in harbor area villages, villages connected to the harbor area by paved road, or in the remote villages of the Tutuila north shore and the Manu'a group. Later areas are labeled: "No differences of blood pressure are found between migrants from the modern area of Samoa and sedentes residing in that area" (1979:469).

James, McGarvey, and Baker (1983) continue to use the tripartite division. But here they refer to the three categories as residence designations, labeling a village rather than an individual as modern, intermediate, or traditional.

Education, commercialization, and the very mobility of its people are increasing the homogeneity of American Samoa. Sources of variability must be sought in factors additional to a simple rural/urban continuum. In regard to individual variability, a wider range of social and cultural variables than education, employment, and languages spoken must be examined.

American Samoa and the port town concept

The consideration of "modernization" in American Samoa would also benefit from another look at the nature of rural and urban areas in relation to exposure to modernizing influences and to migration. Implicit in the focus on the harbor area as a "district center" or "port town" may be the idea that individuals are both influenced by it and drawn to it, either temporarily and diurnally, or more permanently and residentially (Spoehr 1963, 1960).

The draw of the "district center" may be constrained, however, particularly in societies where kin ties are still important in determining access to resources. To the extent that the district center still participates in the traditional polity, access to it may be limited. Contacts with relatives and friends in Hawaii, the U.S. mainland, and elsewhere may also be a more or less independent modernizing influence. Additionally, in American Samoa, the draw of the harbor area may be diminishing due to the establishment of new shopping and entertainment centers, particularly in the Western District. Educational and commercial media have also lent a homogenizing influence, decreasing variability and the significance of the "port town" concept for American Samoa.

Because of the nature of migration in American Samoa, it is difficult to determine whether internal migrants or immigrants from Western Samoa account for population growth and maintenance in the Pago Pago area (Park 1979:51). From 79.0 to 87.9 percent of American Samoa residents were native to their district of residence in 1974, but the Central District was the most heterogeneous.

> In short, although to a certain degree the Central and Western districts have attracted people from other districts, internal migration within the island of Tutuila does not seem to be extensive. This is probably because of the small land area and the improvement in public transportation and roads. Tutuilians may not need to move within the island to be closer to jobs and schools, as all areas now have easy access to these necessities of modern life. (Park 1979:52)[5]

In addition to the factors Park considers (jobs, schools, and transportation facilities) I would add that such necessary "auspices" as kinship ties may be lacking for potential internal migrants to the harbor area. Tilly and Brown (1967:142) define auspices as "the social structures which establish relationships between the migrant and the receiving community before he moves. . . . An individual migrates under the auspices of kinship when his principal connections with the city of destination are through kinsmen, even if he comes desperately seeking a job." Nontraditional housing is also limited and expensive on Tutuila and especially so in the harbor area.

For Western Samoan migrants to New Zealand, Pitt and Macpherson (1974) recognize that earlier migrants tended to gain experience in Apia prior to moving onward, but that many subsequent migrants have been brought directly from rural villages. Pitt and Macpherson imply, however, that virtually all movement from American Samoa to Hawaii and California goes through Pago Pago. While Manu'ans do tend to sojourn in the Pago Pago area, we suspect that many American Samoans move directly from rural villages to Hawaii and California. As Samoans move outside the Samoan islands, additional cultural and social factors complicate the picture of variation. Language environments and cultural expectations for behavior differ more widely. Educational systems, employment, and social assistance programs offer additional possibilities.

Oahu, Hawaii: Samoans in church town and secular city

Samoans have been coming to Hawaii since the 1920s, but the greatest influx began in the 1950s with the end of naval administration in American Samoa (Ala'ilima 1982). Of Samoans in Hawaii covered by the

1980 census, almost equal numbers were born in American Samoa, Western Samoa, and Hawaii (Franco 1984).

A Tongan participant in a conference on migrant adaptation observed that Hawaii confronts a migrant "with a wide range of value systems—there's a Japanese set of values, a Caucasian set, the military, and, finally, a Polynesian value system. You don't know what to adjust to" (Macpherson et al. 1978:34). Samoans in Oahu are also involved with smaller cultural and religious groups such as the Filipinos, Hawaiians, and Mormons.

Adjustment in Hawaii appears to be difficult for Samoans. Many Hawaiian residents cannot distinguish, for example, between Samoans and Tongans. Nevertheless, Samoans are distinctly and negatively stereotyped in Hawaii as big, aggressive, and violent.

Negative stereotypes and lack of job and language skills may help account for the high unemployment rate of Samoans in Hawaii. In 1980, the labor force participation rate of Samoan women in Hawaii was lowest of all reported ethnic groups and that of Samoan males was second lowest. The percentage of employed Samoans in the civilian labor force was by far the lowest of any ethnic group. Only the Vietnamese had a lower male labor force participation rate and lower median income levels than the Samoans (Franco 1984). Caution is advised in interpreting these data, however, because employment figures may show the influence of personal choices and cultural norms. For example, cultural expectations are that most Samoan women will not seek employment (Markoff and Bond 1980). Nevertheless, Samoans have very low incomes, a young population, and large household sizes, and face a high and rapidly rising cost of living in Hawaii.

As of 1980, Samoans were estimated at about 2 percent of the Hawaiian population. More than 98 percent of Samoans in Hawaii lived on Oahu (Franco 1984), concentrated in a relatively small number of communities and census tracts within the city of Honolulu. Prominent areas of concentration are the major public housing developments in Honolulu, communities on the Waianae coast, and the communities of Laie and Hauula on the windward coast of Oahu.

Honolulu

The largest concentrations are in the census tracts of three public housing developments in the Kalihi/Palama area where 23 percent of the 14 073 Samoans enumerated in the state of Hawaii in the 1980 census reside. Samoans comprise 54 percent of the population of the Kuhio Park Terrace census tract, 1428 persons; 44 percent of the population in the Kalihi Valley Homes tract, 1285 persons; and 30 percent of the Mayor Wright Housing tract, 519 persons (Franco 1984). Public housing is inexpensive by Hawaiian standards and Kalihi/Palama is well situated in relation to public and private employment opportunities, as well as health and social services. An additional 681 Samoans (17 percent of the tract population) lived in the public housing area in Palolo Valley at the opposite side of Honolulu (U.S. Bureau of the Census 1983).

The Waianae coast

Along the shoreline west of Honolulu, the life-style "is a little more like home" (Ala'ilima 1982:105). In 1980, 2716 Samoans (14 percent of the Hawaiian Samoan population) lived in the communities of Maili, Makaha, Nanakuli, Waianae, and Waipahu. There is a concentration of Samoans in the Samoan Church village at Nanakuli where traditional village social structure and ceremonial activities are more closely followed than in other areas in Hawaii. "Participation in Samoan groups and village activities was noticeably higher in the Nanakuli Samoan Village than, for example, in Mayor Wright Housing" (Fauolo and Salanoa 1965, reprinted in Young 1974).

Laie and Hauula

Laie and Hauula are neighboring communities on the north east shore of Oahu on the opposite side of the island from Honolulu. As one approaches the Laie/Hauula area, the road becomes narrower and increasingly hugs the coast, much like the road in American Samoa. Bananas, trees, and flowers seem familiar; the houses and yards are witness to a more Polynesian life-style than is possible in most of the Honolulu housing where Samoans live. The main attractions of this area for Samoans are the Mormon Temple and employment at the Polynesian Cultural Center, both in Laie. In addition, about 8 percent of the students at Brigham Young University-Hawaii campus in Laie are Samoan (Graham 1983). For many years, a large proportion of Laie area residents have been Samoan (Ala'ilima 1982; Stanton 1978). In 1980, 1398 or 10 percent of the Samoans in Hawaii lived in Laie and Hauula (U.S. Bureau of the Census 1983).

In contrast to Honolulu, living in the Laie area offers a different mix of value systems to Samoans. Samoans in Laie have exhibited some ambivalence about the matai system, and the Mormon Church has taken an official stand against it, contending "that if a person properly fulfills his religious obligations in accordance with the guidelines provided by the Church that there is no need for the matai system in Hawaii" (Stanton 1978:289-293). The Laie setting combines a particular conciousness of Samoan culture with a potentially clashing set of other values.

While not all Samoan residents in the Laie area are Mormons, according to Stanton, many who are not eventually move elsewhere. Assimilators tend to move from Laie as well. "The transition to a non-Samoan status [usually Hawaiian or Caucasian] is easier in communities not as conscious of Samoan culture" (Stanton 1978:279).

The emphasis on church affiliation in Laie should not obscure the fact that the majority of Samoans elsewhere on Oahu also identify strongly and are identified with their churches. But Ala'ilima notes that except "in Laie and a few Samoan churches [elsewhere on the island] . . . it is difficult to find the traditional sense of community within which Samoan families used to provide mutual assistance and social controls" (1982:107).

Samoan communities in California

While initial Samoan settlers in California tended to come through Hawaii, by 1960, "the mainland communities included many . . . who had migrated directly from Samoa . . ." (Lewthwaite et al. 1973:142-143). Since that time, participants in the "New Neighbors" conferences suggest that California communities differed considerably in the primary source of migrants from the Samoans. For example, Tofaeono suggests that the Oxnard people are largely from Fitiuta and Olosega (Macpherson et al. 1978:29).

Samoans in California reside primarily in communities where military personnel were transferred when the navy closed its American Samoa base in the early 1950s. The original settlements were in Oceanside, National City, San Diego, Long Beach, and San Francisco. Samoan Wards of the Mormon Church were established in Southern California in 1956 and in San Francisco by 1957. The first Samoan Congregational Church congregation was begun in 1957 in San Francisco.

In Southern California, sizable populations are now found also in Santa Ana, Torrance, Compton, and Oxnard. The primary population centers in northern California are Daly City, South San Francisco, Oakland, San Jose, and Monterey. Within San Francisco itself, Samoans live primarily in the southern part of the city where housing is relatively affordable and abundant. According to the California State Data Center, in 1980 approximately 80 percent of all Samoans in California were living in five counties (in order of numbers): Los Angeles, San Diego, Orange, San Francisco, and San Mateo (Janes 1984). There is considerable travel between northern and southern centers of Samoan population concentration in California, especially for weddings and funerals (Lewthwaite et al. 1973).

The functional adaptability of the 'aiga in California is similar to that described by Pitt and Macpherson (1974) for New Zealand. The 'aiga serves in the mobilization of capital and life crisis situations, sometimes in new ways, and takes additional significance in the provision of accommodation and employment. Although the family is less extended in California, the 'aiga functions as a "mutual aid and employment agency" accounting for "the gravitation of Samoan men into ship building, metal-jobbing and construction work and of women into nursing . . ." (Lewthwaite et al. 1973:151). As in New Zealand, the social hierarchy of matai is less significant in California where the "churches quickly became the centers of Samoan life," with some churches more or less clearly perpetuators of fa'aSamoa (Lewthwaite et al. 1973; Ablon 1971c:83).

While the initial settlement of Samoans in California appears to have been more economically successful than in Hawaii (Lewthwaite et al. 1973; Ablon 1971c, but cf. Chen 1973), Samoans in California are becoming increasingly economically stressed and perhaps more like their Hawaiian counterparts. In the mid-1960s, Ablon (1971c) found nearly full employment among Samoans in the San Francisco Bay Area. According to recent estimates, between 10 percent and 40 percent of Samoan adults in California are unemployed (Janes 1984; Literacy and Language Program 1984; Hayes and Levin 1983).

Rolff (1978a, 1978b) suggests that participation in Samoan activities and organizations differs according to factors such as wealth, age, life experiences, non-Samoan ancestry, and interracial marriage. In her view, early migrants were more economically self-sufficient because they were young, had small households, were initially isolated from other Samoan households, and arrived in time to take advantage of favorable military and blue-collar job markets. As they have aged and accrued more dependents, and the Samoan community has expanded, resources have been strained. Moreover, the quality of the community has changed as it has grown.

In contrast to the large urban communities in Los Angeles, Orange County, and San Francisco, adaptation in the relatively small Samoan communities of Oxnard (Rolff 1978a, 1978b) and Seattle, Washington (Kotchek 1978, 1977) appears to be somewhat easier. Kotchek attributes this to less ethnic visibility and a corresponding freedom of choice in adaptive strategies.

Size and age of communities appears to affect cultural variability on the mainland. Rolff's work in Oxnard and Kotchek's research in Seattle illustrate that the pressures for conformity are relatively weak in the very small communities, where families and individuals have opportunities to avoid Samoan social networks and the expensive rituals they entail. In San Francisco and Los Angeles, a certain community density was reached fairly early in the migration. Between 1965 and 1970 descent groups were reconstituted to a level permitting large formal exchanges. The first large Samoan funerals in San Francisco were held, for example, after a 1965 fire in a Catholic parish hall (Ablon 1970). The reestablishment of such Samoan rituals occurred somewhat later in San Diego and Oceanside.

Janes (1984) notes that it is also important to distinguish migrants on the basis of the cohort in which they migrated. He describes three groups of migrants: those migrating under military auspices in the 1950s, kin-linked family migration from the late 1950s to the late 1960s, and a more recent immigration of elderly. Military migrants are, as a rule, much more sophisticated in dealing with American institutions than are those without military experience. Families migrating in the decade of the late 1950s to late 1960s easily found good employment on the West Coast as heavy industry experienced growth during the Vietnam era. More recent migrants have had significant economic problems such as lack of affordable housing and a paucity of jobs for the unskilled. A large number of aging parents are now migrating to be with children or to take advantage of medical care on the mainland. They tend to be a financial drain on their families. Many even live alone in public housing, for the homes of their children are too crowded with other relatives and grandchildren (Janes 1984).

SUMMARY

Classic studies in the anthropology of Samoa have described features of that distinctive culture and society. Important shared principles and familiar structures of family, land, village, and church guide daily life in Samoan communities. But along with cultural unities come different realizations depending on scale, indigenous history, and eventually on encounters with other cultures and their political, social, and economic forms.

In this brief overview of variation between and within areas where Samoans live in the 1980s, we have seen that contrasts in colonial experience and differences in traditional social organization influenced the course of development in Western and American Samoa. In different communities in the Samoan islands, and in Hawaii and California, family affairs, political and community events, and church activities play different roles for communities and individuals. Commercial activities and engagement in wage labor directly affect Samoans in the Apia area and in much of American Samoa. Residential patterns and consequent densities of involvement in local communities vary yet further in the metropolitan destinations of migrant Samoans. Reliance on kin and community is only one of an expanding range of options, sometimes positive, sometimes painful, for individuals.

NOTES

1. The differences and similarities between American Samoa, specifically Manu'a, and Western Samoa have recently arisen as issues of more than parochial interest in anthropology with publication of Freeman's criticism of Mead (Freeman 1983) and the many responses it has provoked (e.g., Brady 1983).
2. Concentrations of Samoan migrants are found also in New Zealand (Pitt

and MacPherson 1974), in Washington State (Kotchek 1977), Utah, Kansas City, and New York City (Hayes and Levin 1983).

3. Students of Samoan culture are not in terminological accord. For example, what Shore refers to as the 'au'aiga, Orans calls the umu 'aiga.

4. At least two airlines now serve the Manu'a group (Scheder 1984, personal communication).

5. Some north shore villages are still not connected by paved road to the south shore of Tutuila.

Chapter 4

The Demography of Samoan Populations

SARAH F. HARBISON

The interaction between ecological and cultural factors and the demographic processes of the last century have resulted in the present population system of Samoans in the Pacific. Western Samoa and American Samoa, close both geographically and culturally, have also been linked historically by migration. Since World War II, these islands have also sent increasing numbers of migrants to Hawaii, to the West Coast of the United States, and to New Zealand. The Samoan migrant population in Hawaii, for example, numbered about 15 000 at the time of the 1980 Census. Many of these Samoan migrants remain closely tied to families in Western or American Samoa, as do Samoans on the West Coast of the United States and in New Zealand. Return visits to the homeland for ritual occasions, the arrival of new migrants, and the remittances sent by migrants to their families at home, all serve to reinforce ties among subpopulations of Samoans.

The focus of this chapter, then, is on the interrelated demographic histories of Western Samoans, American Samoans, and Samoans in Hawaii, with an emphasis on the role of migration as the demographic, cultural, and economic link between the populations. In addition to linking two populations, however, migration may also be viewed as an adaptation of the donor population to existing fertility and mortality patterns. As Baker has pointed out, migration can be seen both as "a process which, by putting individuals in a new environment, necessitates a new adaptation" and "on a different level migration itself may be viewed as an adaptation to population size and structure" (Baker 1981:3). It is clear, for example, that one reason that Western Samoa has not been harder pressed by land and resource shortages is that migration has siphoned off a significant portion of its population growth. Therefore, this discussion will evaluate the impact of migration on the Samoan populations in the Pacific on two levels.

First, on the aggregate level, migration affects the age/sex structure of both the population of origin and destination in several ways. Certain age categories are diminished in the population of origin, and these categories are increased at destination. Furthermore, migration may also affect the fertility and mortality rates of the populations by drawing selectively from age groups with atypically high or low age-specific rates. It may also affect nuptiality by creating shortages of potential mates in certain age groups. Second, on the individual level, migration may, by putting individuals in a totally new environment, change their social, biological, or economic characteristics. These changes may, in turn, affect individual demographic patterns and overall demographic rates. For example, the possibility of increased education or labor force participation of women migrants may change the market values of the woman's time, and consequently the perceived values and costs associated with children.

Even when the focus is on trends in fertility, mortality, and migration at the population level, however, it is on the level of the individual, the household, and the 'aiga that the decisions are made that result in the observed demographic rates. It is through membership in a particular family and 'aiga that an individual gains access to resources and is linked to other members of the village and the larger society. The number of children that a couple has, the decision to migrate, the timing of a marriage, and the selection of a mate are all demographic decisions that can be better understood within the culturally defined context of the family. A schematic representation of the context of demographic decision making is presented in Fig. 4.1. The historical convergences and divergences of the three populations of Samoans living in the Pacific can be interpreted as the sum of individual, household, and village adaptations to differing ecological, social, and demographic environments.

In this framework, the demographic options available to an individual are influenced by his or her place in the family, the demographic structure and economic status of that family, and the place of the family in the

Fig. 4.1 A framework for demographic decision making.

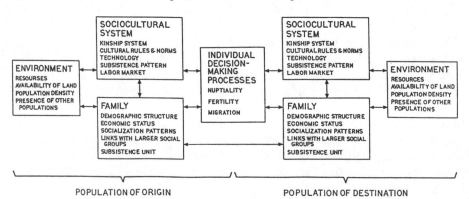

large community. The family serves as the link between the individual and the sociocultural system and environment. Decisions will not only be affected by characteristics of the sociocultural system and environment in which an individual is living, however (i.e., the left side of Fig. 4.1). Information is also received from individuals who have already migrated concerning the new situation (see right side of Fig. 4.1), and this, too, is an input to decision making. These linkages are explored for each of the three populations of Samoans.

A BRIEF REVIEW OF THE DEMOGRAPHIC HISTORY OF THE SAMOAN ISLANDS

The nineteenth century

While it is generally accepted that relatively reliable demographic data are not available before the 1920s for the Samoan islands (Park 1980; McArthur 1967), the reports of early missionaries and administrators provide a general picture of demographic dynamics of the nineteenth century. The period was one of increased contact with Europeans, as missionaries arrived in increasing numbers and western nations, including Germany, Great Britain, and the United States, developed trading interests there (Lockwood 1971). In general, "the reconstruction of Samoa's population from all evidence available for the nineteenth century suggests fluctuations in total numbers within a relatively narrow range of 34 000 to 39 000. In most years the birth rate was probably a little higher than the death rate, but this may have been temporarily reversed by sporadic outbreaks of disease and by intermittent warfare" (McArthur 1967:115). Tensions among external interests as well as internal politics made for frequent disputes, conflicts, and battles. Mortality in the villages was high because of famines resulting from "the prolonged absence of men from their villages, as well as the havoc caused by shelling and burning" (Gilson 1970). Combined with the epidemics of diarrhea, measles, dysentery, and influenza, these factors account for the lack of growth in population of the Samoan islands during the nineteenth century (see Table 4.1). It seems likely that the islands followed the typical "pretransition" demographic pattern of high fertility and intermittently very high mortality offsetting each other and leading to an approximately stable or perhaps very slowly growing population.

Actual data on mortality and fertility rates for this period are severely limited (Park 1980). Park suggests that a crude birth rate (CBR) of something over 40 is a reasonable estimate, with an "intermittent excess of deaths over births." Concerning migration, Park suggests that "although there must have been extensive inter-island movement, records do not indicate any emigration to areas outside the Archipelago," and that it is unlikely that any of the islands were net gainers or losers.

Table 4.1 Estimated Population of the Samoan Islands in the Nineteenth Century

Island	1839	1849	1853	1863	1874	1879	1887	1899	1900
Upolu	25 000		15 587	17 556	16 568	15 000	15 750		17 755 [c]
Manono	1100	20 000	1015			1500	1200	18 000	887
Apolima	500		191						
Savai'i	20 000	12 000	12 444	12 670	12 530	12 500	13 000	10 000	14 022
Tutuila	8000	3700	3389	3450	3746	3700	3500	4000	3923
Manu'a	2000	1300	1275	1421		1400		1759	1756
Total Samoans	56 600	37 000	33 901	35 097	34 265 [a]	34 100	33 450 [b]	33 800	38 494
Foreigners			120		919	2300			1600 [d]

Source: McArthur 1967.

[a] Population reported for Manu'a in 1863 (1421) was included in the total shown.

[b] It was claimed that Manu'a's population was included in this total, but it appears not to have been.

[c] Includes 195 Polynesians from Niue (99), Fiji (48), Rotuma (13), Futna (15), and Uvea (20), but excludes 787 Melanesians from German New Guinea.

[d] German Samoa only: 787 Melanesians and "approximately 800" foreigners.

Table 4.2 The "Samoan" or "Native" Population and Total Population Enumerated in Censuses since 1900, by Sex

Date of Census	"Samoan" Population							Total population of Western Samoa		
	Upolu, Manono, and Apolima		Savai'i		Western Samoa					
	Males	Females	Males	Females	Males	Females	Persons	Males	Females	Persons
Aug.-Oct. 1900	9403	9390	7491	6531	16 894	15 921	NA	NA	NA	NA
July-Sept. 1902	19 411		13 201		NA	NA	32 612	NA	NA	NA
Oct. 1, 1906	10 550	10 112	6598	6218	17 148	16 330	33 478	NA	NA	37 320
Oct. 1, 1911	21 182		12 327		NA	NA	33 554	NA	NA	38 084
July 1, 1917	NA	NA	NA	NA	19 199	16 205	35 404	20 334	16 997	37 331
Apr. 17, 1921	10 660	9972	5908	5982	16 568	15 954	32 522	19 442	16 901	36 343
Jan. 1, 1926	12 228	11 824	6413	6223	18 641	18 047	36 688[b]	20 986[a]	19 245	40 231[a]
Nov. 4, 1936	35 657		16 575		26 468	25 798	52 266[b]	28 727	27 219	55 946[b]
Sept. 25, 1945	22 317	21 451	9517	9137	31 834	30 588	62 422	35 107	33 090	68 197
Sept. 25, 1951	29 165	27 795	11 953	11 240	41 118	39 035	80 153	43 790	41 119	84 909
Sept. 25, 1956	33 484	31 890	13 513	12 946	46 997	44 836	91 833	49 863	47 464	97 327

Source: McArthur 1967.

NA = Not avaliable.

[a] Includes 890 Chinese and 155 Melanesian contract laborers here presumed to be males.

[b] Includes 34 visitors from American Samoa.

The twentieth century

Changes in the political history of the Samoan islands after the treaty of Berlin in 1899 have been reviewed by Gage (see Chapter 2). By 1962, Western Samoa was an independent nation consisting of the islands of Upolu, Savai'i, Manono, and Apolima; the eastern islands of Tutuila, the Manu'a group, Aunu'u, and the Rose Atoll became a territory of the United States in 1899 and are now called American Samoa.

In addition to the political shifts that occurred around the turn of the century, population dynamics changed as well. Although there are some problems with reliability of the data (McArthur 1967), the trend shown in Table 4.2 is clear. The population of Western Samoa increased from about 37 000 in 1906 to about two and a half times that size (97 000) 50 years later. This growth was experienced despite substantial migration from Western to American Samoa, and despite several devastating epidemics. During this same period, the population of the islands of American Samoa grew from about 6000 to 29 190 in 1974 (see Table 4.3).

Figure 4.2 shows the pattern of population growth in Western and American Samoa during the twentieth century. There are contrasts between the two populations in the first half of the century, but the general picture is one of very rapid growth during the period 1940-1970. Although the impact of foreign intervention can be seen more clearly in American Samoa, with a marked increase in population growth in the 1940s and a reduction in the rate of growth in the mid-1950s, the patterns for the two populations are strikingly similar, reflecting the

Table 4.3 Population of American Samoa in Twentieth Century by District and Sex

Year	Total	Tutila Eastern	Western	Manu'a	Swain's Island	Male	Female
1900	5679	2221	1702	1756		NA	NA
1901	5563	2342	1618	1603		NA	NA
1903	5888	2441	1752	1695		NA	NA
1908	6780	3018	1907	1855		3619	3162
1912	7251	3186	2268	1797		3836	3415
1916	7550	3631	2254	1665		3939	3611
1920	8058	3777	2408	1873		4092	3966
1926	8763	4221	2395	2060	87	4494	4269
1930	10 055	5032	2777	2147	99	5208	4847
1940	12 908	6733	3431	2597	147	6612	6296
1945	16 493	9338	4610	2406	139	8565	7928
1950	18 937	10 624	5330	2819	164	9818	9119
1956	20 154	11 405	5902	2767	80	10 107	10 047
1960	20 051	11 137	6113	2695	106	10 164	9887
1970	27 159	15 955	9018	2112	74	13 682	13 477
1974	29 190	16 828	10 520	1842		14 747	14 443

Source: Park 1980

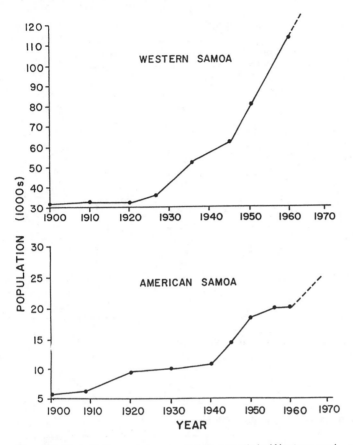

Fig. 4.2 Twentieth century population growth in Western and American Samoa.

continuing close linkage and common cultural background. This growth resulted in a total population in Western Samoa in 1976 of 151 983 (Western Samoa Dept. of Statistics 1980) and a total population in American Samoa in 1980 of 32 297 (U.S Bureau of the Census 1981). The age/sex breakdowns of the populations for Western and American Samoa are presented in Tables 4.4 and 4.5. The population pyramids are presented in Fig. 4.3. The pyramids, representing proportions of the populations of males and females in the various five-year age groups, show some interesting contrasts. Although these differences and their determinants are explored in detail in following sections, it is interesting to note that there is strong evidence of out-migration from Western Samoa, in the ages between 20 and 30, and that the outflow of women in these age groups is almost as large as for men. It is also clear that fertility is considerably higher in Western Samoa with about 48 percent of the population 0-14 years of age. In American Samoa the percentage is 40. The pyramid for American Samoa also shows much less of an indentation for the ages of 20 to 30.

Table 4.4 Population of Western Samoa, 1976[a]

Age	Males	Percent	Females	Percent	
0 - 4	12 995	8.5	11 651	7.6	
5 - 9	13 125	8.6	11 848	7.8	
10 - 14	12 220	8.0	11 407	7.5	
15 - 19	10 398	6.8	9154	6.0	
20 - 24	6457	4.2	5592	3.7	
25 - 29	3892	2.5	3989	2.6	
30 - 34	3294	2.2	3391	2.2	
35 - 39	3321	2.2	3355	2.2	
40 - 44	2885	1.9	2905	1.9	
45 - 49	2647	1.7	2655	1.7	
50 - 54	2294	1.5	2236	1.5	
55 - 59	1732	1.1	1628	1.1	
60 - 64	1240	1.0	1175	1.0	
65 - 69	871	0.6	844	0.6	
70 - 74	593	0.4	595	0.4	
75 +	675	0.4	919	0.6	
Totals	78 639	51.6	71 804[b]	48.4	151 983

[a] Adapted from Western Samoa Dept. of Statistics 1980.
[b] It appears that 1540 females were not listed in this age breakdown.

Table 4.5 Population of American Samoa, 1980[a]

Age	Males	Percent	Females	Percent	
0 - 4	2486	7.7	2300	7.1	
5 - 9	2156	6.7	2062	6.4	
10 - 14	2256	7.0	1947	6.0	
15 - 19	1878	5.8	1971	6.1	
20 - 24	1390	4.3	1667	5.2	
25 - 29	1152	3.6	1236	3.8	
30 - 34	1033	3.2	1033	3.2	
35 - 39	880	2.7	730	2.3	
40 - 44	806	2.5	697	2.2	
45 - 49	638	2.0	546	1.7	
50 - 54	521	1.6	556	1.7	
55 - 59	388	1.2	388	1.2	
60 - 64	333	1.0	302	0.9	
65 - 69	215	0.7	198	0.6	
70 - 74	122	0.4	115	0.4	
75 +	130	0.4	165	0.5	
Totals	16 384	50.8	15 913	49.3	32 297

[a] Adapted from U. S. Bureau of the Census 1982, 1980 Census, unadjusted figures of the Population: Characteristics of the Population, Number of Inhabitants: American Samoa.

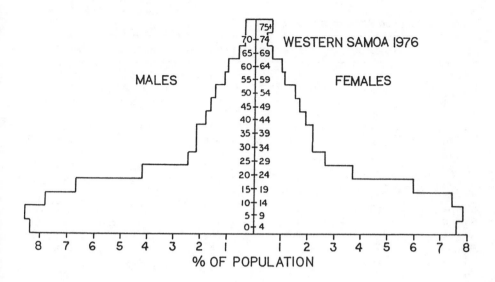

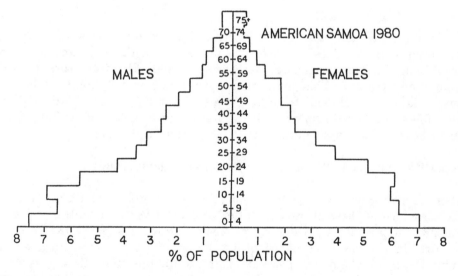

Fig. 4.3 Population pyramids for Western and American Samoa.

Table 4.6 Samoan Population of Hawaii, 1980

Age	Males	Percent	Females	Percent	Total
0 - 4	1096	7.8	1061	7.5	2157
5 - 9	1016	7.2	938	6.6	1954
10 - 14	929	6.6	1011	7.2	1940
15 - 19	892	6.3	862	6.1	1754
20 - 24	593	4.2	664	4.7	1257
25 - 29	569	4.0	598	4.2	1167
30 - 34	456	3.2	461	3.3	917
35 - 39	356	2.5	354	2.5	710
40 - 44	272	1.9	309	2.2	581
45 - 49	216	1.9	244	1.7	460
50 - 54	162	1.2	193	1.4	355
55 - 59	152	1.1	145	1.0	297
60 - 64	101	0.7	108	0.8	209
65 - 69	73	0.5	85	0.6	158
70 - 74	34	0.2	35	0.2	69
75 +	36	0.3	52	0.4	88
Totals	6953	49.2	7120	50.4	14 073

Source: U.S. Bureau of the Census 1983, 1980 Census of the Population, General Social and Economic Characteristics: Hawaii.

To complete the description of the age/sex structures of Samoan populations in the Pacific, Table 4.6 presents the data for age and sex groups of Samoan migrants in Hawaii, and Fig. 4.4 presents the population pyramid. This population represents primarily individuals who have migrated to Hawaii from both Western and American Samoa since the 1950s and their offspring. It is interesting to note the striking similarity between this pyramid and that for American Samoa.

Population dynamics: Western Samoa in the twentieth century

There is somewhat more known about the demographic rates that created the structure of these three populations for the twentieth century than in the preceding century. Data from Western Samoa suggest that during the first half of the twentieth century the population grew at an average annual rate of about 1.9 percent (Tofa 1980), but that there was considerable variability contained within the average. For example, it seems likely that in the early years of the century the growth rate was substantially less than 1.9 percent, perhaps as low as 1 percent due to the force of mortality. In the post-World War II period (1945-1951) the population grew at an average annual rate of 3.7, as fertility remained very high and mortality was substantially reduced. Between 1951 and 1971 population grew, on the average, 2.8 percent annually, and between 1971 and 1976 growth was only 0.7 percent annually as fertility dropped

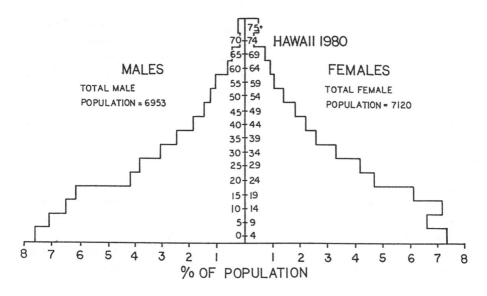

Fig. 4.4 Population pyramids for Samoans in Hawaii.

somewhat and out-migration became a determining factor (McArthur 1967).

Although registration of births and deaths was first introduced into Western Samoa in 1905 by the Germans, the inaccuracies of that system, and indeed most of the systems that followed for the next half of the century, make it difficult to estimate fertility and mortality rates accurately (McArthur 1967). In general, we know that mortality rates continued to fluctuate, as in the previous century, as a result of periodic epidemics. Mortality declined significantly during the first half of the century as a result of the increasing availability of medical care (Born 1970). Banister et al. (1978), using Brass techniques, estimate that in the early 1950s the infant mortality was around 100 per thousand; at the end of the decade it was around 80; and by the early 1960s, it was about 70 per thousand.

Using census data, adjusting for underregistration of adult deaths, and applying life table analysis, Banister et al. estimate that the life expectancy at birth during the early 1960s was about 65 for women and about 61 for men. "The average crude death rate for that five-year period (1962-1966) was about 8.7 per thousand population" (Banister et al. 1978:37). The numbers give an indication of the relatively low levels of mortality at mid-century.

The same data limitations apply to fertility statistics. Nevertheless, fertility must have been exceptionally high during the nineteenth and first half of the twentieth century, perhaps in the range of a total fertility rate (TFR) of 8 to 9, in order to have offset the losses through mortality.

McArthur estimates that in 1951 the general fertility ratio (number of births per 1000 women of reproductive age) was as high as 224.

Banister et al., exploring the results yielded by several different estimation procedures (Palmore regression, own-children), tentatively suggest that the crude birth rate was about 45 in 1961 and the total fertility rate was about 7.5.

These fertility and mortality statistics constitute an explosive demographic picture for Western Samoa, as described by McArthur:

> Even more significant for a population where two-thirds of all males aged 15 years and over in 1956 were engaged in agriculture is the rate at which the work force will increase: in the absence of any marked increase in mortality, only emigration will prevent increments of about 4 per cent a year in the number of males aged between 15 and 59 years, and the 'village' and 'other' agriculture which supported less than 17 000 males in 1958 may be the only means of livelihood for at least twice this number within fifteen years. (McArthur 1967:141)

While there were substantial untapped resources in Western Samoa in the form of arable land (Pirie 1970), the major mechanism for adjustment to population growth and rising expectations has been outmigration.

Although the early decades of the twentieth century were certainly characterized by substantial movement of population among the Samoan islands (McArthur 1967), it seems likely that during the first half of the twentieth century, population loss or gain in Western or American Samoa was negligible (Park 1980). After 1950 permanent out-migration from Western Samoa changed the picture dramatically; this is reflected in the age/sex structure of the population (Fig. 4.3), the reduction in the population growth rate, and in the growing numbers of Western Samoans living outside their country. Pitt (1977), for example, reports that in 1951 there were 1336 Samoans in New Zealand, in 1961 there were 6481, in 1971 there were 22 198, and in 1976 there were 27 400. Likewise, McArthur reports that government statistics show that "between 1951 and 1956, there was a net loss through migration of 1453 males and 704 females" but that "inmigration was probably far more extensive than this, perhaps twice as high" (1967:129). Similarly, Ward and Proctor (1979) point out that although the natural rate of increase (NRI) in the 1970s was about 3 percent, the actual growth was only about 0.7 percent, providing an indication of the volume of out-migration. As is explored in subsequent sections, this migration stream was facilitated by an extremely strong extended family network, close ties culturally with American Samoa, and free access from American Samoa to the United States. In general, the movement was from the more remote villages of Upolu and Savai'i, to the capital city of Apia where papers were processed and arrangements made, and then to New Zealand or American Samoa.

Population dynamics: American Samoa in the twentieth century

Although starting from a much smaller population base than Western Samoa, the population of American Samoa showed substantial growth in the early decades of the twentieth century. Park estimates that during the first two decades the population grew at an annual rate of 1.8 percent, and grew at a steadily increasing rate during the subsequent decades as public health services increased and mortality dropped (Park 1979). By the 1940s and 1950s, he finds that population growth was "equivalent to an annual rate of nearly 4 percent due to both augmented natural increase and tremendous natural influx, mostly from Western Samoa" (Park 1979:15). The population of American Samoa grew from 5679 in 1900 (U.S. Dept. of Navy 1922) to 29 190 in 1974 (American Samoa Development Planning Office 1976).

Shifts in fertility and mortality combined to create a growth peak in the early 1950s. During this period the crude birth rate was about 50, the total fertility rate almost 7, and the child/women ratio about 1000. During the same period, according to McArthur, "deaths to infants and small children were probably halved, and the risks to young adults also lessened" (1967:153). Born (1970) suggests that expectancy of life in 1940 in American Samoa was about 50 years; by the mid-1950s it had risen to over 60 years. These rates, combined with the influx of Western Samoans during the period, created what can accurately be called a population explosion.

It was during World War II and the immediate postwar years that the differing economic and political situations of American and Western Samoa led to very different population dynamics. During World War II Pago Pago, the capital of American Samoa was " . . . a hub of activity. There was almost unlimited demand for labor" (Park 1980:10). Both natural increase and immigrants could be absorbed by a labor market swollen by war-related and service-support jobs. At the end of the war, the labor market shrank. Administrative transfer of American Samoa from the Department of the Navy to the Department of the Interior further depressed the economy. Large-scale emigration occurred during the 1950s, estimated by some to be as high as 3.6 percent annually. Concurrently, fertility began to drop, probably as a result of both modernization of the population and the severe economic conditions. By 1955, the CBR had dropped from 50 to about 42. Park (1980) suggests that this decline was also related to emigration of a large number of individuals in the childbearing years.

It is interesting to note that the flow of migrants from Western to American Samoa in the 1940s was reversed in the 1950s. The poor job market, a severe drought, and other unfavorable economic conditions not only increased the stream of Samoan migrants to Hawaii but also pushed Western Samoans living in American Samoa to return home. During the 1970s return migration to Western Samoa leveled off, and the balance of the flow was again toward American Samoa, and, for many American Samoans, toward Hawaii. By 1980, the population size of American Samoa had reached 33 297 (U.S. Bureau of the Census 1981). Fertility remained relatively high, with a probable TFR of about 5.5. Further reductions in mortality occurred as high-quality health care became more widely available. And migration, as in Western Samoa, played an increasingly dominant role in determining the demographic picture of the country.

THE PRESENT POPULATION SYSTEM OF SAMOANS IN THE PACIFIC

Theoretical introduction

The framework presented in Fig. 4.1 indicates the major inputs to the individual decisions that result in aggregate demographic rates. As Hull (1983) has pointed out, however, the treatment of demographic events as the outcome of individual decisions is a relatively recent development in social demography, and there is still a good deal of controversy about methodology and theoretical interpretation. Since the early 1960s social scientists have explored how people make decisions about fertility, nuptiality, and migration. Although the different disciplines emphasize different points in the decision-making process and use different labels to describe the process, there appears to be a consensus emerging. The individual, rather than having his or her behavior determined by cultural, economic, or household factors, is now seen as maximizing his or her own well-being within the context of those factors. Indeed, cultural values and norms may be seen as providing guidelines or problem-solving routines of the making of certain decisions. Rather than consider a whole universe of options for a particular decision, the individual may consider only a selected range of options. Similarly, some decisions may involve considering a very wide range of options, while others may be resolved by recourse to general, culturally provided solutions (McNicoll 1980).

In general, however, we assume that for any specific decision, the option selected will be that course of behavior for which the benefits exceed the costs by the greatest amount. The ecological, household, and cultural factors listed in Fig. 4.1 both constrain the range of options considered, and influence the perception of costs and benefits associated with various courses of action. As can be seen from this figure, the household has the most direct impact on individual decision making. It is the context of demographic (as well as economic and social) decisions. Household structure, however, is also shaped by those decisions. Figure 4.5 indicates the way in which demographic decisions influence, and are influenced by family structure. For example, the decision of a young person to migrate is clearly influenced by position and status within the family, access to resources, and the perceived financial gains to be achieved by migrating. Once the individual has migrated, however, in

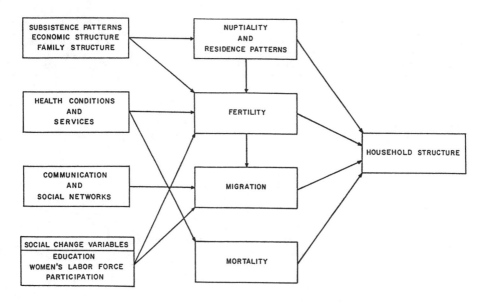

Fig. 4.5 Social and demographic determinants of household structure.

addition to changing his or her own situation, he or she has changed the demographic structure of the family left behind.

The household also provides a unit of ecological adaptation: migration out of the household provides a mechanism for removing surplus labor as well as linking up with a new labor market. On the other hand, absorption of new members via adoption or fosterage of children or shorter term visits of adults may occur where there is a shortfall in the family labor force, or when, as Park (1980) found, in poor economic times, families take in individuals or merge with other families.

The focus then is on the decisions of individuals within a particular family context. The challenge of migration to this kind of analysis is that it frequently puts an individual in a new family situation, a new sociocultural setting, and a new ecological environment simultaneously. Not surprisingly, the demographic decision making process of the migrant is frequently drastically altered. In order to look at changes in demographic decision making, this review examines demographic decisions and demographic rates in a series of environments, along a path frequently followed by Samoan migrants (see Fig. 4.6). It starts in the extremely remote village of Uafato, on the island of Upolu, Western Samoa, moves to Salamumu, a village closely linked to Apia, then to American Samoa, and Hawaii.

Demographic patterns and decisions: Western Samoa

Western Samoans on the islands of Savai'i and Upolu live in a wide range of social and ecological environments. From the geographically remote,

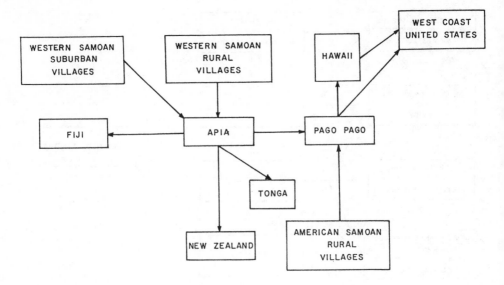

Fig. 4.6 Samoan migration streams.

economically isolated, and socially traditional village of Uafato, on the
island of Upolu to the suburban village of Salamumu, to the capital city of
Apia, differences can be seen in the way social, economic, and demographic decisions are made. Fisk (1962) has suggested that the nature of
the linkage between a village and the market center is crucial to understanding economic, social, and demographic change, and individual
decisions. He developed his theory in the context of subsistence-based
Pacific island economies, and it seems especially well suited to a
comparative study of demographic decision making in Western Samoa.

Many villages on Western Samoa, especially the more remote ones,
have traditional economies, relying on subsistence bush-fallow agriculture
and fishing. Nevertheless, within this generally traditional and relatively
stable situation, important regional differences are starting to emerge.
Within the matai and the 'aiga system, which was described previously,
there is evidence that the extended family as the production and
consumption unit—the unit within which tasks are allocated and resources
are shared—is starting to break down. Indeed many young men report that
one of the advantages of wage labor is that it introduces the possibility of
accumulating some wealth of one's own.

Although most villages in Western Samoa have some degree of contact
with the world beyond the village (they export copra and other
commodities, buy imported goods, go to school and church, visit Apia,
listen to foreign news, and use money) they still maintain "the moral code
of their forefathers, maintaining inherited Samoan institutions, family and
government structures" (Lockwood 1971:xvii). It is this interaction of the

various Samoan villages with the market center of Apia that is a major factor in changing the context of demographic decision making. In his 1971 study Lockwood attempted to test some of Fisk's hypotheses in four Samoan villages. Uafato was the most isolated, least modernized of the four. (All references to Lockwood are to his excellent 1971 study, *Samoan Village Economy*.)

Uafato

Located in a remote coastal area of Upolu, it is isolated from the rest of the island by the mountainous Fagaloa district. Lockwood reported that in 1971 Uafato was the only village on the island of Upolu that could not be reached by road: "whale boats and canoes were the principal means of transport" (1971:50). A road linking Uafato and other nearby villages with the rest of Upolo has since been completed, but it is a very rough road and the sea remains the primary travel route to Apia. In 1980, there were two boats that made relatively frequent trips to Apia, and a significant number of the villagers reported one or two trips per week to the market center. The trip, however, remained time-consuming, with the voyage by sea taking about one and a half hours and the connecting bus ride to Apia taking another hour and a half. (Data for 1980 were collected by the author and D. Pelletier during a brief village survey of household structure and consumption patterns.)

Since Lockwood's measures of strength of economic linkage with the market sector were first *communications* (distance measured by time, distance, and cash cost to Apia) and secondly *marketing services* (facilities available for exchanging village output, as well as the cost and availability of market goods for purchase), it was clear that Uafato was weakly linked to the market sector. The nearest trade store where villagers could exchange their copra and produce for market commodities was over two hours away by foot or canoe and transport to Apia was even more difficult. Another indicator of the linkage of a village with the market center is the pervasiveness of cash in the local economy. Lockwood found that although villagers had been producing and marketing copra since the 1890s, less than 30 percent of the total income of the village was in cash.

Indeed, the economic pattern that Lockwood found and that remained to a great extent in 1980 was one in which men cultivated taro, collected breadfruit, and fished for their families. The women cooked, cared for their large number of children, and did weaving of pandanus mats for personal use and for sale in Apia.

The demographic picture in Uafato that Lockwood reported was a potentially explosive one. The size of the village population was 286 and growing rapidly; more than 56 percent of the population was under 15 years of age. Lockwood estimated that the village had an annual average growth rate of 3.5 percent between 1950 and 1966.

Over a period of about 15 years, the market sector in Apia grew, transportation facilities improved, and radio became widespread. In going back to Uafato in 1980, the similarity to the social, economic, and demographic situation that Lockwood described earlier was surprising. Men were working between 25 and 30 hours a week in subsistence activities, women were still caring for large numbers of children, and most individuals were eating the traditional diet and buying very little in the village store. Trips to Apia were more frequent than 15 years ago, but still somewhat unpredictable, as well as time-consuming and expensive. My impression was that the amount of copra and pandanus mats produced for market had not increased significantly since 1965 and may have actually decreased.

Furthermore, village size was still about 280. The 1976 census also estimated village size at about 280. This gives us a clue as to one possible household adaptive strategy. In a sense, villagers are exporting population rather than cash crops such as copra or other commodities. Nearly every family in Uafato has several members living in New Zealand. Of the households that we surveyed, all reported at least one relative who was a migrant. All reported that they received money from migrant relatives. Some reported that money came on a regular basis; others said the migrant relatives sent money only when asked and for special ceremonial occasions. While we surveyed only about two-thirds of the houseshols of the village, there were no obvious social or geographical divisions in the village that would lead me to expect that the households we covered would be atypical.

In one household that was typical of the ones we surveyed, four of the seven brothers and sisters of the head of the household had migrated. Three had gone to New Zealand, and one to Pago Pago. All sent money regularly. The head of the household reported that he went to Apia about once a week, sometimes to take things to market, sometimes to "see about business." His wife reported only several trips a year. She was 34, had seven children, and was pregnant with an eighth. Their housing was traditional and consisted of a cluster of three fales—a cooking fale, a large living fale, and another small living fale. Their diet was traditional as well, with very little being purchased; the staples were taro and fish.

The feeling in Uafato is one of a quite striking stability that is made possible by exporting population. It is possible for things to remain the same because significant numbers of the population migrate (more or less permanently) to Pago Pago or New Zealand. The relatively few commodities that the villagers do buy are bought with the remittances sent home by migrant relatives. This strategy of household and village adaptation has also provided a fairly stable context for demographic decision making. Population growth has been avoided by substantial out-migration and fertility has remained high. The geographical setting of the village has isolated it from the modernizing influences of involvement with the market sector.

Salamumu

This village of 377 people on the southwest coast of Upolu has a very different economic and demographic structure than Uafato. The village is easily accessible to Apia. Being linked by good-quality roads for most of the 33 miles, the trip takes about an hour and a half. Despite the closeness of ties with the Apia urban center, the village remains traditional in many ways.

There is no doubt, however, that linkages with Apia both socially and economically are increasing. Our most recent survey shows that there is now a reliable daily bus service into Apia in the morning and back in the evening. There are actually a fairly large number of young men from Salamumu who commute on a daily basis or who work in Apia during the week and return home to the village on the weekends. From our 1982 survey of the village, we discovered that approximately one half of the men in the age group 18-35 were working in Apia a significant portion of the time. As a matter of fact, this had led to a new housing pattern in Apia. In our survey work there we found a large number of households of young people between the ages of 18 and 26 or 30 who were all from a single village and lived in a fale in Apia during the week.

In addition to the commuting and short-term residence in Apia, there is also a lot of movement within the village. Genealogies and a household census were completed in 1979 and then again in 1982. There was some demographic change in over 80 percent of the households during this period. The changes included births, deaths, movement to another household in Salamumu or to Apia (on a temporary or permanent basis), and movement to Pago Pago or New Zealand. A large portion of the movement of people from one household to another is the adoption or temporary shifting around of children. In Samoa, as in many societies, the supply of children to a particular household is broader based than the offspring of an individual couple because of the social institution of adoption.

Adoption or fosterage may be used to fill in gaps in the household age/sex structure. Samoans define appropriate behavior for males and females of various ages. For example, the task of collecting fallen coconuts is allocated to children of about 8 to 10 or 12. If a household doesn't have a child of the appropriate age, one can be brought in by adoption or fosterage. It is a social mechanism for adjusting the supply of household labor to the demand for labor.

Economic activities in Salamumu are quite different from those in Uafato. Traditional subsistence activities play a considerably less important role, with many of the men of Salamumu working in Apia. A much larger portion of the diet consists of purchased foods, including both tinned products and fresh meat. Unlike Uafato, flour and rice are both used widely. Most people still eat taro, breadfruit, and fish as staples, but they are no longer the main components of the diet. Fewer men fished, fewer women wove pandanus mats, and there were no roughhewn canoes in the village such as were made in Uafato. Salamumu is a village very closely linked with the market sector, and this linkage has transformed the village.

For example, it is clear that the demand for new goods and services has been escalated by contact with the market. The demand is being met by individuals commuting to wage-paying jobs in Apia. There certainly is no evidence of increased production of the normal subsistence goods for sale in the market. Faced with the option of attempting to increase production by working longer hours in the traditional economy, or by working in Apia, an increasing number of villagers are choosing to work in Apia. The geographic situation of the village, as well as the relatively good quality of the roads make this a feasible option. In contrast, in Uafato, linkage is very weak to the market sector in Apia. The contact with the monetized economy comes via the remittances sent home by migrants. This sort of linkage with the market sector does not imply major socioeconomic or demographic changes in the structure of the village.

In summary, the possibility of paid employment in Apia and residence in Salamumu, has brought the city to the village. In effect, the city of Apia is now part of the decision-making context in Salamumu. It seems likely that the lower fertility that characterized Apia will diffuse to suburban towns like Salamumu.

The theoretical explanation for the difference between Salamumu and Uafato lies in the economic approach to fertility decisions that were discussed earlier. Many of the ecological and sociocultural factors that contribute to the decision to have a larger number of children in Uafato, and to a lesser extent in Salamumu, have been transformed in Apia. In traditional villages such as Uafato, many of the costs of children, including education and child care are diffused throughout the extended family and even the village. This tends to be less true in urban areas. Second, in Uafato there are virtually no labor market opportunities for women, so the opportunity cost of child bearing is low. Third, children in the context of subsistence agriculture make significant contributions to the household economy. The cost-benefit calculation points toward high fertility which is supported by cultural norms and marriage patterns. The consequences for population growth of this pattern thus far have been mediated by the outmigration.

By contrast, in Apia there are numerous employment opportunities for both men and women in the commercial, government, and service sectors. These opportunities serve to pull both men and women into Apia as well as providing employment to lifetime residents and commuters. Table 4.7 shows the flow of women migrants as reported by the 1976 Census of Population and Housing of Western Samoa. It is interesting to note that the inflow of women to Apia, combined with the out-migration of men from Apia to American Samoa or New Zealand has created marked

Table 4.7 "In" and "Out" Female Lifetime Migrants between Place of Residence and Place of Previous Residence, 1971 - 1976

Area	Lifetime in-migrants	Lifetime out-migrants	Net lifetime migration
Apia urban area	1546	537	+ 1009
Northwest Upolu	876	678	+ 198
Rest of Upolu	394	1120	- 726
Savai'i	364	845	- 481

Source: Western Samoa Dept. of Statistics *Census of Population and Housing, 1976* Vol. 2:50.

imbalances in the sex ratio of age groups that could be looking for marriage partners. Wander (1971) points out that women historically have outnumbered men in most of the fertile age groups and that in Apia, the imbalance of potential mates is especially severe as a result of a net outflow of men and a net inflow of women. In Apia in 1966 there were only slightly more than 50 unmarried men between the ages of 20 and 35 for every 100 women between the ages of 15 and 30.

The elevated age at marriage, the increased costs of children in the urban setting, and the decrease in the potential contribution of children to the household economy all serve to change the nature of fertility decisions. It is not surprising that fertility has always been the highest on Savai'i, the next highest in the nonurban areas of Upolu, and the lowest in the Apia urban area.

In summary, the sociocultural and ecological environment of Western Samoa supports high rates of fertility and migration. Just as the extended family diffuses the costs of children, the extended family both at home and in migrant populations shares the costs of migration and supports the adaptation of the migrant to the new situation. Regional differences within Western Samoa reflect the changing nature of fertility and migration decisions, from relatively remote traditional areas with subsistence-based agriculture to highly commercialized urban areas.

Demographic patterns and decisions: American Samoa

The individual decisions of American Samoans that led to the population structure presented in Fig. 4.3 have occurred in a rapidly changing socioeconomic and cultural context. In some ways, American Samoa represents a point intermediate between traditional and modern ways of life. While wage labor is an important part of the economy, transportation and communication systems are relatively well developed, and there is frequent travel to the United States and Hawaii, American Samoans still remain traditional in many ways.

The social structure described for Western Samoa has been

characteristic of American Samoa as well. Hecht et al. in the previous chapter examined the extent to which this traditional system may have been modified since World War II. Certainly there is a major difference between the economies of American and Western Samoa in the relative degree of involvement in a wage economy. Despite the overall higher level of monetization and wage labor in American Samoa than Western Samoa, there are important regional and subgroup differences that are essential in considering the demographic decisions of individuals (Hecht et al., Chapter 3). Not surprisingly, these regional differences have led to differences in demographic decisions.

The importance of migration in determining the overall demographic structure of a population can be seen in Table 4.8, which presents decennial population growth in American Samoa from 1900-1970 (Park 1980). The spurt in population growth that occurred between 1960 and 1970 (average annual growth rate = 3.08) had been reduced by almost 50 percent to 1.8 by 1974. Park suggests that this reduction is not due to a reduction in the natural rate of increase; it is likely that both fertility and mortality fell slightly during the period. Rather, it was due to the substantial acceleration of migration of Samoans to Hawaii and mainland United States. Migration continues to be an option selected by many American Samoans, as indicated by their growing numbers in the United States (Levin and Hayes 1983). Unfortunately, it is not possible to determine these migrants' home villages in American Samoa. It is possible, however, to look at patterns of internal migration within the American Samoan islands.

The 1974 U.S. Bureau of the Census collected data on place of birth and current residence in American Samoa (1974). The pattern of mobility provides an interesting comparison with the patterns described for Uafato, Salamumu, and Apia in Western Samoa. Manu'a, the most remote region in the American Samoan group has had substantial emigration and virtually no immigration. Of all Manu'ans living in American Samoa only

Table 4.8 Decennial Population Growth of American Samoa, 1900 - 1970

Year	Population	Index (1900 = 100)	Average annual growth rate (%)
1900	5679	100	
1912	7251	128	2.06
1920	8058	142	1.33
1930	10 055	177	2.24
1940	12 908	227	2.53
1950	18 937	333	3.91
1960	20 051	353	0.57
1970	27 159	478	3.08
1974	29 190	514	1.80

Source: Park 1980:12.

about half still reside in Manu'a. In contrast, relatively few individuals who are born on Tutuila move to other parts of American Samoa. Like the "commuters" in Salamumu, "Tutuilans may not need to move within the island to be closer to jobs and schools, as they have easy access to these necessities of modern life" (Park 1980:9). On the other hand, large numbers of Manu'ans, like the residents of Uafato, have left their birthplace.

There appear to be significant regional differences in fertility as well as migration. There are many inputs to the fertility decision-making process that underlie differences among regions: economic patterns, educational levels, age at marriage, migration, and household structure. For example, in analyzing fertility data from 744 women living in American Samoa, Harbison (1979) found that fertility was lowest in the Pago Pago area, highest in the rest of Tutuila, and intermediate in the Manu'a group and North Shore region (see Table 4.9 and Fig. 4.7). These results may seem initially puzzling in that it might be expected that Manu'a would have the highest fertility, being the most remote, least modernized, and most traditional in some senses. There are a number of possible explanations. One is that fertility, like a number of other variables influenced by social change (McGarvey and Baker 1979), shows a curvilinear relationship with exposure to modernization. If reduced fertility is considered to be an adaptation to the modern environment, then the relatively high levels of fertility found in the intermediate areas may be interpreted as an initial response to modernizing influences (including increased household income), before the adaptation process has led to the lower levels of fertility seen in the bay areas. It also may reflect a lessening of the budget constraint. According to the economic theory of fertility, if a household has more disposable income, if children do not become more costly, and if values remain constant, fertility will rise.

Like differences in economic structure, differences in educational patterns can affect fertility decision making. The availability of education to women, whether or not it is compulsory, and the average duration of school attendance by women, are all important factors in determining the fertility impact. In a study of the impact of education on fertility, Harbison et al. (1981) found major differences between Western and American Samoa.

In Western Samoa, where education is primarily in the context of the local village, education within the lower and the middle range does not seem to be an indicator of social class. In fact, the traditional 'aiga system that dictates the sharing of food and resources, assures a degree of homogeneity within the society. In the local villages it seems unlikely that there are significant differences in health, access to medical facilities, or nutrition that would lead to systematic differences in the supply of children. Furthermore, since there is not a significant wage labor market for most females, education does not constitute an investment in future earning capacity; therefore, it does not increase the relative value of the

Table 4.9 Age-specific Fertility Rates by Place of Residence in American Samoa

Age	Pago Pago Bay area		Tutuila and other		Manu'a and north shore	
	Rate	PYL[a]	Rate	PYL	Rate	PYL
15 - 19	55.9	1359	85.5	1695	42.3	614
20 - 24	212.0	1231	262.3	1540	228.8	590
25 - 29	242.8	1071	318.6	1331	278.0	554
30 - 34	219.0	918	273.0	1128	252.9	518
35 - 39	170.8	767	182.7	958	211.3	497
40 - 44	80.2	636	100.7	765	132.7	422
Over 45	16.1	497	28.0	607	29.2	343
TFR[b]	5.0		6.3		5.9	

[a] Persons Years Lived.

[b] Total Fertility Rate.

Fig. 4.7 Age-specific fertility rates by place of residence in American Samoa.

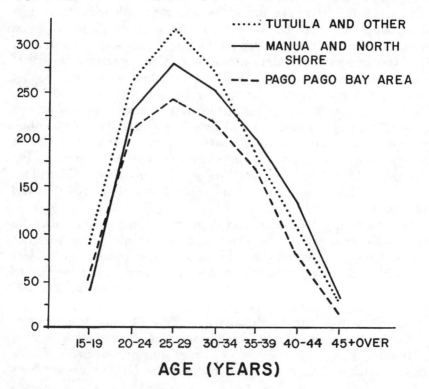

wife's time or the relative cost of children. Finally, education provided within the village context by native Western Samoans is unlikely to reduce the preference for children by changing attitudes and values.

In American Samoa, on the other hand, the prevalence of wage labor, the availability of jobs for women, and the greater heterogeneity of the society, have transformed the nature of the relationship between education and fertility.

Economists have suggested that the education of women increases the relative costs of children by increasing the value of the wife's time, and consequently the relative cost of children. In American Samoa, however, household structure and definition of family roles provide many alternative caretakers of children. This aspect of the social structure reduces the cost of childbearing to the educated women and permits a quick return to the labor market. Since there is an active labor market for women, and education improves a woman's chances in that market, it seems reasonable to assume that the major way in which education is related to the demand for children on American Samoa is through income. If education lessens the budget constraint by providing women with marketable skills and additional income, then we would expect the effect of education on fertility to be positive. If both tastes and costs remain constant while income increases, then fertility will increase as well.

Figure 4.8 presents own-children age-specific fertility estimates for three educational levels for Western and American Samoa. It should be noted that there were very few women in the highest educational level in Western Samoa, so interpretations based on that group should be made cautiously. The 0-6 years of education group and the 7-12 years group contain substantial numbers of women; however, these two groups of women were extremely close in both the age-specific pattern of childbearing and the overall rate. This finding is consistent with the absence of a significant labor market for women. In the context of traditional village economy, the impact of education on fertility is very slight (Lockwood 1971).

In American Samoa, the regional differences in fertility seen in Fig. 4.7 are consistent with the education/fertility relationship. The women with the highest levels of education are resident in the Pago Pago area where fertility is the lowest; similarly, the women with 7-12 years of education tend to be residing on Tutuila. Their fertility is significantly higher than women on Manu'a, where the educational levels are low (see Fig. 4.8). These findings for Tutuila are consistent with an environment that provides a significant wage labor market for women and a monetization of the economy but, at the same time, traditional extended family structure provides alternative caretakers for the children if the mother chooses to work. A reasonable interpretation of the low fertility of the most educated group is that these women are the most modernized, are most likely to participate in the labor force, and most likely to have lower ideal fertility.

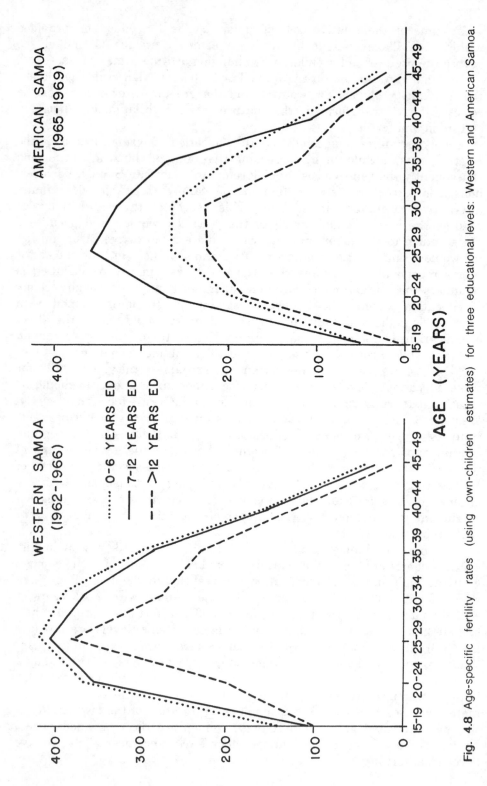

Fig. **4.8** Age-specific fertility rates (using own-children estimates) for three educational levels: Western and American Samoa.

88

In summary, the various environmental and household factors that affect individual fertility decisions appear to be changing at different rates in different regions of Samoa. The extended household still is appropriately viewed as the primary context of decision making, and the link between fertility decisions and migration decisions is especially close. Practically because members of the extended family, living both in American Samoa and Hawaii, provide a very effective facilitating network for migration, as well as providing a pool of caretakers for children, fertility in general remains relatively high. Migrants to Hawaii, however, find themselves in a very different sociocultural and economic environment, and thus have transformed the demographic decision-making process.

Demographic patterns and decisions: Samoans in the United States

For the potential migrant from American Samoa, there are numerous inputs to the decision. The availability of funds for the trip, the family situation at home, and the presence of relatives already resident in Hawaii or on the mainland, all feed into the decision. Both the economic and demographic situation at home and the situation at the destination, perceived by the potential migrant, will be important considerations. Data from the 1980 U.S. Census of Population provide ample support for the assertion that a large number of Samoans have made the decision that they will be better off by migrating to the United States. Preliminary data indicate that in 1980 there were 20 096 Samoans in California, 14 168 in Hawaii, 1830 in Washington, 764 in Utah, and 563 in Texas. A more recent analysis of census data in the context of known patterns of migration suggests that these numbers are probably between 10 and 17 percent too low (Levin and Hayes 1983).

Samoan migrants to the United States, like almost all other migrants cite improved educational opportunities and a better living as the major reasons for migrating. Some also mention the advantages of escaping the cultural requirement that support, financial and otherwise, be provided to members of the 'aiga. The possibility of accumulating some wealth of one's own appears to be an input to the decision to migrate as well.

Ironically, educational, employment, and welfare statistics from both Hawaii and California indicate that many of the improvements in living that were anticipated may not have materialized. In both Hawaii and California, Samoan children have a great deal of difficulty in school, and unemployment and dependence on welfare are extremely high. In California, Oakey reports that "Samoans, like most other groups come to improve their educational and economic opportunities. However, they arrive with so little preparation, no marketable skills, and such weak English language capability that almost 75 percent receive some sort of welfare assistance in order to survive" (Oakey 1980:5).

Not all migrants, however, experience these hardships to the same

Table 4.10 Age-specific Fertility Rates for American Samoans, Samoan Migrants to Hawaii, and Hawaiians[a]

Age	1969-1971 American Samoa	1975 Samoan migrants in Hawaii[b]	PYL	1975 OEO survey of Hawaii	Female population
15 - 19	53.2	62.1	644	53.7	40 821
20 - 24	293.3	225.3	537	126.6	43 017
25 - 29	322.7	261.8	477	131.5	37 731
30 - 34	228.0	204.6	391	73.8	30 795
35 - 39	193.5	122.8	334	28.5	23 323
40 - 44	71.9	52.8	265	5.2	22 430
45 - 50	18.9	41.5	193	0.8	24 622
TFR	5.9	4.9		2.1	

Source: Harbison and Weishaar 1981.

[a] Because the age-specific rates reported in this table are estimated using different methods depending on the data available, and do not refer to precisely identical periods of time, the comparisons and contrasts can only be suggestive. Because the rates based on the 1975 Samoan migrant survey reflect births to women over their reproductive period, the figures do not represent a 1975 rate, but rather a "cohort rate" for migrant women.

[b] The age-specific fertility rates for Samoan migrants in Hawaii were calculated by a team headed by Baker in 1975. Births to women by age were tallied from reproductive histories. This number was divided by person years lived (PYL) or the number of years spent by Samoan migrant females in a particular age category.

extent. Just as there is variability among Western Samoan villages and American Samoan villages, migrants to Hawaii enter different sociocultural and economic situations, depending on what region of Oahu they settle in, and also their own unique socioeconomic and demographic characteristics.

The fact that migrants are not a random sample of the population that they leave makes the analysis of the impact of migration on fertility somewhat complicated. For example, looking at Table 4.10 it can be seen that the fertility of Samoan migrants in Hawaii is intermediate between that of Hawaiians overall and American Samoans, but much closer to that of American Samoans.

As noted previously, however, there are important differences in the types of communities to which Samoans migrate. Of the four communities that were studied by Harbison and Weishaar (1981) two were characterized as traditionally Samoan, and two as more modern. In summarizing the differences between the more and less traditional groups, they suggest:

> The more traditional communities are each linked to a single church and Samoan chief. One of these two communities is, in fact, an incorporated church village, separated physically from other neighborhoods. Here, the traditional Samoan practices of prayer hour, curfew, and chiefly authority

over residents are enforced and maintained. In contrast, the less traditional areas are marked by diverse neighborhoods and public housing.

The residents belong to several different Samoan churches and organizations, but membership in these social groups does not vary with areas of residence within three locales. Thus residents in less traditional communities are not tied to them socially, politically, or religiously. (Harbison and Weishaar, 1981:269)

It is not surprising to find that the total fertility rate for women living in the traditional villages is considerably higher (5.3) than that of women living in the less traditional areas (4.3). Thus most of the reduction in fertility of Samoan migrants as a whole is accounted for by the migrants to the less traditional areas. This still does not tell us, however, whether this reduction in fertility represents an adaptation to the new environment, or whether migrants to the less traditional areas had lower fertility even prior to migration. Selection does seem to be playing an important role, since when age-specific fertility rates are calculated for the premigration period, migrants to both types of villages had TFRs below American Samoa as a whole. Furthermore, migrants to less traditional communities had substantially lower premigration fertility than those Samoans who went to traditional areas.

In summarizing the population dynamics of Samoans in Hawaii, many of the same considerations are relevant as were discussed for the villages in Western and American Samoa. If fertility in the traditional areas of Oahu remains at its present high levels, population pressure will necessitate substantial out-migration in search of jobs. This migration, however, will mean leaving behind the traditional support structures of the extended household, the 'aiga, and the church-related village. These characteristics of the traditional villages may be seen as isolating villagers from modernization; nevertheless, they have proved to be a relatively successful adaptive strategy to a new environment.

The situation of Samoans in California is somewhat different; nevertheless, a similar adaptation model for understanding demographic and other biocultural changes seems to be appropriate (Pawson and Janes 1983). Like Samoans in Hawaii, Samoans in California tend to live together in relatively tightly knit, regionally defined communities. Pawson and Janes point out that many of these areas were associated with U.S. Navy bases where young Samoans "were transferred when the U.S. Navy terminated their operations in American Samoa in the early 1950s" (1984:2), and this migration history has led to certain differences between the California Samoan communities and the Hawaiian Samoan settlements. The traditional values associated with the 'aiga and family-related economic reciprocities remain to some extent, however. As economic pressures increase, and Samoans move into the larger society, these values, as well as the typical demographic patterns, will tend to disappear.

SUMMARY: SAMOAN POPULATION DYNAMICS

Migration, in the context of high fertility, relatively low mortality, and limited natural resources, plays a predominant role in Samoan population dynamics. As an adaptive strategy (viewed on the aggregate level) it has the advantage of drawing off surplus labor and increasing foreign exchange through remittances. The disadvantages, however, are that it leads to a population with a high dependency ratio, from which the relatively more skilled and educated individuals in the most productive age groups may be missing.

There is evidence, however, of the beginning of a fertility decline in Apia, and a moderate decline of fertility in Pago Pago. The factors that have contributed to this decline, that have led individuals to decide that they will be better off if they have fewer children, are increases in labor force participation of both men and women, involvement in a cash economy, higher levels of education, and reductions in the strength of the extended family and the matai. It seems likely that, as in Salamumu, as the modern, wage labor economy of the city becomes part of the life of more and more villages, fertility will continue to drop.

Chapter 5

Mortality Patterns and Some Biological Predictors

PAUL T. BAKER
DOUGLAS E. CREWS

The rapid growth of the total Samoan population in recent decades was primarily due to a decline in crude mortality rates and relatively high fertility rates. Such rapid declines in mortality have been common among traditional populations where food supplies are adequate and effective public health programs are introduced. The reduction in mortality rates is generally greatest among infants, with a progressively lower reduction in rates among older age groups (Preston 1976; Teitlebaum 1975). The shift in the age structure of mortality occurs primarily as a consequence of the declining levels of mortality from infectious disease. Thus, populations even in low per capita income countries often have age- and cause-specific mortality patterns that resemble those of the higher per capita income countries.

While such transitions are similar in many respects, substantial variations exist in the contemporary cause-specific mortality rates of these populations and the rate of transition (Teitlebaum 1975). Some of the current variations in mortality patterns have relatively obvious causes, such as the presence of treatment-resistant diseases like malaria and that in those populations that now utilize machinery, automobiles, and agricultural chemicals, death from accidents and trauma become more common. On the other hand, why the number of people dying from the chronic and degenerative diseases varies is generally obscure. For example, there is no satisfactory explanation of why such groups as the Pima Amerindians of Arizona (Bennett and Knowler 1979; Bennett et al. 1976) and the Micronesians of Nauru (Zimmet 1979) have developed mortality rates from type II diabetes that are many times those reported for other modern groups and populations. Neither is it clear why people such as high-altitude Amerindians who have migrated into the coastal regions of Peru have middle-aged cardiovascular-related death rates that are much lower than the local sedente population (Dutt and Baker 1981).

As may be deduced from data presented in the previous chapter, Samoan subpopulations, whether located in Western Samoa, American Samoa, or California, all appear to have life expectancies approaching those found in the modern industrial countries. Given that these areas differ in wealth from among the lowest per capita income in the world to among the highest and vary greatly in health care facilities, such a similarity in life expectancy might not be anticipated. The frequency of such health-risk factors as obesity, hypertension, and type II diabetes (discussed elsewhere in this volume) also varies substantially among these Samoan groups. Thus, a detailed comparison of the mortality transition and current causes of death for each could provide numerous insights into the causes for particular mortality patterns. Such a comprehensive comparison was not possible as part of the present studies because the available data were incomplete. Since the American Samoan death registry and censuses were quite comprehensive, it was possible to examine in detail the changes in mortality and its causes from 1950 to 1981 for that population (Crews 1985). An analysis of recent Western Samoan mortality (Crawley 1984) and some Hawaiian and Californian death records (Hanna and McGarvey, in press; Pawson and Janes, in press) also permit the presentation of some comparative statistics. Finally, the results of two 1976 health surveys involving approximately one-half of the American Samoans over 30 could be matched with death records through 1981 in order to assess the relationship between selected individual biological, social, and psychological characteristics and mortality (Crews 1985). In the present chapter the results of these various studies are summarized in relation to the changing biological and cultural environment.

DEATH RATES IN AMERICAN SAMOA, 1950-1981

Methods and database

All census data available for American Samoa, during the 32-year period covered by the death records, were used in this analysis. Populations at risk were determined from U.S. Bureau of the Census decennial census publications for the years 1950, 1960, 1970, and 1980 (U.S. Bureau of the Census 1983, 1973, 1963, 1953). Park's (1980, 1972) two reports, one on the 1970 census of American Samoa and one on population change in American Samoa, were also used. Three censuses conducted by the Government of American Samoa, one in 1956 (American Samoa Government 1956a), one in 1974 (Levin and Pirie 1976), and a 10 percent sample census conducted in 1977 (American Samoa Government 1978),

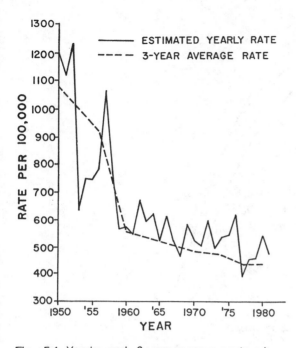

Fig. 5.1 Yearly and 3-year census-centered average crude mortality rates in American Samoa between 1950 and 1981.

Table 5.1 Age-specific Mortality per 100 000 Population of American Samoans, 1950 - 1981

Age	1950	1956	1960	1970	1974	1977	1980
0 - 1	9742.3	6360.0	2946.8	2509.0	2668.1	2394.8	2305.9
1 - 19	406.8	383.6	145.1	113.8	109.1	63.3	87.1
20 - 34	357.7	363.5	251.9	142.6	174.4	262.7	204.2
35 - 49	694.4	592.0	566.1	399.0	330.5	402.7	356.8
50 - 64	1706.0	1588.8	1338.0	1458.1	1130.2	1011.4	1165.6
65 +	8663.7	7273.9	4942.3	5168.0	6731.7	5243.9	4217.8

were also used. Because of the small number of deaths per year, 3-year census-centered average crude death rates were calculated using death registry counts. A regression model was used to determine the population at risk for determining yearly death rates. Univariate linear regression of population size by year was done. The estimated equation ($R^2 = .95$) was used to predict population size during intercensal and census years.

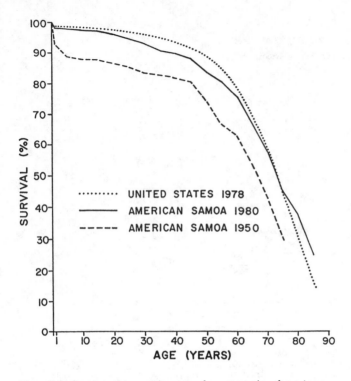

Fig. 5.2 Survivorship estimates for men in American Samoa 1950 and 1980 compared to United States men 1978. (U. S. data from *Demographic Yearbook 1980*)

Results

The results of these calculations are illustrated in Fig. 5.1. As the 3-year averages show, the major decline in mortality had occurred by 1960, but a steady, slow decline continued until the late 1970s. Crews' (1985) analysis of mortality by age groups (see Table 5.1) shows that, as might be expected, the rapid early decline in overall mortality was related to a sharp drop in infant mortality. The later, slower, but continuing decline after 1960 appears, however, to have been produced by declines in all age groups, including the middle-aged population.

The impact of the overall transition is more easily visualized from the survivorship estimates illustrated in Figs. 5.2 and 5.3. Among men the gains were largely the result of better survival during infancy and childhood, although survivorship also improved at all subsequent ages. In 1950 approximately 93 percent of male infants survived to their first birthday; in 1980 almost 99 percent did. At age 30 men showed a gain in survival of approximately 10 percent, while at age 70 approximately 15 percent more men survived in 1980 than did in 1950.

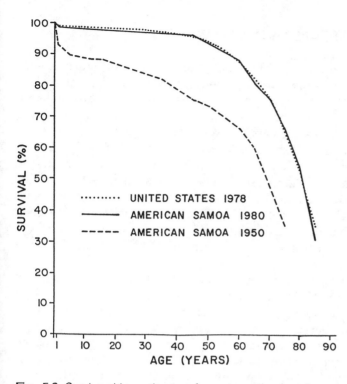

Fig. 5.3 Survivorship estimates for women in American Samoa 1950 and 1980 compared to United States women 1978. (U. S. data from *Demographic Yearbook 1980*)

Survivorship among Samoan women improved more than that of Samoan men. Among women not only did infant and childhood survival improve but there was substantial improvement in survival throughout adulthood. This was particularly true during the childbearing years. Survival to age one increased by 6 percent, as it did among men. During the second and third decades of life, however, women experienced more improvement in survival than men, gaining about 13.5 percent, while at age 35 this gain was 15 percent. Women also achieved 28 percent better survivorship at age 70 in 1980 than in 1950.

A comparison of the 1980 survivorship estimates for American Samoans with those for the total 1978 U.S. population shows that the mortality shifts, while resulting in survivorship for women that is indistinguishable from the general U.S. experience, did not result in comparably high male values. The analysis undertaken did not allow us to identify all of the cause specifics for this sex difference, but the retention of a higher male infant death rate and a continuing higher trauma death rate in boys and young men contributed to the difference.

DEATH RATES IN WESTERN SAMOA 1982 AND 1983

Attempts to calculate accurate birth, death, and migration statistics for Western Samoa have been thwarted by the high rate of underregistration of both births and deaths. Calculations based on census data also indicate serious underenumeration in the census, with rather unpredictable age and sex patterns. For these reasons the Western Samoan Department of Vital Statistics undertook a vital statistics sample survey (Crawley 1984). This stratified survey conducted during 1982 and 1983 permitted the calculation of 1-year age-specific death rates based on a person-years-at-risk sample of 29 994. Age- and sex-specific rates based on the report from this study are given in Table 5.2. The age categories have been adjusted to make the data as comparable as possible to the age classifications used by Crews in his analysis of the American Samoan data.

The number of deaths and person-years-at-risk in this sample is substantially lower than those provided by the American Samoan data (Table 5.1) and the California data (Table 5.3) to be reviewed later. Nevertheless, the numbers are large enough to make a reasonable comparison when both sexes are combined. When the rates are compared to those shown for American Samoa in 1979-1981, the values appear to be considerably higher at all ages in Western Samoa. While the differences in the age groupings make detailed comparisons difficult, the age-specific rates appear to approximate those found in the 1960-centered rates in American Samoa. If this comparison is correct, it suggests that the Western Samoan population, while no longer having very high mortality from infectious disease, continues to have a significantly higher mortality than American Samoa from birth through young adulthood from such causes as infectious and viral diseases. Death rates during middle age also appear to be unusually high, particularly for men, but this specific deviation from the experience in other Samoan populations may be a result of the small number of person-years-at-risk in the sample.

Table 5.2 Estimated Yearly Age-specific Mortality (all causes) from a Western Samoan Sample Drawn in 1982 and 1983[a]

Age	Males	Females	Combined sexes
0 - 1	4148.5	2375.3	3299.2
1 - 14	193.3	177.9	186.4
15 - 24	296.2	100.1	204.0
25 - 44	232.6	222.6	227.5
45 - 64	2720.3	1728.0	2236.0
65 +	6436.8	9295.2	7874.0

[a] Based on information provided in Crawley 1984. Rates are per 100 000, based on 29 994 person-years-at-risk.

Table 5.3 Estimated Yearly Age-specific Mortality (all causes) of California Samoan Population for Period 1978 - 1982[a]

Age	Samoan Population			All California		
	Males	Females	Combined sexes	Males	Females	Combined sexes
0 - 1	1379.3	608.1	969.5	1289.1	1051.7	1087.2
1 - 14	37.6	27.6	32.7	46.0	42.8	44.7
15 - 44	114.2	41.1	77.8	212.0	90.6	152.1
45 - 64	1003.9	646.6	834.6	1115.2	642.6	870.7
65 +	5810.1	3308.3	4314.6	5991.3	4450.4	5078.3

[a] Rates are per 100 000 total population.

DEATH RATES FOR SAMOANS IN CALIFORNIA, 1978-1982

Method and database

Janes and Pawson (in press) in attempting to determine death rates among Samoans in California found it necessary to use a rather unusual method in order to retrieve the Samoan death records in that state. The Samoan deaths were not identified by ethnic category and certificates were catalogued under the very large residual category *other*. They therefore took advantage of the uniqueness of the Polynesian language and the fact that most Polynesian speakers in California were Samoan to use a name-based algorithm for preliminary sorting. From the records sorted they were then able, on the basis of reported place of birth and ethnicity, to verify 255 certificates that were identified as Samoan. The population base they used to determine mortality rates was the 1980 U.S. Census, which separately identified Samoans (U.S. Bureau of the Census 1982). Pawson and Janes noted that their record-sorting method missed Samoan deaths when the individual did not have two names that conformed to their name algorithm or an inappropriate ethnic identification. While these difficulties were likely to cause an underretrieval of death records, the rates estimated may be somewhat more accurate than anticipated, since it is also estimated that the census underenumeration is between 10 percent and 17 percent.

Results

The age-specific mortality rates derived by Pawson and Janes compared to all California are shown in Table 5.3. For ages 1 to 65, the age categories utilized in this study do not exactly coincide with those used by Crews for American Samoa. Even so, a comparison of these results with Table 5.1 suggests that death rate estimates for Samoans in California until about age 50 are much below the estimates for the population of American Samoa. The fact that the estimated California Samoan rates for under 45-year-olds are also below the general California rates strongly suggests that the method for retrieving Samoan death certificates led to substantial underenumeration in these age categories.

The validity of the rate estimates for Samoans over 50 years of age appears to be a separate question, since for these age groups, the rates are in general agreement with those for American Samoa and all of California. They could, of course, still be underestimates if the actual rates were higher than they are in the larger populations. As noted earlier, however, a primary reason for missing death certificates for Samoans in California was the lack of a full Samoan name, and it is probable that older Samoans were more likely to have full Samoan names than younger ones.

Although there have been Samoans in California since the end of World War II, the majority have arrived as migrants since the mid-1960s. This means that almost all of the decedents over 50 were at least over 30 years old when they migrated, while many of the younger decedents were actually second generation migrants. From the experience of other migrants to the United States it can be suggested that there would be a much higher probability that the younger individuals would have non-Samoan names because of intermarriage, identification with the recipient society, and loss of the native language. If this is the situation for the Samoan migrants, then the mortality rate estimates for the older age groups should more closely approximate the correct ones.

CAUSES OF MORTALITY IN AMERICAN SAMOA, 1950-1981

In order to examine how the causes of death changed during this period all of the individual death records were examined and the underlying cause of death classified using the 9th Revised *Manual of the International Statistical Classification of Diseases, Injuries and Causes of Death* (World Health Organization 1977). For some analyses the data were combined into classes suggested by Preston (1976). For the purposes of this summary we will examine, in broad categories, how the causes of death changed over time.

Infectious disease

The large decline in the crude death rate from the 1950-centered rate to the 1960 one was primarily the result of a decline in deaths from infectious disease. As shown in Fig. 5.4, the death rate from major infectious diseases was almost 440/100 000 in the 1950 calculation. The number of such deaths accounted for about 42 percent of the total deaths.

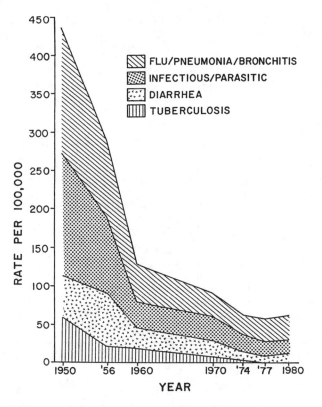

Fig. 5.4 Contribution of different causes to the decline in mortality rates from infectious diseases between 1950 and 1981 in American Samoa.

By 1960 the death rate from these same causes dropped to 128, and the number of such deaths accounted for only 23 percent of all deaths. By the 1970s tuberculosis no longer appeared as a cause of death, and mortality from diarrhea, while still occurring among infants, was uncommon. It is also worth noting that maternal-related deaths, which showed a low but significant rate of 14.1/100 000 in the 1950 calculation, essentially disappeared by the 1970s.

Of possible interest is the persistence of mortality from infectious and parasitic diseases as well as mortality from influenza, pneumonia, and bronchitis. Table 5.4 contains the age-specific rates for these diseases and as can be observed from the data, the deaths in the 1980 calculation associated with each of these causes occurred almost exclusively among infants, the middle-aged, and the old. This pattern of residual deaths associated with infectious disease very closely parallels that found in the economically developed countries for that time. On balance, it appears that by 1980 American Samoa showed only minor variations in death from infectious diseases when compared to other high per capita income populations with adequate health care.

Table 5.4 Changes in Age-specific Mortality per 100 000 from Selected Infectious Diseases: American Samoa, 1950 - 1981

Disease class Age	Year[a]						
	1950	1956	1960	1970	1974	1977	1980
Infectious and Parasitic							
0 - 1	1004.8	766.8	114.8	228.1	173.3	194.2	169.6
1 - 19	124.7	66.7	26.7	17.8	15.3	10.2	8.3
20 - 34	85.5	57.4	18.0	6.2	10.9	11.0	4.4
35 - 49	133.0	32.9	26.3	10.0	8.5	17.1	7.8
50 - 64	65.6	95.3	29.1	18.5	—	—	53.6
65 +	440.5	316.3	164.7	301.5	139.3	40.7	63.0
Influenza, Pneumonia, and Bronchitis							
0 - 1	2228.1	902.1	191.4	32.6	34.7	97.1	33.9
1 - 19	78.7	53.4	24.7	6.7	—	2.0	—
20 - 34	—	9.6	9.0	6.2	—	5.5	—
35 - 49	44.3	16.5	13.2	20.0	42.4	8.6	23.3
50 - 64	262.5	190.7	116.4	166.1	79.6	99.7	160.8
65 +	514.0	695.8	549.2	387.6	603.5	609.8	440.7

[a] 3-year-centered rate.

Trauma

As the mortality from infectious disease declined over time in the American Samoan population, the relative importance of all forms of trauma as a cause of death increased. In the 1950-1952 calculation the rate was 47.5/100 000, and the number of deaths accounted for only 4 percent of the total deaths. In the 1960 calculation the rate was 58.2/100 000 and the deaths accounted for 10 percent of the total deaths. The 1970-centered rate was down to 45.4/100 000, but by numbers still accounted for 9 percent of deaths. During the 1970s the death rates from trauma began a steady rise, while the overall death rates continued a slow decline. Thus by the 1979-1981 calculation, trauma was the second leading cause of death with a rate of 81.6/100 000, and accounted for nearly 19 percent of all deaths. While during the 1970s some part of the rise in deaths from trauma was due to motor vehicle accidents, this did not explain the major part of the increases.

Attempts to obtain from the data a more detailed understanding of the age- and cause-specific trends in death from trauma was limited by the small number of deaths in given categories. Table 5.5 shows one such attempt. The pattern of mortality from motor vehicle related accidents suggests that such deaths during the 1970s occurred primarily to children and young adults. It should be noted, however, that the actual number of

Table 5.5 Changes in Age-specific Mortality per 100 000 from Motor Vehicle Accidents and All Other Traumas: American Samoa, 1950 - 1981

Cause of death / Age	Year[a] 1950	1956	1960	1970	1974	1977	1980
Motor vehicle accidents							
0 - 1[b]							
1 - 19	3.3	3.3	3.0	2.2	6.6	10.2	6.2
20 - 34	—	—	—	12.4	10.9	11.0	8.9
35 - 49	14.8	—	13.2	29.9	—	34.3	7.8
50 - 64	—	—	—	18.5	31.8	—	26.8
65 +	—	—	—	—	—	40.7	—
Other traumas							
0 - 1[b]	43.7	45.1	76.6	32.6	34.7	—	—
1 - 19	29.5	60.0	32.6	33.5	45.8	28.6	39.4
20 - 34	77.8	105.2	72.0	37.2	81.8	136.4	133.1
35 - 49	29.6	82.2	118.5	39.9	33.9	120.0	93.1
50 - 64	65.6	158.9	58.2	55.4	31.8	57.0	93.8
65 +	73.4	189.8	—	43.1	46.4	122.0	94.4

[a] 3-year centerd rate.

[b] For this age group motor vehicle and all other trauma death rates were not segregated, because actual number of deaths per year never exceeded two in the combined categories.

deaths per year in any of the age categories does not exceed five in 3 years.

The deaths from other traumas are more numerous, and some confidence can be placed in the age-specific rate trends. From these data it appears that death rates among adults over 50 and children did not change in a systematic manner over the 32 years. While the trauma-related death rates in the 20- to 25-year-olds also fail to show a sustained long-term trend, this age group did show substantially higher rates from 1976 to 1981. In his more detailed analysis Crews (1985) noted that the 1980 trauma-related death rates for the 20- to 34-year-old Samoans were very much higher than the rates for the United States during the same period. A comparison of the rates of older adults and children with U.S. rates also suggested that the Samoan rates were greater during the late 1970s, although the differences were not as great as those found for young adults.

Because of the recent controversy over the extent of latent and manifest aggression among Samoans, Crews examined the specific deaths attributed to suicide and homicide in the 32 years of death records. In order to ensure that the results applied only to Samoans, he excluded those individuals whose death certificates indicated other than a Samoan or Polynesian ethnic origin. The percentage of such individuals has varied slightly over time but, for example, in the 1974 census this group was about 5 percent of the population and was almost equally divided between Caucasians and others. The results of this analysis, including deaths from trauma of undetermined causes, are shown in Table 5.6.

There is a substantial sex difference in the death rates as death by homicide or suicide appears to be quite rare among women. Even if some of the deaths from trauma of undetermined origin involved personal violence, there is no suggestion of time trends or indications of high frequencies of aggression-related deaths among women.

The data do suggest that both suicide and homicide reached substantial levels among Samoan men during the late 1970s and early 1980s. Even if a significant percentage of the deaths from trauma of undetermined

Table 5.6 Mortality Rates[a] from Selected Traumatic Causes among Persons Identified as Samoan or Polynesian on their Death Records, per 100 000 Population 15 Years Old and Older[b]

Cause of death	Year						
	1950	1956	1960	1970	1974	1977	1980
Men							
Suicide							
Rate	19.2	23.1	6.7	4.6	12.7	30.0	24.5
N[c]	3	3	1	1	3	7	7
Homicide							
Rate	—	—	26.6	4.6	12.7	12.9	24.5
N	—	—	4	1	3	3	7
Trauma (undetermined cause)							
Rate	38.5	69.3	59.9	32.2	42.3	25.7	38.5
N	6	9	9	7	10	6	11
Women							
Suicide							
Rate	—	14.3	6.5	4.7	—	—	6.9
N	—	2	1	1	—	—	2
Homicide							
Rate	—	—	—	—	—	3.8	3.5
N	—	—	—	—	—	1	1
Trauma (undetermined cause)							
Rate	33.5	28.5	12.9	14.0	4.1	7.5	3.5
N	5	4	2	3	1	2	1

[a] 3-year-centered rate.

[b] The total census population of American Samoa was used as the denominator in these calculations.

[c] N's are for 3-year period.

cause did not involve personal violence, it is apparent that aggression-related mortality has risen among the men.

In this analysis Crews also notes that the non-Caucasian minority on American Samoa, many of whom are crew members on the fishing boats for the American Samoan fish canneries, has an unusually high death rate from trauma in all forms. These deaths contribute significantly to the trauma-related death rates found for 20- to 34-year-olds in the total American Samoan death rates.

Degenerative diseases

When the total crude death rates of populations decline as a consequence of an increasing control over mortality from infectious diseases, the leading causes of death generally shift to cardiovascular, neoplastic, degenerative, and chronic diseases. Such a shift is to be expected, since

Fig. 5.5 Contributions of selected desease categories to the mortality rates in American Samoa from 1950 to 1981.

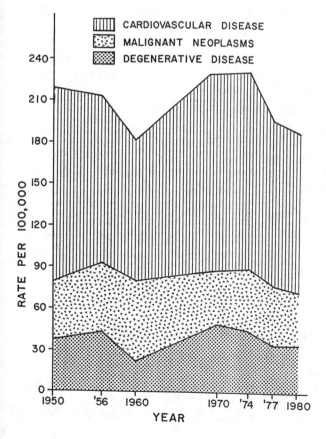

the majority of deaths then occur in middle and old age and these causes are generally anticipated in the older age groups. To some extent, this shift occurred in American Samoa during the 1950s through the 1980s, although death from cardiovascular disease (CVD) was already the third ranked cause of death in the early 1950s. While CVD became the first ranked cause of death thereafter, trauma, because of the youth of the population, remained more important than neoplasms as a cause of death into the late 1970s and early 1980s.

The increasing fatness and apparently rising blood pressure of the Samoan population leads one to expect that death rates from CVD and certain degenerative diseases should actually have risen. As reflected in Fig. 5.5, however, total population rates showed no consistent trends in CVD, neoplasms, or degenerative and chronic diseases. The lack of trends could, of course, have been a result of the population age structure, so it is useful to examine the age-specific rates as presented in Table 5.7. The age-specific rates also fail to show any specific trends in the deaths from neoplasms, although for 35- to 49-year-olds they are generally lower in the 1970s.

The death rates from degenerative and chronic diseases appear to be higher for the old during the 1970s than they were previously, but the establishment of the LBJ Tropical Medical Center in American Samoa during the 1960s may have led to a more careful diagnosis of cause of death in the old, thereby leading to increasing the number of deaths assigned to these causes. As is discussed later, diabetes mellitus was listed as the underlying cause of death for a significant number of individuals from the 1970 census period onward and the number of such deaths may have significantly increased the death rates in the category of degenerative and chronic diseases.

Some interesting trends do appear in the age-specific death rates from CVD. These rates appear to be substantially higher in young adults and children from 1950 to 1961 than they are in more recent years. This may be caused by a relatively high prevalence of rheumatic heart disease during the early period that declined sharply during later years. Although the data we have available do not allow proof of this hypothesis, it seems reasonable in light of the continuing death rate from infectious diseases during the 1950s. If this is indeed the case, then a certain percentage of the older adult and middle-aged cardiovascular mortality must also be attributable during this earlier time period to residual heart-valve damage from rheumatic fever. Our analysis also did not allow us to distinguish such deaths if they occurred.

In order to examine in more detail the trends of mortality from selected degenerative diseases among American Samoans, Crews calculated 3-year-centered rates for the Samoan- and Polynesian-identified indi-viduals over age 15. The results of these calculations (see Table 5.8) reveal a number of sex and time differences in the death rates. While the rates from all neoplasms show only slight differences between the sexes,

Table 5.7 Changes in Age-specific Mortality per 100 000 from Degenerative Diseases: American Samoa 1950 - 1981

Disease class Age	Year[a]						
	1950	1956	1960	1970	1974	1977	1980
Cardiovascular (390 - 459)							
1 - 19	19.8	30.0	5.9	8.9	6.6	—	2.1
20 - 34	46.7	47.8	63.0	18.6	32.7	38.3	13.3
35 - 49	133.0	131.6	171.1	179.5	110.2	94.3	124.1
50 - 64	459.3	413.1	436.3	646.0	509.4	384.6	375.1
65 +	3450.8	1897.5	1318.0	2282.5	3244.8	2601.6	1983.0
Neoplasms (140 - 239)							
1 - 19	—	10.0	—	11.2	2.2	4.1	2.1
20 - 34	31.1	19.1	7.0	6.2	—	5.5	4.4
35 - 49	88.7	115.1	92.2	49.9	50.9	68.6	23.3
50 - 64	196.9	63.6	349.0	166.1	175.1	185.2	268.0
65 +	587.4	822.3	713.9	430.7	974.9	650.4	409.2
Degenerative and Chronic (250; 530 - 534; 570 - 573; 580 - 589)							
1 - 19	13.1	13.3	3.0	—	2.2	—	4.2
20 - 34	31.1	28.7	—	4.8	10.9	16.4	4.4
35 - 49	44.3	16.5	26.3	39.9	42.4	17.1	31.0
50 - 64	196.9	286.0	174.5	276.9	206.9	242.2	134.0
65 +	220.3	316.3	219.7	732.1	835.7	365.9	472.1

[a] 3-year-centered rate.

the deaths from lung neoplasms among men were, from 1973 to 1981, almost triple the number reported for women. This parallels the findings in industrial countries and probably reflects a higher frequency of male smokers. Even so, the rates must be considered relatively low, and there was no obvious tendency for the rates to rise for either sex.

Death rates from CVDs are also consistently higher for men than for women from 1970 to 1981. The male data suggest that there may have been a generally rising rate of mortality from CVD over time, although there are substantial fluctuations in the rate from one time period to another. The women do not show a similar trend and indeed, the highest rates occur in the early 1950s.

The information on deaths from diabetes mellitus deserves special mention, both because the high relative body weights of Samoans suggest they should have high prevalence of type II diabetes and because the data suggest a very rapid and recent increase in deaths associated with diabetes. The rates are somewhat higher among men than women, but the actual number of deaths is so small that the apparent differences may

Table 5.8 Mortality Rates[a] per 100 000 from Selected Chronic Diseases: American Samoans, 15 Years Old and Older, Identified as Samoan or Polynesian 1950 - 1981

Cause of death	Year						
	1950	1956	1960	1970	1974	1977	1980
Men							
All neoplasms (except lung)							
Rate	83.3	69.3	119.9	36.8	67.7	55.7	56.0
N [b]	13	9	18	8	16	13	16
Lung neoplasms							
Rate	—	23.1	26.6	13.8	21.1	38.6	21.0
N	—	3	4	3	5	9	6
Cardiovascular disease							
Rate	179.5	169.3	206.4	326.9	274.8	274.2	192.3
N	28	22	31	71	65	64	55
Diabetes mellitus							
Rate	—	23.1	13.3	55.3	46.5	51.4	31.5
N	—	3	2	12	11	12	9
Women							
All neoplasms (except lung)							
Rate	73.7	85.5	58.1	51.3	57.5	48.7	44.8
N	11	12	9	11	14	13	13
Lung neoplasms							
Rate	—	7.1	12.9	23.3	12.3	3.8	10.3
N	—	1	2	5	3	1	3
Cardiovascular disease							
Rate	274.8	199.6	116.1	167.9	213.4	116.2	158.6
N	41	28	18	36	52	31	46
Diabetes mellitus							
Rate	6.7	7.1	—	37.3	36.9	22.5	34.5
N	1	1	—	8	9	6	10

[a] The total census population of American Samoa was used as the denominator in these calculations.
[b] All N's are the total sample for the 3-year-centered rate.

be a statistical artifact. The difference between the rates before and after the 1970-centered estimate is quite obvious, but it is questionable whether or not this reflects a real difference in the importance of diabetes as a cause of death. The presence of diabetes was undoubtedly more frequently recognized after the establishment of the LBJ Medical Center and, in addition, diabetes is increasingly being recognized as an under-lying cause of death. Nevertheless, the mortality ascribed to diabetes is quite high and, in conjunction with prevalance data from other studies,

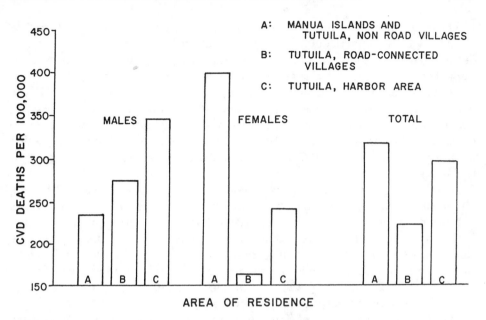

Fig. 5.6 American Samoan cardiovascular disease death rates by area of residence for age 30 and over (1962-1974 inclusive). (From Crews and MacKeen, 1982)

suggests that Samoans may have unusually high rates of type II diabetes.

In an earlier analysis Crews and MacKeen (1982) examined in greater detail the cardiovascular and diabetes mellitus mortality from 1963 to 1974. In this study it was their intent to examine whether the changing life styles of the Samoans and their increasing body weights and blood pressures were affecting death rates from these diseases. In order to calculate the rates the 1970 and the 1974 censuses were extrapolated to estimate the population at risk and to examine time trends. Four-year average mortalities were used instead of the 3-year average Crews used in his later analysis. Deaths of non-Samoans and non-Polynesians were excluded from the analysis, as were the deaths of individuals whose customary residence was outside American Samoa.

Age-adjusted rates for the overall time period showed a CVD death rate of 242.5 per 100 000. This rate was substantially below the 368.6 rate reported for the United States in 1962 (Shryock and Siegel 1975), but higher than the rates reported for some other countries such as France (177.2), Japan (121.7), and Chile (198.6). They also found that the age-adjusted death rate from diabetes, at 32.3 per 100 000 was far greater than that reported for 38 national populations (Kurihara et al. 1970). The rate exceeded a reported U.S. rate of 13.4 (Marks 1966) by more than twofold.

Of more particular interest are the results that suggest that life-style variations in American Samoa may be related to mortality from CVD. Figure 5.6 shows the rates by residence areas in American Samoa. As

Hecht et al. discussed in Chapter 3, these areas do show some differences in life-style, and particularly life on the Manu'a islands differs in many details from that on Tutuila. Both body weights and blood pressures appear to vary some by region. The results by area strongly suggest regional differences in CVD mortality, although the high rate for the women from Manu'a breaks an otherwise organized pattern. Again, because of the relatively small number of deaths, interpretations must be cautious. This is particularly the case for the Manu'a island data which included only 27 out of 287 CVD-related deaths reported in this part of the study.

The authors found that the occupations listed on the death certificates for most of the males also made it possible to classify the decedents according to whether their occupations had been active or sedentary. While older men and women at all ages could not be so categorized, the classification provided some information on how the residential areas compared by occupation and how the occupations related to causes of male mortality. While only 42 percent of the deceased men from the Pago Pago harbor area were in active occupations, 53 percent of those from villages on paved roads in the rest of Tutuila were in this classification. For those decedents who had been residents of the Manu'a islands and the villages on the unpaved roads of Tutuila, 65 percent had been in physically active occupations. Only 29 percent of the deaths of those in the active occupations were CVD related, while 43 percent of the deaths of men in the sedentary occupations were from CVD.

CARDIOVASCULAR DEATHS AMONG SAMOANS IN HAWAII

McGarvey and Hanna (in press) obtained from the Hawaii Department of Health, Vital Statistics Division, a complete listing of Samoan deaths from 1968 to 1978 for individuals aged 35 and older. Caution must be used in interpreting the data, since no census of Samoans was available during this time period and because there may be a number of reasons why all Samoan deaths may not have been identified. If the Samoan population over age 34 in Hawaii as recorded in the 1980 census is used as the population at risk, then, as shown in Table 5.9, the death rate from all causes in this age group from 1974 to 1978 was about 922 per 100 000. This is lower than the 1167/100 000 rate found in American Samoa for the 1977-centered census analysis on this same age group. Since in-migration to Hawaii was at a high rate during the mid-1970s, however, the Samoan population in Hawaii was probably smaller than recorded in 1980. Therefore, it appears that most if not all of the Samoan deaths in Hawaii were recorded in the death certificate summaries.

The percentages of deaths from CVD for Samoan Hawaiians during the 1974-1978 time period is similar to those found in American Samoa in the 1976-1978 period, while the CVD death rates are essentially identical.

Table 5.9 Mortality of Samoans in Hawaii Compared to American Samoans: Both Sexes, over Age 34

Cause of death	Hawaii		American Samoa[a]	
	1968-1973	1974-1978	1973-1975	1976-1978
All deaths/year				
Rate/100 000	—	901.9[b]	1260.4	1167.4
N [c]	21.4	26.4	85.0	82.3
CVD deaths/year				
Rate/100 000	—	464.6[b]	567.9	482.3
N	6.8	13.6	38.3	34.0
Percent of Deaths from CVD	32.0	52.0	45.1	41.3

[a] Based on data provided by Crews 1985.
[b] Based on 1980 U. S. Census estimations of population at risk and number of deaths reported by Hanna and McGarvey (in press).
[c] *N*'s are average number of deaths per year on 5-year data in Hawaii; 3-year data in American Samoa.

This does not necessarily mean similar age- and sex-specific death rates for the two groups of Samoans, but such a presumption is supported by the comparable age and sex structure of the two populations at risk (see Harbison, this book, Figure 4.3). During the 1968-1973 period, Hanna and McGarvey found a much lower percentage of deaths from CVD. Since the age and sex composition of the Samoans in Hawaii may at that time have been much different than it was by 1980, however, it is impossible to know what this finding purports for time trends in CVD death rates. The analysis by Crews (1985) does not suggest a comparably lower risk of CVD deaths at this time among older Samoans in American Samoa.

DEATHS AMONG OLDER SAMOANS IN CALIFORNIA

Pawson and Janes (1982), using the record recovery system described earlier in this chapter, also examined the death rates from cardiovascular disease in the older Samoans in California during the period from 1978 to 1982. Table 5.10 shows the results of their analysis, including rates from malignant neoplasms and a comparison with all-California rates. While the 45- to 65-year-old age category varies from the 50- to 64-year-old age category used by Crews, some comparisons can be made between the California Samoans and American Samoans. When the data in Table 5.10 are contrasted to those shown in Table 5.7, it can be observed that death rates from malignant neoplasms are higher in the California Samoans from ages 45 to 64 than they are in American Samoa for 50- to

Table 5.10 Estimated Age-specific Mortality Rates per 100 000 from Selected Causes of Death: California Samoans and All Californians 1978 - 1982

| Cause of death | Samoans | | | All California | | |
Age	Males	Females	M/F	Males	Females	M/F
Malignant neoplasms						
45 - 64	386.1	258.6	325.7	288.9	290.4	289.6
65 +	1229.1	526.3	809.0	1335.9	852.4	1019.3
Heart disease						
45 - 64	347.5	194.0	274.8	392.5	153.2	277.0
65 +	3128.5	1578.9	2202.2	2516.1	1910.2	2157.1
Cerebrovascular disease						
45 - 64	96.5	21.6	61.1	37.8	38.5	38.2
65 +	335.2	300.8	314.6	547.6	643.8	604.8

64-year-olds, even though the California population must be somewhat younger. The difference in death rates from cancer are even greater in the over-65 age category. While the causes for this difference cannot be specified from the available analysis, the large sex discrepancy in the California rates compared to the lower difference in sex-specific rates in American Samoa (Table 5.8) suggests that smoking and consequently lung cancer may be more common among Samoan men in California.

The combined death rates from heart disease and cerebrovascular disease in California Samoans is somewhat lower than the rate for American Samoan men aged 50-64. This is to be expected given the inclusion of the 45- to 50- year-olds in the California study and suggests that the frequency of death from cardiovascular disease in these ages is similar in these two Samoan populations. The combined heart and cerebrovascular disease death rates for the California Samoans approximate the rates found in American Samoa. While the combined rate in California is higher than the one calculated for 1978-1981 in American Samoa, the rates in these islands varied widely during the 1970s, and the California rate fits well within this variation.

The comparison of the death rates of the Samoan population with all of California produces no striking contrasts, although the death rates from cerebrovascular disease appear to be lower in Samoans over age 65. The most obvious difference is that Samoan women for almost all diseases and age categories have lower rates than all California women. While an underenumeration of deaths due to an incomplete recovery of Samoan death records in these age categories is possible, it does not seem probable that the recovery method would seriously alter the relative percentages of CVD versus malignant neoplasm deaths recorded. Since the relationship between cancer and CVD related death in Samoans and all California is similar, we agree with Pawson and Janes's conclusion that there is no reason to believe Samoan death rates from CVD exceed the general California rate.

PREDICTORS OF MORTALITY IN AMERICAN SAMOA

The age- and cause-specific death rates in the late 1970s suggests that Samoans still had a mortality-cause pattern that differed from the general U.S. pattern. The rates also indicated that at least the CVD-related deaths were at a somewhat lower rate than might be expected in a population with such high body weights and frequencies of hypertension. In order to examine the problem further in American Samoa the data obtained from two health surveys conducted in 1976 were linked to the American Samoan death records from that year through 1981. One of the surveys was conducted by the Public Health Department of American Samoa, while the other was part of the Penn State Samoan Project studies.

Sample Sources

During 1976, 4693 residents in all parts of American Samoa were included in a Body Build and Blood Pressure Survey (BBBP) conducted by the Public Health Service. In this survey the name, age, sex, village of residence, weight, height, systolic, and diastolic blood pressure of each participant was recorded. The BBBP participants represented slightly over 40 percent of American Samoa's total population aged 30 or older in 1976. During 1982 and 1983, Crews linked—by name, age, sex, and village of residence—the 1976 through 1981 death records with the records of participants in the BBBP survey. One hundred thirty-five linkages were found. These 135 decedents represent 3.0 percent of all BBBP participants and suggest a death rate for the total sample of 30 per 1000 during the 6-year follow-up period.

During 1976, investigators from the Penn State Samoan Project also interviewed and measured a geographically dispersed sample of 1732 residents of American Samoa age 20 and over. After linking the records of participants in this survey with the 1976-1981 mortality records, 52 linkages were found. These 52 decedents represent 3.0 percent of the total Penn State sample and 6-year death rate equal to that of the BBBP sample. In the Penn State study the same variables were measured as for participants in the BBBP survey. In addition, three skinfolds—triceps, mid-axillary, and subscapular—upper arm circumference, response to a social questionnaire, and the Cornell Medical Index (CMI) (Brodmen et al. 1949) were also recorded. This sample is more completely described in McGarvey and Baker (1979) and McGarvey (1983).

The total decedent sample thus included 135 from the BBBP survey and 52 decedents from the Penn State study. There was some overlap between the two surveys. Data were available from both surveys for 19 decedents, suggesting an overlap of about 37 percent with the Penn State sample. If this proportion is assumed to apply across the data base, then of the 6242 persons measured and interviewed, there were approximately 5784 different American Samoans represented. These 5784 persons would represent 20 percent of the total population of the islands and approximately half of the population aged 30 and older in 1974. After removing the overlap between the samples, the 158 different decedents represent a mortality rate of about 2.7 percent among the 5784 individuals who participated in these two surveys. Complete data were available for only 77 deceased men and 44 deceased women from the BBBP sample (121 decedents) and 34 men and 13 women from the Penn State sample (47 decedents).

Results

Table 5.11 shows for the integrated samples the contrast in biological characteristics between the survivors and decedents. Age, at the time of the survey was, of course, substantially greater for the subsequent decedents than survivors. Of potentially greater interest is the fact that the blood pressures of the decedents were significantly higher than the survivors, while body weights and relative weights were not.

This pattern of biological differences between survivors and decedents was similar when the decedents from CVD were compared to survivors, but varied some when those who had died from malignant neoplasms or diabetes were compared to survivors. For neoplasm-related deaths only women had significantly higher blood pressures than survivors; body weights were lower in the decedents, although not enough lower to be statistically significant. Both blood pressures and weights were high in the individuals dying from diabetes, but only the systolic blood pressure and body mass index of the males showed a statistically significant difference between the decedents and survivors. It must be noted that there were for this analysis only a small number of deaths from diabetes and that the sample of women was particularly small.

As another method of examining how biological risk factors may have affected deaths in the sample, the integrated sample was divided into individuals who showed hypertension or obesity compared to those who did not, and a chi-square test of relative mortality was calculated. The resulting significance of the chi-square tests is provided in Table 5.12. Hypertension, whether based on the systolic or diastolic value, increased both the overall and most of the cause-specific risks of death. In this aspect the results parallel the previous findings of higher blood pressures in decedents than in survivors. The method did yield a slightly different result in relation to the body mass index. This variable continued to show, for women, no statistically significant relationship to probability of death from any cause. Among men, however, the obese individuals

Table 5.11 Mean Values of Selected Biological Characteristics among American Samoans Who Remained Alive or Died During a 6-Year Follow-up

Biological characteristics	Survivors		American Samoans who died from:							
			All causes		Cardiovascular disease		Neoplasms		Diabetes	
	Men	Women	Men	Women	Men	Women	Men	Women	Men	Women
Age (yr)	44.23	43.41	58.51[a]	57.53[a]	62.73[a]	56.17[a]	56.94[a]	58.67[a]	56.10[a]	54.50
Weight (kg)	86.85	85.20	86.40	83.22	87.91	81.83	84.41	80.85	91.70	97.35
Height (cm)	169.97	159.97	169.34	158.71	169.42	156.89[b]	170.13	160.35	168.19	165.10
Systolic blood pressure (mm)	129.07	127.71	140.11[a]	154.77[a]	149.18[a]	159.65[a]	125.06	153.00[a]	152.70[a]	140.00
Diastolic blood pressure (mm)	84.42	81.58	88.62[a]	92.23[a]	92.82[a]	97.96[a]	85.24	88.89[b]	90.00	95.00
Body mass index[c]	30.01	33.25	32.01	32.78	30.48	33.07	29.10	31.22	32.31	35.72

[a] Here p < .01.
[b] Here p < .05 for a two-tailed t-test of the means comparing survivors and descendants.
[b] Body mass index = (weight/height^2).

Table 5.12 Hypertension,[a] Obesity,[b] and the Relative Risk of Death from Selected Causes in American Samoa: Statistical Significance of a Chi-square Test of the Differences in Relative Risk Reported

	All causes		Cardiovascular disease		Neoplasms		Diabetes Mellitus	
	Men	Women	Men	Women	Men	Women	Men	Women
Systolic hypertension	<.01	<.01	<.01	<.01	—	<.01	<.01	ns
Diastolic hypertension	<.05	<.01	<.01	<.01	<.01	<.01	<.01	ns
Obesity	ns	ns	ns	ns	<.05[c]	ns	ns	ns

[a] Hypertension was determined as a systolic blood pressure of 160 mm or a diastolic blood pressure of 95 mm or greater.

[b] Obesity was determined as a body mass index (weight/height^2) of 30 or greater.

[c] Obesity showed an inverse relationship with mortality from neoplasms.

showed a lower risk of death from neoplasms than did the nonobese, while the obesity category of the individual was not significantly related to the probability of death from diabetes.

Logistic regression models
Predictions based on the previous methods can be misleading because of the statistical interaction of the biological variables with each other. Such methods were also not suitable for some of the analyses, since only the small Penn State study sample included many of the biological measurements plus social and Cornell Medical Index (CMI) measures. To increase the accuracy and power of the analysis, Crews (1985) estimated nine logistic regression models for predicting death and cause of death among American Samoans. The reader is referred to Walker and Duncan (1967) for a fuller discussion of logistic regression methods. The strategy of this analysis was first to estimate a prediction model using six biological measures—age, weight, height, systolic and diastolic blood pressure, and the body mass index; called Model I, it contained only the main effects of these six predictors. Second, Model II was estimated. Model II contained the same six predictors as Model I, plus five terms representing interactions between them—age by weight, systolic and diastolic blood pressures, and weight by the two blood pressures. Model II can be called a full model and Model I a main-effects model. Next, Crews tested whether the additional predictors, the interactions, improved the prediction of death. This was done by examining the residual chi-square when the chi-square for Model I was subtracted from the chi-square from Model II. Since these were MLE chi-squares, the residual provided a test statistic to determine whether the two models were significantly different.

If they were, then at least one of the interactions had to be included in any prediction model. If they were not, then no interaction term improved the prediction of death and none need be included in later models.

Third, a reduced model was estimated. In Model III only sufficient predictors were retained to ensure that the reduced model was not a significantly poorer predictor of death than either Model I or Model II. At this step several models were estimated to ensure that different combinations of predictors were examined and that a "best" model was presented as Model III. Not all possible models could be examined. Model III was then tested against both Model I and Model II using the difference in chi-squares between models.

Fourth, the sample was reduced to include only subjects who participated in the Penn State survey. Then Model III was reestimated for this reduced sample. Other possible models were estimated to ensure that Model IV was the best reduced model. Fifth, the effects of four additional predictors were estimated along with those retained in Model IV. This was Model V, it included the three skinfold measures and upper arm circumference as predictors. If any of these four predictors had significant estimated regression coefficients, they were retained, the rest dropped and the model reestimated. Either this reestimated model or the one with all four additional predictors included was Model V.

Sixth, the sample was reduced to include only subjects for whom social predictors were available. Several models were estimated using this small sample and the best one retained as Model IVa. Next, the social predictors, which an exploratory analysis had suggested might be predictive of death, were added to Model IVa. These included marital status, years employed and educational attainment. Several models were estimated using different combinations of these social predictors, and the best model was presented as Model VI. Model VI was tested against Model IVa to determine whether social predictors improved the prediction of death.

Seventh, the sample was reduced to include only subjects where CMI responses were available. Several reduced models were estimated using this sample and the best one retained as Model IVb. Then those CMI section scores, which an exploratory analysis had suggested might be predictive of death, were included with Model IV. These include the illness, eye/ear, cardiovascular, and fatigability section scores. Several models were estimated using different combinations of these predictors, the best model was presented as Model VII. Model VII was tested against Model IVb to determine whether CMI section scores improved the prediction of death. Last, prediction Model III was used to examine the relationships between a given rise in either systolic or diastolic blood pressure and the probability of death.

To exemplify the results of this modeling strategy the analysis of predicted mortality from all causes for Samoan men is provided in Tables 5.13 and 5.14. Based on these and all of the other models generated,

Table 5.13 Estimated Logistic Regression Models Using Biological Characteristics to Predict Mortality among American Samoan Men

| | | Estimated Regression Coefficients | | | | | | | | | | |
| | | Biological characteristics | | | | | | Interactions | | | | |
Models[b]	Intercept	Age	Weight	Height	Systolic blood pressure	Diastolic blood pressure	Body Mass Index[a]	Age × systolic blood pressure	Age × diastolic blood pressure	Systolic blood pressure × weight	Diastolic blood pressure × weight	Age × weight
I. Main effects of six biological characteristics	-2.938	.090[d]	.050	-4.169	.006	.014[f]	-.138					
II. Full model with interactions	1.222	.001	.059	-3.441	.030	-.104	-.122	.00003	.00131[d]	-.00030	.0052	-.00030
III. Reduced model	-9.820[d]	.089[d]			.014	.015[f]						
IV. Reduced model for Penn State Samoan Project sample[c]	-10.378[d]	.097[d]			.014	.002						
V. Model IV plus four additional variables	-8.172[d]	.081[e]			.021[e]	.006						

Chi-square comparisons between models: $\chi^2_{II} - \chi^2_{I} = 4.29(5)$[g] $\chi^2_{II} - \chi^2_{III} = 5.30(8)$[g] $\chi^2_{II} - \chi^2_{III} = 1.01(3)$[g] $\chi^2_{V} - \chi^2_{IV} = 4.19(4)$[g]

[a] Body mass index = (weight / height^2).

[b] All models are based on the logistic regression where $Y_i = \log (P_i / 1 - P_i) = \beta_0 + \beta_1 \chi_{1i} + \ldots = \beta_k \chi_{ki}$.

[c] The Penn State Samoan Project sample included only subjects for whom upper arm circumference and three skinfolds were known.

[d] Here $p < .01$ significance levels for a MLE chi-square test that the estimated coefficients are different from zero or that the model is significant.

[e] Here $p < .05$.

[f] Here $p < .10$.

[g] Here $p > .10$, not statistically significant.

Table 5.14 Estimated Logistic Regression Models Using Selected Biological and Social Characteristics and CMI Section Scores to Predict Mortality among American Samoan Men

| | Estimated Regression Coefficients | | | | | | | | | | |
| | | | Biological characteristics | | Social characteristics | | CMI sections[a] | | | | |
Models[b]	Intercept	Age	Systolic blood pressure	Diastolic blood pressure	Education level	Widow-hood	Eye/ear	Illness	Respiratory	χ^2(df)	R(c.p.)
VI. Reduced model plus selected social variables	-13.022[c]	.091[c]	.013	.013	.870[c]	.681[c]				34.37(5)	.361(.78)
IVa. Reduced model for men without social data	-8.749[c]	.078[c]	.016	-.002						24.81(3)[c]	.317(.76)
VII. Reduced model plus selected CMI section scores	-9.909[c]	.090[c]	.011	.009			-.020	-.022	.065	37.61(6)[c]	.357(.81)
IVb. Reduced model for men with CMI data	-9.520[c]	.089[c]	.008	.011						36.68(3)[c]	.391(.81)

Chi-square comparisons between models: $\chi^2{}_{VI} - \chi^2{}_{IVa} = 9.56(2)^c$ $\chi^2{}_{VIII} - \chi^2{}_{IVb} = .93(3)^d$

[a] CMI section scores are variables representing the total of positive responses to different sections of the CMI.

[b] All models are based on the logistic regression model where $Y_i = \log(P_i / 1 - P_i) = \beta_0 + \beta_1 \chi_{1i} + \ldots + \beta_k \chi_{ki}$.

[c] Here $p < .01$ significance levels for MLE chi-square test that the estimated coefficients are different from zero and that each model is statistically significant.

[d] Here $p > .10$, not statistically significant.

Crews concluded that elevated blood pressures independently of all other variables were predictive of death not only from all causes but also for most of the sex- and cause-specific categories. While this conclusion agrees with those from a number of studies (Holtzman et al. 1983; Keys 1980a, 1980b; Lew and Garfinkel 1979), the American Samoan models appeared to show that elevations in diastolic blood pressure increased the risk of dying somewhat more than elevations in systolic. This was a finding somewhat different than the conclusions reached by Kannel et al. (1971) reporting from the Framingham Study, since they find systolic pressure to be a slightly better predictor.

Body weight, relative body weight, and skinfold measurements failed to show predictive values in any of the models despite the fact that the cruder comparative techniques had suggested some possible relationships between relative weight and deaths in men from such causes as neoplasms and diabetes. Prospective studies in other populations have also failed to find a consistent link between relative weight and later mortality (Andres 1980; Keys 1980a, 1980b; Sorlie, et al. 1980; Kannel and Gordon 1970).

Among the social variables examined, only the loss of a spouse was a significant predictor of mortality for both sexes. This association has been found in a number of U.S. studies, for example, Goldfarb (1971) and Palmore (1971). The fact that Samoans generally live in larger and more socially integrated groups than most older individuals in the United States, however, make this finding one that would not necessarily be anticipated. A more unusual finding was that for men an increased educational level was predictive of death from all causes and from cardiovascular disease. This is the opposite of the reported associations found in studies in the United States (Jenkins 1976; Palmore 1971; Kannel and Gordon 1970). The cause of this association cannot be determined from the data in the mortality studies, but, from other data reported in this book, it can be suggested that the more educated men in this population may also have been more exposed to some of the life-style changes that are associated with an increase in psychological stress.

The association of both total and section-specific CMI scores with mortality was found to be inconsistent and failed to yield a logical pattern. The sample size that included the CMI was quite small and associations would have necessarily been rather strong to be statistically significant in the models. It should also be noted that as discussed in Chapter 8 the responses to the CMI questions may have been strongly influenced by culturally conditioned attitudes toward reporting illness.

OVERVIEW OF MORTALITY — CAUSES AND PREDICTIONS

The lack of accurate cause of death data for Western Samoa, the high rates of immigration and emigration in American Samoa, and the difficulties of recovering the death certificates for Samoans in Hawaii and California, all make it difficult to assess mortality and its causes accurately in Samoan populations. With some careful analysis the mortality causes in Western Samoa could probably be clarified but, since none of the studies conducted as part of this overall project on change in the Samoan population included such an analysis, we decided it was better to exclude the topic from this chapter. Of the studies that were undertaken, the most extensive was the one of American Samoa based on the frequent censuses and the relatively reliable record of death certificates from 1950 to 1981. Although the analysis of records from Hawaii and California is based on less reliable recovery systems, these records provided an essential supplement of information about the constant outflow of migrants from both American and Western Samoa.

From the American Samoan analyses we may conclude that the transition from a mortality based almost entirely on infectious disease to one based on the degenerative and traumatic causes was already well underway by the early 1950s and was essentially complete by the early 1960s, even though mortality from infectious diseases continued to fall slowly thereafter. The reliable sample survey from Western Samoa suggests a current age-specific mortality resembling American Samoa's in the early 1960s, but one that may differ in some particulars. Despite the early transition in American Samoa the pattern of cause of death still remained somewhat unique in the late 1970s and early 1980s. Relatively low death rates from cardiovascular disease persisted, while relatively high death rates from trauma and diabetes mellitus were found. Exactly how unique these rates were is hard to determine, since the age structure of the Samoan population stayed young not only in American Samoa but also in the migrant locations of Hawaii and California. It appears from the age- and sex-specific rates that the high trauma death rates can particularly be ascribed to deaths among the young adult males. These deaths can be partially attributed to the increase in the use of motor vehicles, but deaths from other causes including homicide and suicide are also high, as are deaths from unspecified trauma. The high trauma-related death rate of the non-Samoan males in American Samoa was a significant contributor to this rate, but there was also an increased rate of trauma-related death among young Samoan men in the 1970s and early 1980s compared to previous years.

While the high blood pressures and obesity of the Samoans suggest that the death rate from cardiovascular disease may rise in the future, the rate for middle- and old-aged individuals remained relatively low as late as 1980 not only in American Samoa but also among Hawaiian and Californian Samoans. Because of the youthful age structure of these groups it is difficult to compare them with published data on other populations, but in their analysis of Samoan rates for the 1963-1974 period Crews and MacKeen (1982) found the age-adjusted rates for American Samoans substantially below U.S. rates. Pawson and Janes (in press) conclude that the 1978-1982 rates for Samoans in California must be no higher than all California rates.

Because of improving diagnostic capabilities, the rate of death from diabetes mellitus in American Samoa may not have risen as substantially over time as the rates suggested. Nevertheless, the rates throughout the 1970s were very high. In his 1985 analysis of the rates Crews showed that the crude mortality rate reported for American Samoans was nearly 25 percent greater than the U.S. crude rate, but given the different age structures of the populations in the comparison, the rate differences in the older age groups must be much greater.

Finally, the biological and social characteristics of up to 50 percent of the adult American Samoan population was compared to their 6-year mortality in the hopes of understanding how the relatively high blood pressures and obesity of this population contributed to their deaths. As might be anticipated from other studies, the elevated blood pressures did contribute to mortality, but somewhat surprisingly the very high relative body weights of American Samoans in the 1970s did not seem to do so. Of course the Samoans may have, as a group, been so fat that the effect was not apparent or the contribution of increased relative weight to increased mortality may have occurred only through effects on blood pressure. In any case, the analysis suggested that currently the extreme relative weights of Samoans contributed only indirectly, if at all, to middle- and old-age risk of death.

The sample of deaths for which social measures and CMI information was available was quite small. Even so, the models suggested that widowhood may increase the risk of death, as may a greater amount of formal education. Elsewhere in this book it is argued that the transition from the traditional life in a Samoan village to participation in more modern urban life-styles may be damaging to selected aspects of health and welfare. A more extensive study of prospective mortality, which includes detailed and better data on the social and psychological stress characteristics of a large sample is necessary before we can determine how the change is affecting cause-specific mortality.

Chapter 6

Life Histories

FAY ALA'ILIMA
MARY LIANA STOVER

During the past 30 years, thousands of Samoans have moved from small traditional subsistence villages into the heart of American cities. Today there are over 14 000 Samoans in Honolulu and over 20 000 in California. Such a drastic change in physical and cultural environments requires untold changes in physiology, and individual and social behavior on the part of the migrants. Human biologists, psychological anthropologists, and demographers have studied hundreds of these migrants. In this volume, they report on many of these changes. We can from these studies develop an objective view of the migration process and the stresses involved for the migrant population.

In this chapter, we present a complementary view of the migration process. With the thought that subjective analyses can sometimes give us insights that may enhance the understanding provided by objective measures, five individuals were interviewed by the senior author and associates about their migration experiences. Individuals were selected who were born in Samoa, had lived in the United States for over 10 years, and who came from a range of occupational backgrounds. All were interviewed in Honolulu and chose to talk of their experiences in English.

The interviews were deliberately as unstructured as possible. The interviewer tried to guide the conversation to include four topics: (1) life in Samoa; (2) first impressions after migration; (3) long-term adjustment; and (4) summary and evaluation. These topics were presented as the general questions which follow:

1. What was your life like growing up in Samoa? Why did you decide to leave, and how did you arrange it?
2. What surprised or troubled you most when you first arrived?
3. What has happened to you since?

4. What has been the hardest to get used to over the years? What do you feel you gained or lost by moving? Do you want to move back?

Interviews lasted as long as the narrator felt he or she had something to say. The shortest was 3 hours at a single sitting, the longest involved six interviews spread over several weeks. All of the sessions were taped, and the written versions edited by the interviewer with the help of the informant to eliminate repetitions, grammatical errors, and irrelevancies, and to improve descriptiveness and sequencing. The complete version of the migrant life histories is being published elsewhere. Here we present highlights of each of the five interviews, followed by a final section summarizing the conclusions of the senior author.

TAU (MALE, BORN 1951)

Samoa

Tau was born in the village of Leusoali'i in Western Samoa, the second child of 10 children. His father was a Methodist minister, and so the family moved frequently. As the oldest of five boys, Tau was responsible for the household tasks: building the earth oven, killing pigs, serving the food, washing clothes in the river, sewing, minding babies, pulling weeds, planting taro, and carrying bunches of bananas and heavy baskets down from the plantation. He attended school as often as he could, but the duties required of him in a pastor's household held him back and he was considered a slow learner. He completed form V, however, and passed the entrance requirements for the Teachers Training College in Apia.

The minister's household was very strict. Family prayers were held at dawn and dusk each day. The most important commandment was "honor thy father and thy mother," which meant to him "do what you're told." Any second requests were followed by the belt.

When Tau was 18 he was sent to American Samoa to get a job to earn money for a contribution to a church dedication. He was not paid in cash by the aunt that employed him. Instead she promised to send barrels of beef to his father for the church contribution, and to buy him a ticket to America if he stayed after that. Although the foodstuffs were sent to the dedication, the ticket did not materialize. In the meantime he was exposed to the new life of Pago Pago where the young men wore pants and shoes instead of lavalavas and bare feet, and actually got to earn money to keep for themselves.

He went back and forth several times from then on between the two islands. His aunt in American Samoa promised him the proceeds from the taro patch to pay for a ticket to America. He worked for her for over a year, but he found that the more taro he grew the more people he fed, and the more his aunt contributed to ceremonies. After he had given up, she promised him a position on a dance team going to California if he would plant 1000 taro and build a house. This he did, and was ultimately allowed to go.

America: First impressions

When the team returned to Samoa, however, Tau stayed in California to study electronics on a student visa. Since his father was born in American Samoa, Tau was able to apply for papers as a "national," and was thus able to work. He had a variety of manual labor jobs, living with different uncles in different parts of the United States, and always relinquishing his entire paycheck to them. It was after he decided to keep a bonus for himself and was violently accused by an uncle of cheating that he decided to move out on his own with a sister. He signed his paycheck over to his sister at that point, with the understanding that she would send some of it home to his parents. This she did not do, and when he realized that he signed up at a recruitment office for the U.S. Army. This, he says, was the best decision he ever made.

Boot camp was easy for him since he was strong and the drills were no problem. He ate better than he had in his life and the sergeant would tell him what to do. No one asked for his paycheck. Of the $280 he earned the first month, he kept $80 and sent $200 to his parents.

Europe

His first assignment was Europe, but as he could not understand his orders, he seldom left the barracks to prove he was a good soldier. His roommate was convinced there was something wrong with him. It was not until he met some other Samoans who could read his orders that he discovered what his duties were. During this time the church was of real help to him. He attended every service. Prayer calmed him down, and the Protestant chaplain was very kind to him.

Hawaii

After three years he was sent to Hawaii, where he met a well-educated Samoan girl and was married. He continued to support his parents in Samoa, contributing to ceremonies and paying for school fees, as well as travel expenses to and from Hawaii. The army gave him everything he needed, and after a few years he was assigned to be a recruiter.

Summary

The one thing that bothers him is that young Americans may talk back to superiors. This seems like disrespect to him, and his reaction is to lash out at them. He respected his parents when he was a child. Now he expects younger people to respect him.

Family troubles have also been occurring because of the constant necessity to support his family in Samoa. Medical bills come frequently, as do travel fares and church contributions. Various family members have come to stay with them. His wife has threatened to leave him and he is deeply in debt. He wants to be a good son, however.

He also wants to be a good father and husband. For this reason he thinks he will stay with the army. The sergeants know how to give orders and enforce discipline. Besides there is plenty to eat and a paycheck of one's own.

EDDIE (MALE, BORN 1930)

Samoa

Eddie was born in Tafitoala in the Safata district of Western Samoa, the first son and the oldest of six children. His father, a village orator, was determined that his sons should have an education, since he himself had dropped out of primary school. Eddie was sent to a Catholic boarding school once he had completed primary school in the village.

The boarding school was reestablishing itself on Upolu after operating during the war years on Savai'i. Two months were needed for the students to clear the vines from the old school buildings and to plant the taro for their food. Many of the boys started dropping out not only because of the hard work and the small rations but because they weren't doing anything different from what they would be doing at home. Of 800 students, only 300 remained that first year. The reason he didn't quit, he said, is that he knew his father would beat him if he did, and send him right back again.

At age 14, Eddie was the youngest, and as such at the constant beck and call of the other students. When the New Zealand principal saw what was happening, he removed Eddie from the situation by assigning him to be his houseboy. Eddie's life was made easier and he was able to learn English faster than the other students. The following years were better for him, for now he could start ordering the new boys around. By the time he graduated, he was a school prefect. He also held Samoa's interschool record for the hundred-yard dash.

New Zealand

Under a newly established scholarship scheme with the New Zealand government, Eddie was selected to attend St. Patrick's College, a boy's preparatory school in Wellington. He was the only Samoan at the school, living in a dormitory with 30 New Zealand boys. The boys made fun of him, calling him "black boy." Again, he felt like quitting, but he knew his father would be waiting for him with a stick if he returned to Samoa.

The sports master noticed that Eddie was a fast runner and invited him to join the football team. The rest of the team was older and bigger than he was and wouldn't let Eddie join in football practice. One day during lunch when the team was passing the ball around, Eddie was excluded from the circle. Frustrated, he jumped into the circle, grabbing the ball and sitting on it. When the biggest footballer tried to force him to give up the ball, Eddie sprang up with a upper cut to the fellow's jaw. The principal, trying to diminish racism in the school, was prepared to give the maximum number of cane strokes to the older footballer. Feeling sorry for the fellow, however, Eddie volunteered that the problem was partly his fault. So the penalty was reduced. The footballer thanked Eddie once they were outside the office and the two became friends. The next year Eddie was made captain of the team. He also broke the New Zealand record for the intercollegiate hundred-yard dash.

After graduation from St. Patrick's, Eddie passed the government exams and entered the University of Auckland. He wanted to be a teacher, but the government insisted Samoa needed electrical engineers. Eddie thought for a while of becoming a wrestler along with his friend Pita Maivia, but after being whirled across the room by the wrestling instructor, he decided to go back to engineering. During his second year of university, he met and married a girl on a Samoan scholarship studying nursing. When he finished his degree in 1956, they returned to Samoa with three children.

Samoa

Since Eddie was the first Samoan trained to design and install electrical systems, he advanced up quickly in the public works department. There was much work to be done in the early days after independence since many projects were being financed and most of them required electrical wiring. Eddie was bonded to work for the government for 7 years to repay his scholarship. When his obligations were complete, however, he decided to move his family to American Samoa. Not only was he offered a job with very high pay, but he wanted to be near a hospital where his wife, now suffering from breast cancer, could be treated.

Jobs and money came easy for him in American Samoa. So did the alcohol and the women. His wife's cancer became progressively worse, however, to the point that blindness was setting in. She couldn't talk and was ultimately too weak to move. His son announced that he was going to church to have the congregation and pastor pray for her. Eddie went out drinking. The next morning his son called him at work announcing that his mother was cured; that she could see and was walking in the yard. Eddie thought he was lying in order to get him to come home. But his wife was cured. And the doctor at the hospital said he had never seen anything like it. A glaze had come over the wound. The cancer never returned.

That evening Eddie went with his wife and son to church, and after listening to the missionary he went forward to be forgiven. Eddie sees that his life was turned around at that point.

He became involved in the Assembly of God church, conducting Bible studies and organizing a new church building. He began a correspondence course to become a pastor, and, once ordained, he built another church and became its first pastor. He continued starting new congregations and erecting new buildings, until he received a letter asking him to come to Hawaii to organize a Samoan congregation there. He was very reluctant to go, however. Not only did he have his church work but he was making a high salary working for the government and was happy with his work. The governor did not want him to leave, either. It was only after being reassured that his mission field was the world that Eddie decided to go.

Hawaii

In Hawaii, Eddie's energies and ingenuity were really called upon. Not only did his Samoan congregation not have a church, but they had no money with which to build one. Most of his parishioners were on welfare. Within a few months, however, the church had its own church building, and within a few months after that yet another larger building was found to accommodate the growing congregation. Members were asked to make large sacrifices so the purchase could be made, and Eddie insisted that if anyone felt they were being asked too much, they should leave. The membership declined from 500 to 75. By asking as much as he could from the Samoan families, by borrowing as much as he could from the bank, and finally through radio pleas to the community, Eddie was able to collect enough not only for final payment on the land, but also to build the building and pave the parking lot. Membership was gradually increased again, and Eddie is in the process now of developing 11 more Samoan congregations on Oahu and on the outer islands.

Summary

Eddie sees many problems for his congregation in Honolulu. Poverty is one of these. In Samoa everyone has land on which to grow food and build a house, but in Hawaii, food is expensive and rents are high. It is usually a long wait to get into public housing. Most of his congregation is dependent on welfare. Prejudice is another problem, making jobs difficult to obtain, and exploitation is often experienced on the job. Temptation is great in the city and the family less in control. Older men run off with younger girls, food money is used for alcohol, and there is no one to discipline the children. The teenagers end up in jail for lack of strong guidance.

He and other pastors have organized a "hot line" where people may call when they are in trouble. Recently, a principal at a large high school called him when nine Samoan boys got themselves into trouble by threatening other students to get their lunch money. The parents of the boys were in the office threatening the principal. Eddie convinced the parents that if the principal wasn't allowed to hold the line, the children would not listen to the parents either. One of the fathers complained that indeed his son had threatened him with a two-by-four.

Eddie firmly believes in corporal punishment. He feels that more Samoan teachers, using Samoan methods of discipline, are needed in the schools. That was how his teachers were. Also, if Eddie had run away from school, his father would have beat him with a stick. Now he's grateful for the discipline, because if his father hadn't threatened him, Eddie would have quit before he knew what he was doing.

ATO (FEMALE, BORN 1944)

Samoa

Ato was born in the village of Tafatafa on the south coast of Upolu in Western Samoa. Her father held a title in that village and, like the other families, they lived off the taro, bananas, coconuts, pigs, and chickens that they grew on family land. Ato was the third of ten children and with many hands to do the work there was always enough to eat.

Ato particularly remembers how in the evening around the kerosene lamp her grandparents used to tell stories and sing songs about the ancient heroes and warrior. The children were made to memorize their genealogy. Her family was well known, she was told, because of its high Samoan title, and the family was clever and hardworking because of its German and Chinese blood. It made her feel proud. She was determined to work hard and keep up the family's reputation.

Her parents were strong Methodists. Almost every evening the children were sent to the pastor's house where they were taught to sing hymns, read the Bible, and pray. This was Ato's first school. Her family sent the pastor the first basket of food every time the earth oven was uncovered. Most of the cash that was earned was also given for church contributions.

When Ato was nine years old, she was adopted, Samoan fashion, by an aunt who was a teacher at a Methodist girls' boarding school. Life at the school was very difficult. Discipline was harsh and the diet poor, so the girls were often sick. Ato often thought of the ancient heroes and how life was not easy for them either. But they persisted and that's why her family was so proud of them.

The basic object of this boarding school was to train girls to be wives of Methodist ministers. The routine was rigid and discipline was strict.

Beyond their studies, which began in the early morning, the girls worked on the plantations each afternoon not only to raise food for themselves but for the pastors of the mission. Bedtime hours were strictly enforced. Ato began to realize that being the wife of a minister was not what she wanted. She wanted to be a teacher. When she graduated, her aunt permitted her to enter the Teachers' Training College in Apia. Two years later she began teaching at her aunt's boarding school, the youngest teacher there.

Ato was not happy teaching, however. As the youngest she had to take orders from all the older teachers, even though she was the only one with training. She wanted to teach science, but instead was given Bible and weaving classes, which were considered more important for pastors' wives. She tried to buy meat for the girls, but was reprimanded by the mission. She also objected to reporting in every night now that she was a teacher. It was finally decided that it would be easier for everyone if she got married. An elderly pastor whose wife had died was chosen for her. Ato did not want him, but she didn't know how to get out of the situation without disappointing her family.

In the summer of 1969 a group of students from Hawaii visited the school as part of an enrichment program. The team was led by Mr. Apa, a Samoan who had been educated in the United States, and his American wife. Ato helped orient the students and find Samoan families for them to live with. At the end of their stay, the Apas invited her to return with them to Hawaii. She could live with them while furthering her education. They thought she might be able to get a scholarship at the liberal arts college where Mrs. Apa was teaching. Ato could hardly believe her good fortune: she could attend college and get out of marrying the pastor.

Hawaii: First impressions

Ato was again adopted in the Samoan way, which now made her the eldest of six children in the Apa family. Since Mrs. Apa was teaching, Ato helped with the housework and with the children. The college gave her free tuition and a grant for books.

Even with all the help, adjustments were not easy. She found it scary to take the bus across the island to school. It was difficult understanding the teachers because they talked so fast, and Mrs. Apa's children studied with the television, which made it hard to concentrate. She was also appalled at the way American students argue with their teachers and talk back to their parents. She felt insecure because people kept asking her what she wanted instead of telling her what to do. Also, everything cost money in America, including medical care. She might have given up and gone back if it had not been for those stories her grandmother had told her about determined warriors and family pride.

One thing that particularly helped her was that as a lay preacher, Mr. Apa had a Samoan congregation. Ato could therefore speak Samoan and eat Samoan food with the church members at least once a week. She also discovered that while America seemed strange to her, she was getting acquainted with American customs faster than other Samoans, even though she had just arrived. This was not because she was smarter, she claims, but because she lived with a part-American family, which showed her how American families lived. Also, her teachers and fellow students were interested in where she came from and were willing to compare and explain. She had many American friends, while most Samoans hardly mixed at all.

Hawaii: Long-term adjustment

After the first year, with the consent of her adopted parents, Ato lived with other families. The first was the family of an Air Force officer, the second was that of a professor. She had to speak English in these families and to eat American food all the time, and while the situation was difficult for her, she gradually adjusted. It was during the last year of college that she met her husband, a retired military man, who was studying auto mechanics at a community college. His parents were well-to-do members of the black community on the East Coast. This was Ato's first introduction to black people, and yet another culture to adapt to. Ato was married shortly after her graduation, much to the displeasure of the teachers back in Samoa, who claimed that the old pastor was still waiting.

Like other American couples, Ato and her husband rented a small apartment in a large building where they knew no one. It would have been unbearably lonely for Ato, who was used to living with a big family and having vegetables and chickens in the yard, but another Samoan custom helped her out. The Apa's children and nieces and nephews were getting married by this time, so everyone met often to plan weddings and picnics and welcome each other's babies. Though the families lived separately, they were always in touch with each other.

Ato realized that while she had been planning to teach, her real interest was in helping immigrant families adjust to life in America. She received a minorities grant to study social work at the University of Hawaii. Her daughter was born during her first semester, and Ato might have had to withdraw from school, except for another Samoan custom. She asked her younger sister to come to Hawaii and take care of the baby and house while Ato studied. As soon as she finished her degree, Ato arranged for her sister to enter school.

Ato's first job was as a social worker at a children's hospital. She worked with all kinds of people, but was very thankful she spoke Samoan. When Samoan children are sick, their relatives want to live at the hospital, cook food, and take care of them. They want to sleep under the child's bed to be on call. American doctors and nurses insist that quiet, diet, and hygiene are more important. The result was angry people on both sides, and very little communication. Ato spent a lot of time explaining to both sides and trying to work out compromises. Payment for medical services was another area of tension. In Samoa medical care involved only a token fee, so she had to explain the bills to her Samoan clients and that Americans meant them to be paid.

When that hospital closed, Ato moved to a new children's hospital as part of a team to assess children with learning disabilities. With immigrant children it is not often easy to tell whether a child is really disabled or just hasn't learned the language and culture yet. She was able to rescue a few normal Samoans who might otherwise have been erroneously classified. She also began to work with families of children with emotional problems, learning that it was not only immigrants who felt they were losing control of their children. Many of her clients were middle-class American parents.

Her favorite extracurricular interest was organizing a group of young educated Samoans who wanted to do something for their own people. She found quite a few people who were on their way to becoming doctors, lawyers, teachers, and members of the helping professions. They called themselves *Fetuao* (Morning Star). One of their first projects was to organize a series of meetings with Samoan parents to discuss the confusion between Samoan discipline and the American concept of child abuse. Another program was established for Samoan children who were confused about their cultural identity. Ato introduced them to her grandparents' songs and stories about Samoan heroes.

Two years ago Ato was appointed social worker for an intensive care unit for infants. She is on call day and night to help parents of very young children in life and death situations. While her predecessors have termed this a "burn out" job, Ato has not "burned out." What has helped her through, she sees, is the tough discipline of school in Samoa, and living with different kinds of families in Hawaii. She thinks that the strains of her own life have made her more able to sense problems and develop practical compromises for others.

Summary

Ato still has many family problems of her own, in spite of her profession of helping others. Her parents still live in Tafatafa, but as they are not well and medical facilities are limited, she brings them to Hawaii quite often. Since her parents have no medical coverage, the cost is quite high for Ato. She wonders how her children are going to eat. Relatives in Samoa also count on her to help out with weddings and funeral contributions, and school fees. Her pregnant niece came to live with them as well. Twins were born who would not be alive today if Ato had not worked for an advanced children's hospital. They continued to need care, so Ato and her husband adopted both them and their mother. Ato understands that her husband has been very patient. Their house has only two bedrooms, and with all these dependents and visitors, it is sometimes hard to find a place to lie down in the living room. They finally got an unlisted telephone number to keep people from calling. It wouldn't be so hard if they were living from the land, Ato says, but she feels she looks like a millionaire to her relatives who have no idea what things cost. Also, since Ato is working she doesn't have long evenings to tell stories to her children. She finds it hard to be a good American mother and a loyal member of a Samoan family. She feels uncomfortable about returning to Samoa. Her future is in America where social workers are needed who know the problems through their own experience.

LIA (FEMALE, BORN 1931)

Samoa

Lia was born in Lalomanu, a small village on Upolu. Her father had the Leifi title, which was an important one in that area. Lia was the only girl of six children, which, she says, is how she learned to be a fighter. She planted taro, killed pigs, and made earth ovens just like the boys. Lia attended village primary school and was considered "cheeky," getting into fights with both boys and girls. When she was ten, the teacher whipped her in front of the class. Lia never returned to school.

For the next four years she worked at home, weeding, cooking, and washing. In the evenings she walked over to the pastor's house, where the children were taught to sing hymns and read the Bible. Like the other girls, Lia would spend the night there. She enjoyed her life, and hoped to marry a pastor some day.

New Zealand

When she was 14 her father sent her and one of her brothers to New Zealand to earn money and establish a place for the family there. A family friend was there to help them settle in. Lia immediately got work in a shoe factory. The work was easy. Although she knew only two words in English, "hello" and "good-by," she mimicked what she saw others doing, and managed to get by. In several months she and her brother had earned enough for the rest of the family to join them. With most of them working, the family was soon able to make a down payment on a house. There weren't many Samoans in Auckland in the 1940s, but the Fanene's from Saleilua moved into their neighborhood. Included in the extended family was a 10-year-old boy, Pita.

Pita was an unusually strong boy. He hated school, spoke almost no English, and loved fighting. He wasn't afraid of his teachers or policemen. He also hated work. By the time he was 15, when he needed money he would challenge men in the bars to fight and take bets. He always won. He would also fight if anyone said anything bad about

Samoans in his hearing. Lia thought he was wonderful and fell in love with him.

Lia says she was pretty wild in those days, as well. She'd lose her job every month or so, usually because she couldn't understand instruction. If someone laughed at her, she would grab the person by the hair. No one approved of that except Pita. Her family did not like Pita, and they threatened to throw her out of the family if she continued seeing him. Although the thought of living alone frightened her, she remained defiant. They lied about their ages when the got their marriage license at city hall. Pita was 15; Lia was 17.

Neither of them had ever lived alone before, and they found everything much more expensive than they had thought. Now Lia didn't dare fight for fear of losing her job and not being able to pay the rent. Pita stopped cruising the street and became a doorman at a billiard parlor. They thought things might be easier if they started going to church. Maybe God would help them. Within a short time Pita became a deacon. Then with two other Samoan couples they organized their own church. Three years later the church building was finished, the same year their only child, a daughter, Ata, was born.

For two years Pita tried his hand at boxing at a local boxing school. But whenever he lost he would start a fight in the streets, so the school let him go. Then Pita heard about a wrestling school that was opening in Auckland, and decided to try that. As soon as Pita started wrestling in the ring, he stopped fighting in the streets. He even began to help other Samoans who were in trouble. Lia didn't have to bail him out at the police station any more; instead, he began to earn a little money in competitions. Now it was Lia's turn to get in trouble. Sitting in the front row by the ring, she would fight anyone who called out anything derogatory about Pita.

In 1963 Pita accepted an invitation to wrestle in Europe. Lia was very unhappy to have the family split up, but two months later Pita flew back to New Zealand to get Lia and the baby. He had a 3-year contract and they were going to live in London Lia had never heard of London, but she thought she might as well see it.

Europe

The promoter found them a luxury apartment, and told them they would have to change their name from Fanene. He suggested Maivia because it would remind people of a Spanish bullfighter. Pita like bullfights and the name sounded good in Samoan, so they agreed.

For once Lia didn't have to work, but because her daughter was in school she couldn't travel with Pita. Only after a friend volunteered to look after Ata could she join Pita. It was then that her ringside problems started again. Lia couldn't sit still when she heard people calling out negative things about Pita or Samoans. After a championship fight in

Scotland, when people yelled for the "barbarian" to go home, Lia grabbed a chair and hit a man, who lashed back at her with a knife. The police were called. Lia was firmly told by the promoter that if she wanted her husband to succeed, she should never do that again.

Back in London, Pita was offered the role of a Japanese sumo wrestler in a James Bond film. He couldn't speak Japanese, but they told him to speak Samoan, because when they speeded up the tape it would sound like Japanese. The film was a success. The Queen and her husband attended its premier. Pita and Lia were also invited.

The promotor was pleased with Pita, but he wanted Lia to be more photogenic, but she refused to wear the makeup the beauticians put on her. The promoter decided to hire a beautiful girl each time to give Pita a lei and walk down the aisle for the television cameras. The first time Lia saw the stunt, she jumped into the aisle and grabbed the girl by the hair. "I'll take care of that pretty face of yours," she hissed. The girl never came back and the promoter was angry, threatening to send them back to New Zealand if it happened again. Lia was ready to leave, because she felt Europeans had no sense of morality. But the manager's wife took her aside and explained that TV girls are just part of a wrestler's life. It was the hardest western custom Lia had to adjust to, she maintains.

Pita made another movie and was becoming increasingly popular all over Europe. When his London contract was finished, the family moved to Paris for three years. Their daughter learned French and attended school there. Lia was still having trouble with English, however, so when the opportunity came to move to the United States, they decided to accept it.

The United States

In America they became celebrities, and in San Francisco they were invited out to one of the finest restaurants by the local politicians. They were refused service, however, and a brawl started. Both Pita and his wife were arrested, but were released after a few hours. The police knew the community would be gathered the next day, and many Samoans would be upset if they heard Pita was in jail for standing up for the politicians.

The family moved around the country, spending several years in different cities. While they were in New York, their daughter turned 21. She could speak English, French, Spanish, but not much Samoan. She also didn't think of Samoa as home, which disturbed her parents. Lia and Pita wanted Ata to marry a Samoan, but Ata was in love with a black named Rocky who worked out in Pita's gym in San Francisco. Lia was relieved when the family moved to New York. But when they returned to San Francisco for a short tour, Ata secretly married Rocky in Las Vegas. Pita and Lia were furious, and Pita beat Rocky. Ata begged and cried for forgiveness. Remembering what she and Pita had done to her parents and they were even younger than Rocky and Ata, Lia convinced Pita to let them go. Pita would have nothing to do with the young couple, until Ata had a baby. Then he rushed to the hospital to see his new grandson, and the family has been close ever since.

In 1979 Pita accepted an offer to become a wrestling promoter, and was given the wrestling franchise for Hawaii. Lia became his business partner. Since her responsibilities included scheduling, finding crews, and borrowing money, she wished she could write better. But one thing she knew thoroughly was wrestling.

Pita enjoyed training the local boys in Hawaii, but he dreamed of training boys from the Pacific islands. As his business manager, Lia discouraged the idea. It was expensive to get to the islands, people don't reply when you write them, and they don't stick to schedules. But Pita decided to try. Their first tour to arouse interest was to American and Western Samoa in 1979. The response was overwhelming. Pita wanted to expand to Tonga, but the king would not give permission. Other promoters had tried and failed to enter the country. Pita decided he should talk personally with the king to convince him of the opportunities for the young people. He succeeded, and a fight was scheduled in Tonga for February 1982 between Pita and the champion of Italy. As soon as Pita entered the ring, Lia could tell something was wrong. He won the fight, but collapsed as soon as he entered the dressing room. The diagnosis upon their return to Hawaii was advanced cancer, with three months to live.

Pita and Lia decided to spend the last three months talking with each other, so Lia moved into the hospital. Pita had known for a year that he was sick, but felt he had to go on being champion as long as he could. Lia promised that she would carry on his work when he died, and Pita promised to help from the other side.

Pita was buried on Diamond Head, contrary to the wishes of his family in New Zealand and Samoa. After his death, family members tried to dig up his body and take it back to New Zealand. Lia resisted them. She needed Pita close by to help her. Every day she visits his grave to receive his guidance.

A week after he died, Lia signed the formal papers taking over the National Wrestling Alliance franchise for Hawaii and the Pacific, becoming the only woman wrestling promoter in the world. Since then her life has become filled with wrestling preparations and negotiations. The main problem she has is in finding men who will work with her and not try to take over the operation, but she has no intention of being a figurehead. This is her business and Pita's. Some promoters are embarrassed to negotiate with a woman. When she attends the international conventions, she always packs five fresh leis and five Samoan dresses, one for each day of the week. The other promoters bring their wives, but Lia is the only woman allowed in the meetings.

Summary

Lia sometimes thinks of going back to Samoa, but the last time she went she didn't get farther than Apia. Relatives came in asking for money for ceremonies, school fees, medical bills, and church contributions. She couldn't say no, and within two weeks she was $10 000 in debt. Her mainstay is the local kids she is working with. She is very close to the boys she promotes, and they watch her apartment and drive her around. As she says, they're not the kind with whom you'd want to argue. They come to her as well when they're in trouble, for advice on money, police, and girlfriends. And she explains TV girls to their wives.

TUTUILA (FEMALE, BORN 1926)

Samoa

Tutuila was born in Lau village in American Samoa, in a family of seven girls and one boy. When Tutuila was one week old, her father's sister came from Upolu to take Tutuila's twin sister to her father's family in Western Samoa. The twin was named Upolu; the one which stayed was called Tutuila.

The title in Tutuila's family, Fei, was held by her mother's brother. Her parents, who served him, slept in the kitchen house. Tutuila was allowed to sleep in the larger *fale afolau* (oval Samoan house) with Fei. As a child she weeded taro, cooked, and served her uncle. When there was less work to do she would attend school. She was also allowed to attend the pastor's school, where she learned to read the Bible. She finished the fifth grade.

When her mother died, her father returned to his family on Upolu and Tutuila went with him in order to see her twin. Her father's family wanted her to stay, but she did not like the school where her father's brother was teaching. There were no desks, so the children sat on mats. When a church *malaga* (visiting group) came from American Samoa, Tutuila returned with them. Later, when she was 17 years old, she tried to return to Upolu, but Fei insisted she stay with him, because he had a young wife who needed help with the children. Since there were no boys left in the family, Tutuila became the principal food gatherer as well. She climbed the mountain for taro, breadfruit, and yams, and caught squid, octopus, eel, and fish on the reef.

When the war came, Fei joined the marines, and even though he was away from the village, he still considered himself Tutuila's protector. Once when, the *aualuma* (village womens' group) sponsored a dance for the Samoan marines, Fei attended with his fellow marines. Seeing Tutuila watching the dance, he beat her. Tutuila never went to another dance. When one of the village boys wanted to marry her, Fei chased him away and beat him. In the face of Tutuila's tears, however, the old man finally gave his consent. But he didn't like the young man, and so refused to attend the wedding ceremony.

Lafo and Tutuila stayed in Lau for the first years of a their marriage, serving Lafo's father. After some time, however, Lafo began working for the police and went to live in Fagatogo, where he lived with another woman. When Tutuila and Lafo's sister went into town to get his paycheck, they found no money, only the woman, who considered herself to be Lafo's wife. Tutuila declared that she had had enough and informed Lafo's father that she was going home to Fei.

So Tutuila returned with her two children to serve her uncle. Again the burden of the household was on her shoulders. Although she refused to get a divorce from Lafo, she bore three boys by a man of the village. Eventually, when Fei and his wife moved because of illness, Tutuila was left alone in Lau with her five children. Because she had no man and no money, she continued to support the family, by planting taro, catching fish, and making the *umu* (earth oven). She also, along with other women of the village, earned money by weaving mats and hats and selling them along the road, and by dancing for tourists.

A friend, Toa, with whom Tutuila had been working, was corresponding with a Filipino in Hawaii. Eventually, Toa went to Hawaii to get married, writing back to Tutuila that her husband had a friend who also wanted to marry a Samoan girl. If Tutuila would come to Hawaii for a visit, he would pay the plane fare. Tutuila accepted the offer. She decided to get officially divorced before she left, however. Since Lafo had moved to Hawaii and now was living with several girl friends, Tutuila wanted to have legal custody of her children before leaving Samoa.

Hawaii: First impressions

Tutuila flew alone to Hawaii, leaving her children in Uncle Fei's care. She was very frightened. Her English was not good and she had no money. But Toa and her new husband were there to meet her. The husband was small, noted Tutuila, but rich; he had a car, a cheap plantation house, and an icebox full of food. Toa and her husband introduced Tutuila to their friend, Rico. He was small, too, but he informed Tutuila that he loved big Samoan women and wanted to get married right away. Rico had worked in the sugar plantation, but now he was a carpenter. He took Tutuila to his house in Waimanalo, which felt just like Samoa to her. Five days later they were married at the Samoan LMS church (Congregational) in Maunalua.

Rico was a very jealous husband, but a good one, in Tutuila's opinion. Every payday he would turn his paycheck over to her. Within the first year they saved enough to pay the airfare for Tutuila's five children. This was over the objections of Uncle Fei, who wanted the children to remain in Samoa to serve him. Tutuila insisted the children be brought to Hawaii so they could learn English, go to school, and get good jobs. With the money they would be able to serve the family and eventually gain titles for themselves. If they stayed in Samoa with Fei, he would work them hard, planting taro and cooking food. There would be no time for school and they would stay "stupid" like she was.

Rico's small, two-bedroom house was soon filled with children, including two more that Rico and Tutuila had together. The family was never wanting for food, however. Tutuila planted taro, banana, and manioc, and after work the whole family would go fishing, Tutuila and her daughter collecting sea cucumbers, sea urchins, and octopus from the reef.

Hawaii: Long-term adjustment

Eight other Filipinos in Waimanalo had also taken Samoan wives. The families all went to mass on Sundays. But after some time, they decided to organize a Samoan church for themselves. A Samoan pastor was contacted and services were held in Samoan. Tutuila particularly enjoyed the opportunity to sing Samoan hymns and eat Samoan food, just like at home.

Her enjoyment of the church was counteracted, however, by her shame at being duped by a man selling life insurance. He convinced her that the insurance would protect her and her children should Rico die. If she paid him only $7 a month, she would get $5000 upon Rico's death. She didn't tell Rico what she was doing.

Rico died in 1969 but the insurance man never came to pay the $5000. All Tutuila got each month was $120 from Rico's social security benefits. She tried selling hats for a while and made $30 to $50 on a weekend. But when the government tried to regulate roadside selling, she could not make any money, so she gave up her business.

All her children finished high school. The boys then joined the armed services, and sent their checks back to Tutuila. Her daughter married a Samoan, but three of the boys married *palagi* (whites). This angered Tutuila, as she sees that palagis don't serve the family, but only care for their own babies. Also they don't like their husbands to send money home. She had promised Fei that she would bring back smart children to serve the family, but the non-Samoan wives would let neither their husbands nor the money go. Only one son married a Samoan girl. Tutuila hoped this couple would go back to Samoa to get a title and make her proud, but they didn't understand the custom. This discouraged Tutuila. Why send the children to school at all?

But her real trouble was with Rico's boy, Minko. He was truant, never obeyed her, and smoked marijuana with his friends. After the social worker came to see her about him, Tutuila decided to send the boy to Samoa where she knew Fei would make him obey. Indeed, Fei made the boy work in the plantation, make the oven, and serve food. The boy had never experienced this kind of hard work, but he also had no friends in Samoa he could relax with. Tutuila knew he was unhappy, but when she allowed him to come back to Hawaii, he got into trouble again. Therefore when her daughter married and moved to Texas, Tutuila insisted that Minko go with them. The daughter was one of the few people who could discipline him. The arrangement seemed to work out.

Summary

Now, all of Tutuila's children are grown and she lives alone. She is too weak to plant taro and catch fish, so she is grateful for welfare and the food stamps she gets. Even though she has raised many children, she finds she has no one to serve her now. She had her daughter's two girls sent from Texas to look after her, but the girls are still too young to be of much help. Nevertheless, they are easier to handle than boys. Tutuila instructs them to study hard, to obey, and take care of their grandmother. She wants them to get smart, and go back to Samoa to serve the Fei title. But Tutuila thinks now it might not happen. Maybe by bringing her family to Hawaii she didn't make the family strong. Maybe she broke it.

The doctors have told Tutuila she has high blood pressure. She was hospitalized once, but did not like the place. When she tried to leave, however, she could not find her clothes. She still goes to Samoa three or four times a year for *fa'alavelaves* in spite of her poor health. No one can stop serving the title, she says.

But she always returns to Waimanalo where Rico is. He was her best husband, she says. The welfare is good to her now. They pay her hospital bills, but she has told them not to send her any more food stamps, as she is going to teach the children how to garden and to catch fish.

IMPRESSIONS

In reviewing the life histories we were impressed first with the uniqueness of each of the narrator's adjustments. While we tried initially to posit contradictions between Samoan and American institutions and culture, we realized that such an oversimplification was not meaningful, since Samoan traditions and American urban life were experienced quite differently by each narrator. First we discuss the Samoan experience; then the response to American culture and the migration process; and finally, the long-term adjustments of a Samoan in the United States.

Experience with Samoan cultural institutions

All the narrators mention three Samoan practices as important in their upbringing: strict discipline, church attendance, and service to their extended family. They coped with these, however, in different ways.

Discipline

Lia recalls fighting with students who called her names, and being "cheeky" to the teacher. She was whipped for her impertinence in front of her classmates in the second grade. Her response was not obedience but prompt rebellion. She walked out, which ended her formal education. This pattern of response to authority probably affected her response to the police in San Francisco later on.

Tau responded to strict discipline in a different manner. He made sincere efforts to obey the commands of his parents, even to the point of feeling heavily overburdened. This pattern made it difficult for him to say "no" to requests later in America.

When Ato felt the discipline of the Methodist school unduly restricted her activities, her response was neither obedience nor physical violence. She found she could manipulate the situation by going to the school down the road. One of her most useful gifts today in handling conflicts between medical personnel and patients is an ability to think up ingenious accommodations.

Of all the narrators Tutuila seems to have experienced the harshest discipline and the least opportunity to pursue her own development. Service to the family came before school. Fei, the family head, beat her for watching a dance, and left her as sole provider for a group of children. She did not respond with defiance, but developed ways of handling him. She got to marry the man of her choice by simple persistence. She persuaded Fei to send her his "boys" and used him when she wanted to discipline Minko. She was educating her children to succeed to his position of authority.

The church

All mentioned a Samoan church in their upbringing, but religion was experienced differently and played a different role in each adjustment process. Eddie mentions his family's Catholicism only as determining where he went to school. It did not deter him from experimenting with "sin." His later conversion, however, was obviously the most significant event of his life. It provided a resolution to great stress, firm guidelines for behavior, and release for his charismatic personality and extraordinary energies. Jumping ethnic cultures was irrelevant after that.

In Tau's family, religion shaped everything. His roots were in a church and not a village. It was something impossible to question, and he never has. When his parents used divine as well as cultural and physical sanctions to enforce their demands, he left unable to rebel or even manipulate this situation. He could only obey until the burden reached explosive proportions. After he found relief in the army, the church played a supportive role in his life. In Germany, the chapel was the one place he felt secure, and prayer and chaplains helped him quiet violent impulses. He feels it has been his most important support in the adjustment process and still attends church regularly wherever he goes.

Lia and her friends built a church in New Zealand despite their poverty. To her, it meant a way of getting God to "bless" them in tangible rather than psychological ways. That was the year she had her only child.

Tutuila was also an organizer of her church in Waimanalo. The church enabled the Samoan wives to get together regularly, enjoy each other's company, and support each other's needs. They began to put on Samoan dances for Honolulu festivals and for other churches. As a group they felt able to venture into the wider community. To them the Samoan church was important socially.

Ato recalls the shock of meeting young Americans at her college who were questioning the existence of God. It was a stressful experience for her at that time. A few years later, however, she realized that the heavy demands of a Samoan church conflicted with her family responsibililties, and she deliberately reduced her financial and time commitment without a great sense of guilt. Thus, each migrant shows a different response to church options.

Family obligation

A third cultural phenomenon was a strong sense of obligation to the family of origin. Except for Eddie, who has transferred his primary obligations to the church, they all find contributing to such Samoan ceremonial activities as weddings, funerals, church dedications, and helping with housing, travel, and medical care burdensome. As Lia said, her family in Samoa considers her a "millionaire" even though she is struggling to make ends meet. In Tau's case, these demands threaten his marriage. Tutuila returns to Samoa for special occasions even at serious risk to her health. Lia wants to go "home" but cannot because the financial demands become so high. Ato cannot say "no" even though it strains her family life in Honolulu. All of them claim the heavy contributions cause them stress and retard their ability to "get ahead" in America.

Why do they continue to do it? Ato says that if she did not respond she would no longer "feel like a Samoan." It is important to her sense of identity to continue to be an active member of her family of origin even though she may never return. Even in Honolulu, it makes clear who she is, and gives her roots. She hopes it will do the same for her children. Lia took her daughter to visit Samoa and paid heavily to keep the connection alive. She wanted to ensure that the family back home remembered them. Tutuila constantly visited and contributed in hopes that her children would be considered for the family title. Much of her current stress relates to the discovery that none of them wants to go back.

It should be noted that while all five grew up in extended families in

Samoan villages, there were differences in their backgrounds. Eddie, Ato, and Lia, for example, had titled fathers with authority over their family members and status in the village hierarchy. Tau's father was a pastor and was thus a privileged person within the village hierarchy. On the other hand, Tutuila's father was an untitled man serving a chief in this wife's family. He was unable to make decisions even over the fate of his own children. Today in Honolulu, Eddie, Ato, Lia, and Tau are busy building their careers in their respective professions. Tutuila's aspiration has not changed. She is still struggling to help her children get a Samoan title.

Response to American society

Because of its broadness and the variability, defining American culture is difficult. In Honolulu alone there are many subcultures, with institutions and practices of their own. Each of the narrators has created their lives in a different subculture. Tau is adjusting to life in the military; Ato is coping with the problems of becoming a professional in a medical setting; Eddie is making his way up the hierarchy of a charismatic church; Tutuila is at home in an ethnic community; and Lia is becoming an established figure in the wrestling world.

We noted that it is critical for the immigrant, as for an American adolescent, to find a subculture with the least stress and the most opportunity for self-realization, where energies can be released. Tau was tremendously relieved when he finally found the army. It gave him the authority figures he needed, a clear line of ascent, and a paycheck of his own. In this subculture he was quite well equipped to be successful. Lia, with her penchant to lash out physically against restriction, would probably not have made it through boot camp. But she was quite comfortable in the wrestling world and has made a successful career there. Tutuila recognized Waimanalo as a community in which she could feel comfortable with little change of life-style. While it did not have a Samoan church, neither did it have the demanding chief. She helped organize a church, and was quite happy to be without the chief.

For Eddie, the church was the foundation of his life. He felt he had been "lost" and was now "found." In this particular church he could release his charismatic personality and energies. The same church might have proved impossible for Ato, who found release for her analytical skills in an academic environment.

Viewed from this subcultural perspective, once the language barrier is mastered, Samoans do not have greater problems than anyone else who might change subcultures. A person bred in an upper-class community in Honolulu would experience more stress if that person tried to live in a plantation shack than Tutuila did after she moved to Hawaii. And an

American church woman would be very uncomfortable in Lia's world of syndicates and physical force.

Exposure and adjustment

Finally, we noted differences in how these narrators were exposed to their new culture. These differences affected their speed of readjustment and their ability to handle a variety of new situations. We can trace the following stages of adjustment: initial and long-term.
Initial adjustments included:

1. *Shock.* Feelings of helplessness, confusion, fear of getting lost.
2. *Language barrier.* Tension learning the language well enough to understand directions and make needs known.
3. *Survival skills.* Tension until one is sure that one can obtain the basic necessities of life.
4. *Social skills.* Tension until one knows enough of new social practices to make friends with people in the new place and to defend oneself.

Long-term adjustments included:

1. *Finding a niche.* Locating some facet of the new environment in which adjustment is possible with least stress and greatest opportunity for self-development.
2. *Depth.* A feeling one can cope with the niche, knows where one is going in it, and that one has the capabilities to get ahead even though there may be unresolved tensions.
3. *Breadth.* Ability to see a variety of niches without feeling confused or threatened. Ability to help others make adjustments to other places.

All five narrators had been through the initial adjustment stages, so the problems that had to be solved were about the same for all of them. But the opportunities they had for learning skills varied considerably and determined both how fast they could go and how far they could get in the long-term adjustments. We call these opportunities "exposure."
In this area Eddie had an advantage. He was exposed to English and a New Zealand family while he was still in his teens. He attended a college in New Zealand and soon discovered how to stand up for himself and make friends. And later at the university, he learned the engineering skills that assured him a living.
Ato was not exposed to the new culture until she was an adult so the language barrier was more difficult for her. She had unusually good exposure opportunities, however, and these speeded her progress through the initial stages and helped her discover a prestigious niche in the new

society. This exposure enabled her to acquire not only depth but breadth in it, as being the only Samoan in a small college, she could immediately make American friends who were willing to explain things to her. She also found living with middle-class American families to be an important experience. These experiences and her academic opportunities enabled her to feel comfortable in the middle-class professional world. The great variety of her Samoan and American family connections has given her resources for helping families from a wide variety of subcultures.

Lia, on the other hand, had very limited academic exposure. In Auckland she had to learn her English, social, and economic skills from observation and often in situations in which people laughed at her. Her friends in the new society were largely street people. Her English is as well adapted to the streets and wrestling arenas as Ato's is to the professional world, as are her social and survival skills. Geographically, she has circled the globe, but her exposure in these places has been limited to the wrestling world. She fits her niche well, but it has been hard for her to get much breadth.

Tau had minimal language and cultural exposure before coming to America as an adult. His English improved in the army, as did his self-confidence, to the point where he can communicate with all groups through his role as recruiter. Tau knows his niche thoroughly, and has the depth needed for long-term adjustment. While he has some breadth and can deal with other subcultures through his role as recruiter, he is most comfortable in that of the army.

Tutuila had the most limited opportunities for exposure, both in Samoa and Hawaii. Her education is limited, she has seen little of America except Waimanalo, and most of her friends are Samoan. Her widest cross-cultural exposure has been to her Filipino husband, and the pidgin language. She has adjusted to her niche, but it will probably be impossible for her to get much depth.

These observations are not presented as definitive conclusions either about the narrators or the adjustment process, but are intended to illuminate the data provided by other studies in this volume.

Chapter 7

Changing Socialization Patterns of Contemporary Samoans

THELMA S. BAKER

Cultural transmission and socialization are processes that provide for th
biological and cultural continuity of a population. The form and content o
these processes provide individuals with an ideational system, an array o
age-related behavioral patterns and technical skills necessary to functio
effectively in a particular environment and society (Williams 1983
Howard 1976; Spindler 1974b, 1973). It can also be argued that th
socialization process can serve as a societal mechanism for stress contro
by providing individuals with behavioral guidelines that facilitate th
ability to respond predictably to a variety of situations in a know
environment.

  In traditional societies, where the natural and social environmen
changes are incremental, the socialization process is stable and the form
institutionalized. With due regard for individual variability and choice, fo
each stage in the life cycle, the process consists of: (1) the *agenda*, th
implicit or explicit rules for what is to be transmitted; (2) the *agents*, thos
responsible for the transmission process; (3) the *participants*, those wh
are being socialized; and (4) the *techniques*, the cultural rules for teachin
and learning. Without sudden alterations in subsistence or populatio
resources, individuals in traditional societies can make behaviora
decisions with a minimum of conflict because the rules for behavior an
the range of choices for most situations have been learned either durin
the early stages of the life cycle in childhood or through continuou
learning in adulthood. In societies where new variables, such as epidemi
disease, drought, crop failure, or culture contact, are introduced into th
natural or social environment the traditional complement of behaviors
skills, and personnel is tested as a template for decisions and adaptation
to the new environment (Spindler 1977; Spindler 1974a).

The environmental changes we have been investigating in the Samoan Studies Project, migration and modernization, have modified the traditional processes of socialization and cultural transmission. The problems to be explored in this chapter are focused on the changes in the socialization process that occur when Samoan villages modernize or when Samoans move from traditional villages to urban centers within the Samoas or migrate to Hawaii and mainland United States.

The changes in the sociocultural environment that Samoans have been experiencing we have called modernization in this volume. As P. Baker indicated in Chapter 1, participants in the Samoan project have interpreted the term in a variety of ways. In the present review I have followed the usage of anthropologists who conceptualize modernization as a subset of culture change (Spindler 1977; Hawthorn 1944), and focus on those aspects of modernization at the societal level that have been shown to affect the socialization process—that is, the centralization of political power, monetization of the economy, institution of formal education systems, introduction of complex technology and communication systems, and rapid demographic change (Wilcox 1982; Hansen 1979; Spindler 1974a).

The cultural agenda for traditional societies is customarily set by family or kin networks for a particular environmental niche. The centralization of political power, either by colonial rule or the establishment of nation states, may change the locus of control over the content of the agenda in order to change ideational systems and values (Wallace 1961). Economic change, resulting from new technology and wage labor, not only introduces different kinds of skills into the agenda but also the social distribution of knowledge, affecting both agents and participants (Gearing 1975; Gearing et al. 1975). New communication systems, particularly radio and television, may modify three components of the process by serving as agents of the transmission of information, by enlarging the classes of participants and by changing traditional transmission techniques (Hansen 1979).

Rapid demographic change, through population redistribution or increased life expectancy, for example, may modify the role behavior expected of young adults or the elderly. Finally, formal education systems, may be used by other components of the sociocultural system to modify the transmission agenda (Carnoy 1974), or become the principal agents of knowledge transmission, define classes of participants, and institute different socialization techniques.

In this chapter I explore the Samoan socialization process among traditional, migrant, and modernizing Samoan communities. I first describe cultural transmission during the stages of the life cycle in Western Samoa. Parallel information is provided for villages in American Samoa and comparative material presented from Hawaiian Samoan communities. The rural villages of Western Samoa serve as the base line for contrasting to the other settings. For each area, I describe the salient environmental variables and the constraints that the particular environment places on the socialization process. For each stage of the life cycle, I describe the agenda or the behavioral, ideational, and technical skills to be transmitted, the agents of and participants in the process, and the methods and techniques of socialization.

THE LIFE CYCLE

The life cycle in traditional Polynesian societies was not measured by chronological age but was recognized as changes in biology or functional capacity (Ritchie and Ritchie 1979). This was true in precontact Samoa, where three distinct stages in the life cycle were recognized, infancy, childhood, and adulthood. The Christianization of the Samoas in the mid-nineteenth century led to the first changes in life cycle categories. The use of the Western calendar to chronicle events introduced chronological age as an additional marker of life cycle stages. Formal education systems differentiated the childhood stage of the cycle. And with the establishment of pension-eligible retirement ages in the 1960s, adulthood also was subdivided. While the three traditional stages still serve as major markers of the life cycle in contemporary Western Samoa, each of the stages has been modified and here identified by chronological age.

Since American Samoa has become tied to a great many age-associated social programs in the United States, the life cycle there is similar to that in the United States and therefore can be defined chronologically. In order to enhance the comparability of information from the three major areas under review, I have designated somewhat arbitrary chronological ages to each of the stages of the life cycle:

Infancy—birth through 2 years of age
Childhood
 Preschool: 2-5 years
 School-age: 5-18 years
Adulthood
 Adults: 19-55 years
 Older adults: 55+ years

WESTERN SAMOA

In this section I describe traditional socialization in the villages of Western Samoa, and then discuss the effect of the modern sociopolitical and economic institutions that are affecting the process.

The analysis of socialization in the villages is based primarily on data gathered in the village of Salamumu, collected initially by Orans (1981), and then by the 1979 and 1982 Samoan Studies Project field teams.

Pelletier in the 1982 field session also collected data on the village of Salea'aumua (1984). Data for the study of the effect of modern institutions on traditional cultural transmission is derived from a number of sources. Data on demographic change, the work force, formal education, and household characteristics were collected during the 1979 and 1982 field sessions. I have relied on published sources for background material describing the centralization of political power (Gilson 1970; Davidson 1967) and economic change (Lockwood 1971).

The natural and social environment of traditional villages

On the large volcanic islands of the Samoan archipelago, the predictable characteristics of the tropical island environment, such as the ocean, a hot-wet climate, and volcanic soil provided the general context for socialization and cultural transmission. The ocean presents two challenges for socialization. The first is the transmission of skills for individual and group survival in a potentially hazardous environment, and the second the transfer of subsistence information and skills. The hot-wet climate of the tropics, on the other hand, is an environmental factor that imposes a minimum demand for cultural skills for biological survival, other than methods for controlling disease. The volcanic soil and the potential productivity of the land for subsistence agriculture necessitate the development of a variety of behaviors and subsistence techniques to provide resources (Beckerman 1977). And finally, the combination of the comparatively large size of the islands in the archipelago, the infrequency of cyclonic storms and volcanic eruptions do not present an environmental stress significant enough for the population to have developed the extensive disaster repertoire necessary for survival on coral atolls, for example (Maude 1981).

The social environment

Not only does socialization serve at the individual and population levels to enhance biological survival, but the process also provides for cultural continuity by ensuring that appropriate sex- and age-related role behavior and ideology are transmitted (Kimball 1982; Dobbert 1975; LeVine 1975; Spindler 1974b). As others have suggested (Freeman 1983; Shore 1983; Mead 1930), critical demands of the Samoan social environment at the population level are a function of the lineage and descent systems, the 'aiga, and the status differentiation associated with the matai system. At the individual level, are the parent/child relationships and the sibling and peer relationships, which both Gerber (1975) and Sutter (1980) have shown are important stressors in the Samoan social environment. And it is on these important relationships that socialization is focused throughout the life cycle.

The physical arrangement of Salamumu and the sociopolitical and economic structure of the village have been examined in detail by Orans (1981) and reviewed in Chapters 3 and 4 of this volume. At the time of our 1979 village census, there were 500 residents of the village in two locations, Salamumu tai (on the sea) and uta (in the bush). Seventy-nine percent (37) of the 47 village households contained extended families of three or more generations. These households ranged in size from 6 to 23 persons, with a mode of 13 people per unit. Twenty-two percent of the households contained only two generations, and they were located for the most part in Salamumu uta, the newer area of Salamumu. These households ranged in size from one household with just three people to one with nine; the other eight households contained an average of seven people. Forty-three percent of all village households included individuals who were over 60 years old, but on the average the two-generational households of Salamumu uta contained younger families than the rest of the village. The age and sex distribution of the village is shown in Table 7.1. These data approximate those collected by Pelletier on Salea'aumua in 1982, where he found similar household size (12 per household), approximately the same pattern of three- or more generational households, and a similar age/sex distribution of the population. The villages on Savai'i studied by Shore (1982) and Sutter (1980) show similar household composition and age/sex distributions.

Infancy

From birth to two years, Samoan infants are raised in a climate of indulgence, warmth, and continuous body contact, a typical Polynesian

Table 7.1 Age and Sex Distribution in Salamumu, 1979 Project Census

	Males		Females	
	N	Percent	N	Percent
Infancy Birth - 2	16	3	17	3
Preschool 2.1 - 5	27	5	24	5
School-age 5.1 - 13	79	16	74	15
Adults 13.1 - 59	129	26	103	20
Older adults 60 +	13	2	18	4
Total	264	53	236	47

socialization pattern (Ritchie and Ritchie 1979; Howard 1970). In Sala-mumu households we observed a pattern of feeding, carrying, and nuzzling of infants that was shared by parents, siblings, and grand-parents through the time of weaning. Mothers are primary caretakers for the first few months of life in that they breast-feed on demand and usually take responsibility for the baby's schedule. During the first months of nighttime feeding, infants sleep on the mats next to their mothers. Carrying the baby, diapering, and washing infant and clothing, however, are tasks shared by all members of the household. When bottle-feeding supplements are used or replace breast-feeding, mothers or other young women in the household ordinarily prepare the formula. After the fragility of the early months have passed, all members of the households share in the child care (Nardi 1983; Sutter 1980). The techniques of socialization in this early period are freely expressed warmth, nurturing, and tactile stimulation by both male and female caretakers of all ages. The young infant learns to respond to a variety of adults, and is accustomed from his or her earliest social interactions to a large group of caretakers, who help him/her develop motor and cognitive skills.

Childhood

During early childhood, or the preschool years, the skills to be taught and learned are the techniques for physical survival in a particular environment, the societal rules for fulfilling biological needs, and the rules for social responsibilities and relationships. In a seaside village, physical survival outside the home depends on learning maritime skills such as water safety and swimming. Youngsters just learning to walk accompany their siblings at the beach and are continuously supervised if they stray toward the beach alone. The responsibility for teaching water skills and overseeing water safety is a concern of the whole village, but is primarily under the supervision of older siblings. Loud voices, stone throwing, and corporal punishment are effective behavioral reinforcing techniques.

Samoan housing and dress, physiologically effective responses to the hot-wet climate of the littoral village, also play a role in the socialization agenda. The fale, a thatched, open-sided structure, built on a raised platform provides adequate cross-ventilation in the heat and protection from the rain. Although fales are functionally differentiated, they all are characterized by undifferentiated space; even sleeping fales consist of one great room. For youngsters the sleeping fales where they spend much of their time pose very few safety hazards, and since there are few possessions in a great deal of open space, supervision is simple. For preschool children, dress of any kind is at the discretion of the household. The traditional Samoan garment, a multipurpose cotton wrap, the lavalava, requires minimal care. For church and other ritual occasions preschoolers wear the lavalava and occasionally, for young girls, western dress. The dressing skills that are learned at such a slow pace in colder climates are easily learned in the early years because of the simplicity of the dress, and thus elaborate training for competence is unnecessary. Feeding and elimination also do not require extensive training periods, since eating utensils are simple and the bush, the sea, or latrines are immediately accessible.

Perhaps the most critical aspect of socialization for preschoolers is the learning of social roles and responsibilities. In addition to the supervision of younger siblings, preschool children are expected to be capable of dressing, eating, and eliminating without supervision. They are also allocated household tasks by the age of three and are expected to fulfill them satisfactorily. These include fetching and carrying for siblings and elders, and assisting in cooking tasks and fale cleaning. By the time the children are of school age, they contribute significantly to the labor force in the household.

As has been observed by almost every ethnographer working in Samoa, status relationships and cooperation are among the most important social rules that have to be observed and learned. Young children learn, through a variety of techniques, that they are the bottom of the status hierarchy in the household as well as the village. One of the areas where this is manifest is in eating precedence. In every household, food allocation and serving are hierarchically organized, and young children usually do not sit at meals or sit at the perimeter of the fale waiting for elders to complete the meal. Even among siblings, there is an order of precedence and in general the little ones eat last. The hierarchy is impressed on the child quite early by the number of people that can control and discipline her or him. If we look at the data for an average household in Salamumu, a youngster of three years may expect to have one younger sibling, and at least five older siblings in addition to parents and grandparents. The siblings provide the socialization group for the preschooler. Sutter (1980), Ritchie and Ritchie (1979), Gerber (1975), and Mead (1930), among others have emphasized the primary importance of this group in establishing the rules of hierarchy by teaching technical skills, work responsibilities, and appropriate behaviors of respect and deference, both verbal and nonverbal. Since all children, regardless of age, have very low status in Samoan society, preschoolers also develop strong affective ties with their age peers, and learn cooperative behavior in this setting.

In order to understand how these socialization patterns might be a cause of stress in a modern setting, it is most productive to look not only at the content of the socialization agenda and the agents of instruction but also at the techniques of instruction. As I have written in another context, child training in traditional societies "is accomplished not by formal methods of instruction by specially designated agents but by observational learning on the part of the children. This learning occurs in the course of all the mundane activities of daily family life in which the young take part. Since very little technological change has been introduced into the cultures, the parents and older siblings are competent behavioral models for the child. Fortes (1970) has suggested that this type of observational learning is accomplished by imitation, identification, and cooperation" (Baker 1976:94). This instructional model is also valid for socialization in Samoan villages where learning takes place at the side of a sibling or adult (Gerber 1975).

Positive reinforcement even for tasks exceptionally well done is usually confined to nonverbal cues. Sutter (1980) has reported that Samoan techniques at home or in the village are based on a system of orders, usually arbitrary, given by one or more members of a household to one child or a group of children. Failure to comply is disciplined by shouting at the child, slapping, and in some cases beating (Gerber 1975).

Through this process by about 5 years of age, the Samoan child has become an obedient member of the family. Before the age of school entry a village child is socialized to be communal in outlook and behavior and is severely disciplined when he disobeys or is "cheeky." The affective details of this are explored in the life histories reported in the previous chapter.

Socialization patterns in traditional Samoan life have been reviewed by many authors, and have been characterized as an initial period of indulgence in infancy, followed by moderate rejection, withdrawal of physical affection, and individual attention at about the time the child learns to walk and talk competently. Ritchie and Ritchie (1979) have suggested that this socialization pattern, common to all Polynesian societies, is directed toward independence training, and that the techniques used rest on sharing the responsibilities of socialization among a large group of caretakers using disciplinary techniques that contrast with the early uncritical nurturance of infancy. In addition to independence training, it seems clear that these techniques reinforce status relationships and cooperative behavior.

The school-age child

In this age group, the tasks the child learns and is expected to perform increase with age and physical capacity. Children learn to use machetes, collect shellfish, carry coconuts from the plantation, and assist substantially in household and village tasks. These are usually but not necessarily sex-specific at that age. If there is a fairly even sex ratio among siblings in a family, adult sex-specific tasks are learned.

With the introduction of formal schooling into Samoa at the time of European contact, the socialization agenda for this age group was affected first by missionaries, then by colonial powers, and later by the national government. When the islands were first evangelized by John Williams in the mid-nineteenth century, pastor's schools were established in the villages, taught by Samoan ministers in Samoan (Ma'ia'i 1957). At first these were the only kinds of formal education in the village and were not primarily aimed at educating children, but rather at teaching Samoan adults to read the Christian Bible. By 1900, several missionary groups had established formal schooling for children, and under German rule, the missions were allowed to maintain the educational system they had developed. The educational system formally became the responsibility of the New Zealand government in 1914 (Beaglehole 1947), then of the Government of Western Samoa in 1962. Today, mission-sponsored schools coexist with secular institutions. At present, 80 percent of Samoan students attend government schools and 20 percent attend mission sponsored institutions (Western Samoa Ministry of Education 1981).

The present government school system, a derivative of the New Zealand system, offers primary, secondary, and tertiary education. Primary schools consist of a 9-year course of study. After completion of the nine years, a student has a choice of secondary systems—either a 3-year Junior High School, a 4-year government college, or the 5-year Samoa College. (Attention of U.S. readers is directed toward the difference in nomenclature of administrative units between New Zealand and the United States.) Mission-sponsored secondary schools have a comparable structure. By 1980 there were 135 primary government schools in the country and almost every village had its own primary school. In Salamumu in 1979, 75 percent of the children ages 5 to 13 were attending the village school.

There are two patterns that young people may follow after the age of 13 when the village no longer can provide appropriate schooling. These decisions are usually made at the 'aiga level. Support for students beyond the primary level is a decision based on meritocratic as well as economic criteria. Since resources for education are in short supply, the 'aiga usually decides which children are most likely to benefit from further education. While at an early age, recognition for outstanding achievement is not publicly accorded, the assessment of an individual's potential contribution to the 'aiga is ongoing. Informants report that because of economic considerations as well as crowding and staffing problems at the village level, only the most capable children are encouraged even in primary school. Traditionally, youngsters of 13 and older are considered part of the labor force and begin the process of practicing and perfecting the sex-specific household, subsistence, and ritual skills expected of adults in traditional society. The traditional sex division of labor is presented in Table 7.2, and may be divided into household activities, communal ones, agriculture, and fishing. Youngsters are expected to assist in all of the plantation work, fishing, food preparation and serving, laundry, care of the house, including mat making and fale repair. Forty percent of the 13- to 18-year-olds in Salamumu were not attending school and either worked in the village or in Apia. For those remaining in the village, canoe building, net making, fishing, housebuilding, and fine-mat making are learned by apprenticeship under the guidance of skilled craftsmen (Franco 1985a). Recent studies indicate, however, that the boundaries of the sexual division of labor are not clearly delineated (Orans 1981; Lockwood

Table 7.2 Traditional Samoan Age/Sex Division of Labor

Task	School-age children		Adults	
	Boys	Girls	Men	Women
Domestic Activities				
Child care and training	O	F	O	F
Cooking				
Traditional umu food	A		A	
European food		F		F
Serving and assisting in food preparation	F	F	F	F
Household cleaning, laundry	F	F		F
Animal care	F	F	A	
Community Activities				
House and church construction			A	
Canoe and boat building			A	
Tatooing			A	
Participating in sociopolitical affairs			A	
Village sanitation, cleanliness, and health		F	A	
Weaving, craft production		F	A	
Hosting village guests			A	
Church maintenance		F		F
Agriculture				
Heavy plantation work	F		A	
Weeding plantations	F	F		F
Collecting and harvesting fruit crops	A	A	A	A
Harvesting root crops	F		A	
Fishing				
Outer reef trolling			F	
Lagoon collecting of shellfish and fish	F	F		F

O = occasionally; F = frequently; A = always.

1971; Hirsh 1958), and as Lockwood points out "this division of labor may once have been strictly observed but the edges have become somewhat blurred" (1971:48).

Starting at this time increasing attention is paid to the acquisition and perfection of individual skills, whether in the traditional domain or in school. At first glance, recognition of individual achievement seems to conflict with the concept of tautua, or service to the family and chiefs, that is inculcated into young men and women at an early age, since it is service to the parents, the matai and 'aiga, and the village that is the dominant theme of socialization in early adulthood. While traditionally tautua meant service in subsistence and ritual activities, it is now conceived of as extending to schooling, so that children excelling at school are relieved of their plantation and village duties because they are contributing to the household through their studies and school achievements.

Adulthood

Samoans regard young men and women as adults from the time of adolescence, although the complete transition comes at marriage. By this time Samoans have learned the social and behavioral skills necessary to function in adult roles; the new behaviors to be learned now are marriage related. Gerber has noted that marriage does not confer independence from parental control.

> Samoan parents never give up trying to tell their children what do do, even after they are married. Marriages do not seem to stabilize until the couple gradually moves into the roles of heads of the family. Once the husband and wife have achieved these senior positions, they each have their own sphere of authority. The woman is concerned with child rearing and household affairs. Men are concerned with matters relating to economic activities, the extended family and the village. (Gerber 1975:155)

(See Chapter 3 for a more complete description of the roles of men and women in Western Samoa.)

The goal of most Samoan men is to receive a chief's title, although not every man receives one. In Salamumu there were 27 titleholders out of an eligible total of 76 (36%) men aged 19 and above. The acquisition of matai status and the ritual and behaviors associated with this status are clearly understood within the village. Every titleholder is regarded as an elder, regardless of chronological age, although as Orans suggests, in Salamumu "Higher titles tend to be given to older people; in addition younger holders of a title are almost always regarded as lower than older holders of the same title and especially if they received the title later" (Orans 1978:20). Traditionally old age is functionally defined so that when a chief suffered from a chronic illness or could no longer walk to the gardens he could voluntarily retire or have his title removed by his 'aiga (Pitt 1970). In the household older parents remain in authority unless seriously debilitated by illness. Heavy subsistence work is done by young, capable adults, and the behavioral requirements of older persons shift from an emphasis on subsistence skills to a greater emphasis on sociopolitical skills, in the village, in the church hierarchy, and at the national governmental level.

In summary, we find that through the stages of the life cycle socialization prepares a Samoan to become a participant in a community, at first by learning to be a cooperative, hardworking, obedient young person. Severe disciplinary techniques are used by multiple caretakers; training begins in the household, and then the village as the child becomes an adult. At adulthood, signalled by marriage and usually the acquisition of a title, the community orientation continues—men take responsibility and leadership for others in their domain, and women fulfill household and community roles that usually reflect the status of their husbands. Community responsibility does not diminish in old age and is only relinquished on the basis of functional disability.

CHANGES IN THE SOCIALIZATION PROCESS

The Apia area has served as the entry point for western institutions ever since Europeans established a beach community there early in the nineteenth century. By the end of the century a small urban center had developed containing European, part-Samoan, and Samoan residents (Ralston 1976). European missionaries, traders, and colonial officials instituted many of the processes and institutions we designate as modern—centralization of political power, monetization of the economy, formal education, and complex technology and communications. Apia today is the only urban center in Western Samoa, containing the seat of national government and its agencies, headquarters of the religious denominations, educational institutions, major commercial and industrial establishments, the central agricultural market, hospital, hotels, movie houses, and restaurants. In this section I examine how many of these institutions have modified socialization through the life cycle.

Changes in the natural and social environment

Since Apia has been the initial locus of modernization, I will focus on changes in the Apia area because of the greater variability in behavior that occurs there, but I do not mean to suggest that fa'aSamoa has disappeared in Apia or that socialization in the rural villages has not been affected by modernization.

By the end of the nineteenth century customary land tenure had been modified in Western Samoa and 40 000 acres of freehold land were sold to

Europeans, mostly in the Apia area (Thomas 1981). The juxtaposition of a permanent European population with the part-Samoan and Samoan population led to a variety of changes in the Apia area. In addition to the nucleated littoral village, inland villages were established, and now traditional villages based on customary landholding coexist with unincorporated extended-family household clusters as well as single-family holdings (Sutter 1980). House types range from the thatched fale to a variety of cement block palagi-style houses. Household composition also varies, and Pelletier (1984) found that in the Apia households he sampled, the average number of generations per household was reduced compared to the villagers of Salea'aumua, but household size remained the same. He reports that the major differences occurred in social composition. "Village households generally contain more lineal relatives and fewer affinal or lineage kin that do Apia households. Patterns of large extended families is present in all samples" (Pelletier 1984:129). There is also evidence that in some Apia villages the sociopolitical structure operates differently than the traditional one and that the power of the fono is considerably reduced, being replaced by pastor and church (Sutter 1980).

The socialization patterns in the modern areas of Apia and Western Samoa for infant, child, and adult have developed partially in response to these and other changes from traditional life. In the following sections the socialization patterns that are described are those that differ from the traditional village pattern. There remain in the Apia area, families living in traditional households, following agendas not very different from traditional rural villagers.

Infancy

Daily household composition and household construction are two factors that can be shown to modify socialization in infancy. The modern pattern of early child-rearing differs from the traditional in that the mother is the primary caretaker for most of the infancy period, since many household members are either out of the household during the day in wage employment or at school. The multiple caretaking pattern is atypical for the modern urban village household, not only because of the few people at home during the day, but because the European house type in effect isolates the infant from this traditional practice. While infants are more often left on their own as mothers become responsible for more of the daily tasks of the household, the overall pattern of initial indulgence and warmth characteristic of traditional socialization has not been found to change (Sutter 1980).

Childhood

The pattern continues with the preschooler. Due to the reduced daily population of the household, the preschooler is socialized by fewer people and develops a fairly nonauthoritarian relationship with adults.   The techniques of socialization also differ from the traditional in that pre-schoolers are frequently given suggestions rather than orders.   Where the rural preschooler is socialized as much by siblings as parents and becomes independent at an early age, in Apia the preschooler remains dependent on parents for a much longer period of time (Sutter 1980).   In the modern setting different categories of behavior must be taught to preschoolers. Living in a nonhomogeneous urban environment, children must learn a dress code, appropriate elimination behavior, and the rules for physical safety.   The preschooler in the modern setting is not expected to participate in household tasks to any great extent nor does the child have community responsibilities.   There are fewer opportunities for the preschool child to learn the rules of status relationships and cooperation outside of the household since the focus of socialization is the household and the 'aiga rather than the community (Sutter 1980).

The school-age child

Formal educational systems have modified the agenda, the techniques, and the agents of socialization in modern Western Samoa, since the majority of the school-age population was attending school in 1981 (see Table 7.3).   Both the government- and mission-operated schools have at the national level, and certainly in the Apia area, become significant socialization agents for school-age youngsters, through the age of 13. Only a small proportion of youngsters (65%) after that age attend school, mainly in the Apia area (see Table 7.3).

Although the Ministry of Education is making significant efforts to change the philosophy of education from the present orientation toward western objectives to goals more appropriate to a developing country, and from New Zealand-centered curricula to Samoa-centered ones, these efforts have not so far been totally successful (Western Samoa Ministry of

Table 7.3 Numbers and Percentages of Persons in Given Age-groups Attending School, Western Samoa, 1981

Age	Total population	Number of children attending school	Percent of children attending school
5 - 9	22 848	18 500	81
10 - 14	23 525	22 900	97
15 - 19	20 896	13 600	65
Total	67 267	55 000	82

Source: Adapted from Western Samoa Government Report of the Census of Population and Housing 1981.

Education 1981). The information and skills transmitted to children have no immediate relevance in a traditional village environment other than teaching literacy. While there is specialization in traditional work setting in Samoan villages, there is not the occupational specialization generally recognized in modern societies, as blue-collar and white-collar jobs. Each of these categories of labor exist in the Apia area and requires different kinds of training. In some blue-collar jobs, traditional apprenticeship training is the most usual form of training and the Government of Western Samoa has recently (1978) instituted formal vocational and technical training to upgrade the level of skills of young people in apprenticeship programs (Western Samoa Ministry of Education 1981). The varieties of wage employment available in the Apia area have been described by Pelletier (1984) and by James (1984), and show that for most categories of adult urban labor, the men had completed at least nine years of schooling on average. For white-collar workers, in the James sample 97 percent had more than ten years of schooling. Thus if Samoans expect to participate in the labor force, they remain in school to acquire those skills and behaviors that will allow them to compete successfully. The description of the traditional work by men provided by Greksa et al. in Chapter 13 has shown that heavy manual labor was performed in short bursts of time followed by rest periods. Young men worked under the direction of a matai or the fono as a member of the 'aumaga, but it was a cooperative effort supported by a peer group. Wage labor demands different patterns of endurance and individual effort and the initial learning of these behaviors may take place in the schoolroom (Franco 1985).

In the primary schools, the teacher serves as the primary agent of socialization, replacing parents and peers for this age group. Learning specific skills and behaviors thus becomes a process divorced from traditional contextual learning. As the teacher also is usually not a member of the village, socialization in the village school provides a strong contrast with socialization methods and techniques youngsters are accustomed to at home or in the traditional setting of the pastor's school.

Sutter concludes from his study of schooling (1980) that the socialization techniques employed in traditional households and in school are in direct conflict. Prior to school attendance, socialization is communal and individual attainment discouraged. In school, individual achievement is praised and individual attainment encouraged. Additionally, the demands of the teacher are rationalized, and explanations given for orders. The cooperation and sharing expected in the home is punished at school. Disciplinary techniques, however, seemed not to vary between school and home and "cheeky" children are often caned in the classroom, although corporal punishment is officially discouraged by the Ministry of Education. Nancy Graves (1978) studied similar discontinuities in the social context of learning among Cook Island children.

The results of the discontinuity between traditional socialization at home and school may be stressful for these children, and as Hanna et al. reported in Chapter 9, this stress was reflected to a certain extent in the catecholamine hormone levels tested in two small samples of children attending school in a traditional village and a more modern one in the Apia area. The results suggest that most children attending school initially experienced increased levels of stress, but judging by subsequent performance seemed to level out and adapt to the stress. The children who were most stressed came from traditional households that emphasized communal socialization; these children exhibited inappropriate communal behavior in the classroom and were severely disciplined as a consequence. By studying the children who later dropped out of school, he found that these children showed the highest norepinepherine levels of the group just prior to dropping out of school.

For those children who lived in less traditional households, Sutter suggests that there was not the discontinuity between school and home experienced by school children from traditional settings, since both school and home were more individually oriented and ego encouraging and socialization in class was a continuation of that at home. The results of the norepinepherine tests seemed to support this view.

Adults

The major changes in adult socialization from traditional ones are the changes in subsistence activities and the occupational opportunities that develop from an increasing participation in the wage economy. As Pelletier has described, there is employment in industry, plantation agriculture, commerce, the professions, and the public sector (1984). While some forms of training are available in Apia for these occupations, training for the professions takes place for the most part in New Zealand universities, the University of the South Pacific in Suva, Fiji, in Apia, and at the University of Western Samoa. University entrance courses are offered in Samoa College, Avele College, as well as at many of the mission schools. Since kinship connections play an important role in hiring, however, training can take place on the job as well as prior to employment.

Older adults

A demographic change that has had an effect on the role of adults in Western Samoa is the increase in the proportion of the population over 55 years of age. This has increased from about 5.4 percent in 1951 to 6.8 percent in 1976, while the proportion of the population over 65 years has increased from 2.3 to 3.0 percent during the same period (Crews & Baker 1982). A second change of interest is the disparity in educational attainment between the older population and the younger adults in Western Samoa. Among Western Samoans today, the young and the middle-aged have the highest level of education. On average, among the oldest men surveyed in the project, 92 percent had less than four years of formal education, while among the youngest groups only 5 percent had less than four years of education. Similarly, while 28 percent of men 40-54 years old had five or more years of formal education, 92 percent of the young men between 15 and 24 years had five or more years of education. For those over 45 years of age, the majority of their education was in the village pastor's schools rather than in the government and mission schools in Samoa and New Zealand. As the aged population has increased, changes have occurred in the economic system—older Samoans participate less in wage employment than do other age groups. Of the population over 60 years old, only 3.7 percent worked in the wage economy in 1976, compared to 44 percent of the total working population (Crews and Baker 1982).

In the past, older Samoans were respected teachers of subsistence practices, ritual behavior, and ideology. Now, because of the relationship between formal educational training and successful participation in the wage economy, this role may be changing at the 'aiga level. At the national level, the political power of the elders was codified into national law, and thus they still remain in control of national policy, which in a nation state directs most formal socialization agendas (Davidson 1967). A study of attitudes toward the elderly in Salamumu was conducted by Crews and Baker (1982), who found that there was still very strong support in Salamumu for all aspects of the traditional Samoan way of life and particularly the matai system that is correlated so closely with age. The controlling social behaviors of respect, service, and discipline that are part of the socialization agenda of Samoans living in traditional villages, extends also to the elderly, who in contrast to former times are living longer, and in Salamumu constituted over 6 percent of the population.

AMERICAN SAMOA

In this section I review the effect of modernization on cultural transmission and socialization in American Samoa, with special emphasis on the consequences of formal education, new technology and communication, and demographic change.

The ethnographies of Mead (1930), Holmes (1974), Keesing and Keesing (1956) and Ember (1964) provide the historical base line for this section of our review. Ethnographic information was collected during the Samoan Migration Project 1977 field session. The data provided by Franco (1985a), based on field work in American Samoa and Hawaii, also contributed to the interpretation.

Natural and social environment

The physical environment, history, sociopolitical organization, and demography of American Samoa have been described in previous chapters and comparisons made with Western Samoa. Two significant physical environmental contrasts leading to changes in the rate of modernization and through it changes in the socialization agenda are the uniqueness of the harbor characteristics of Pago Pago and the comparative landmass of the two island groups. As Hecht et al. have suggested in Chapter 3, the American presence in Eastern Samoa was based on the strategic importance of a deep water harbor in 1899 and thus led to the beginning of the American colonial period. The small size of the island group (197 sq km) compared to the 2831 sq km of area in Western Samoa put pressure on natural resources in American Samoa, and the rate of population increase relative to the available landmass has resulted in the restructuring of traditional subsistence patterns and in the migration pattern described by Harbison in Chapter 4. It should be pointed out, however, that the restructuring has modified but not destroyed the cornerstones of Samoan social structure, the 'aiga, the matai, and the village.

American Samoa is increasingly becoming a central point in an international migration system that extends from New Zealand through Samoa to Hawaii and the U.S. mainland, and the sociocultural environment is being modified *in situ* by this new geographic and political context. Socialization processes are attempting to provide individuals with the behavioral patterns and skills necessary to function in this expanded environment (Franco 1985b).

Spindler (1974a) has suggested that formal schooling may serve as a catalyst for culture change or reflect the process of change. Wallace (1961) pointed out that school systems have been used by colonial powers, both secular and religious, and by national governments to serve as agents of culture change. By introducing a formal curriculum, an official or alternative language, different participants, and changed learning environments, the socialization process can become an agent of sociocultural change. On the other hand, Kelley (1977) in his study of Chiapas, Mexico, has shown that the imposition of formal schooling does not necessarily lead to culture change, but is sometimes structured to reinforce traditional patterns of behavior.

Formal schooling was introduced in American Samoa by nineteenth century Christian missionaries (Embree 1934). Secular schooling was instituted by the U.S. Navy, then administered by a non-Samoan governor appointed by the U.S. Department of the Interior (Reid 1941). The system was based on a U.S. model, modified for colonial conditions. The 1957 Report of the Governor of American Samoa on Educational Conditions states that the goals of the system were "to conserve the best

of Samoan culture, and at the same time to guide them in developing
social changes which are to their advantage, and to bring them to a better
understanding of what is best for their well-being, health and general
improvement" (American Samoa Govt. 1957:43). As the result of wide
publicity about the inadequacy of the curriculum, teacher training, and
facilities, the educational system was completely reorganized in 1962
under the auspices of the governor. An educational television network
was introduced as the main source of instruction in the island. By 1964,
educational television was operating in all elementary schools, and by
1966, in the secondary school system. The positive and negative
educational effects of this change in instructional delivery system have
been extensively evaluated (see, for example, Schramm et al. 1981;
Thomas 1980; Masland and Masland 1975). By 1971, in response to
continued criticism, "the system was modified and television became a
supplement to classroom instruction instead of the sole method of
instruction" (Runeborg 1980:42-3). One ancillary effect of the 1971
educational reorganization was the introduction of commercial television
into the islands. When the full capabilities of the television station were
no longer required for educational programming, commercial and public
television programs from mainland United States were purchased and
rebroadcast. A 1976 survey found that 96 percent of the population had
access to a television set (Siegal 1979). Thus in addition to the values,
behaviors, and roles being taught in the classroom, American television
has become an active participant in the socialization of Samoan children
and adults.

The present structure is controlled by the Department of Education
(DOE) of the Government of American Samoa, with responsibility
delegated to a Director of Education and a Samoan Board of Regents. The
system includes 125-130 Early Childhood Education Centers in the
villages, 24 elementary schools, 4 high schools, a center for special
education, and one post-secondary institution, The America Samoa
Community College. Education is compulsory from grades 1-12, or ages
6-18, and is provided free of charge by the Government of American
Samoa. (There are also eight private religious schools on the islands.)
Reliable enrollment estimates suggest that over 90 percent of the 14- to
15-year-olds were still attending school in 1976 and that over half of adult
Samoans have now received education beyond the elementary school level.

Now under local control, the goals of the system have been changed
and, as the DOE states:

The goal of the Department of Education is to help each person reach his
fullest potential as a unique individual.

To achieve this goal the Department of Education is committed to the concept
and practice of individualized instruction. This commitment requires that
each learner's rights, needs and aspirations be treated on an individual basis.

Inherent in this commitment is the obligation to preserve and foster the economic well-being of American Samoa, while at the same time, to prepare each individual for a personally satisfying and socially useful life wherever he chooses to live. (American Samoa Dept. of Education 1980:8)

The emphasis on the fulfillment of individual rather than communal needs provides a striking contrast with traditional culture and conflicts with the hierarchical tradition and that of tautua, or service to families, that are still important components of contemporary Samoan culture.

Analysis of educational programming shows that both children and adults are now being taught their societal roles as well as technical skills in the classroom (American Samoa Dept. of Education 1976:77,80). Although significant efforts are being made to recruit and train native Samoans, there still are a large number of non-Samoan professionals in the DOE in teaching and administrative positions (American Samoa Dept. of Education 1980), thus continuing the influence of the U.S. educational establishment on the functioning of the department.

Infancy

The Obstetrical/Gynecological (OB/GYN) and the Well Baby Clinics of the LBJ Center for Tropical Medicine have significantly contributed to child care and training practices in infancy. (Information on infant mortality is found in Chapter 5; on neonatal health care and feeding practices, in Chapter 10). The high frequency of hospital births, the declining rate of infant mortality, the familiar sight of cartons of disposable diapers being off-loaded in the Pago Pago airport, and public concern about the change from breast- to bottle-feeding indicate that some areas of infant care have changed in response to the modernization that has occurred in American Samoa. The rapid acceptance of modern medicine that reduces the hazards of neonatal and infant life as well as new technology which relieves caretakers of some of the routine tasks of childcare do not necessarily modify the nurturing techniques, the primary caretakers, or the content of the socialization process. The traditional socialization in infancy in the rural villages in Western Samoa describes the patterns in Fiti'uta in the Manu'a group. The conservatism of Samoan society in this regard is evidenced by the fact that even in the villages in Tutuila that have been subjected to the most intense and rapid *in situ* modernization, the socialization process described in most details by Mead (1930) and Holmes (1974) still is the dominant pattern, although there is some variability due to the size of the at-home population in the households. The locus of socialization for infants is the fale, traditional or palagi style, and parents and siblings are the responsible agents, using the traditional disciplinary techniques described for Western Samoan villages.

Childhood

The significant environment for the child living in Tutuila is the Samoan littoral village, changed by vehicular traffic, electricity, and television, but still operating within the traditional social structure. The repertoire of skills needed to function in a traditional village has been broadened to include the new skills emerging from technological change, only now they are being taught in the classroom rather than at home (American Samoa Govt. 1980).

The changing focus on the child as an individual rather than as the least significant member of a family group begins in the preschool ages at the Early Childhood Education centers. Children under the age of six attend the centers in their home villages, for a 5-day 2½-hour educational program. The centers are staffed by Samoan parents or adults who have been trained by DOE Early Childhood Education specialists in a mandated curriculum. In this program, parents teach and children learn, in a formal setting, technical skills formerly learned by observation. They also develop cognitive skills not necessarily acquired in traditional socialization. From the specific goals statement of the Early Childhood Education program, we can conclude that the DOE is setting a socialization agenda that the adults implement. Under the auspices of this program, parents are also being taught how to fulfill their roles, for as the DOE suggests, they "seek ways of helping young adults and parents understand their crucial role in nurturing the child from birth through adolescence" (American Samoa Dept. of Education 1980:14).

An examination of health and safety education programs in the curriculum for grades 1-4 (ages 6-10), illuminates how much of the socialization of Samoan children is planned for in the classroom. There are eight instructional units in the program, one each on Personal Health, Nutrition, Mental Health, Health and Safety, Family Life Education, Consumer Health, Preventing Disease, and Community Resources. Educational objectives for each unit are to teach children a combination of traditional and modern information. In the unit on personal health, not only is the child taught "why physical activity is necessary for good muscular development" but also "why traditional physical chores around the taro and banana plantation, making an umu, and working around the sea stimulate body growth" (American Samoa Dept. of Education 1977-1978:10). In the unit on community resources, the child learns about "health agencies in American Samoa and their functions" as well as "traditional Samoan ways of respecting and caring for grandparents" (American Samoa Dept. of Education 1977-1978:23).

The curriculum reflects the goals of the Department of Education, to train school-age children and adolescents in the behaviors and skills necessary to function in a modernizing American Samoa, or to prepare them for migration. The structure of the curriculum for the elementary and secondary schools is similar to those in Hawaii and mainland United

States, but the implementation of the curriculum is not. The curriculum, the teacher, the DOE, and the television set have not replaced their counterparts in the 'aiga, the village, or the church. They have rather added another dimension to the socialization process that in some cases is congruent with fa'aSamoa and in some cases antithetical. What has resulted is a transitional system that partially mediates the differences between traditional and modern values, but also sets up potential conflict situations between traditional expectations and more modern ones.

The American tradition of continuous, compulsory, daily school attendance has not been wholly accepted by DOE administrators. Absences arising from ritual and kin obligations, while officially deplored, are considered to be valid, and sanctions are often not generally imposed on youngsters absent for these reasons. According to Franco,

> there is the time tension for Samoan youth between village and church responsibilities and school work and extracurricular activities, most prominently, the opportunity for Samoan boys to play football. As this tension has developed membership in village men's and women's groups has become increasingly optional, at least until fa'alavelave events occur . . . . The fa'alavelave support system places great pressure on American Samoan youth, as frequently parents are working forty hours a week in the relatively strong American Samoan economy . . . when a family crisis occurs much of the labor burden falls on students who must be pulled out of school to work preparing food, cleaning the village and guesthouses, and even working part-time to earn money to contribute to family matai. (Franco 1985b:6-7)

It is worthwhile to emphasize that the DOE has assumed the responsibility of preparing students for their sex-specific roles in the labor force and the family. They provide training for young men, not only in modern trade and industrial skills, such as small engine repair and drafting, but also in 1981 instituted a program in vocational agriculture to teach traditional subsistence skills (American Samoa Dept. of Education 1980:14). Young women of high school age participate in the consumer homemaking program, which provides instruction in nontraditional areas, such as family finance and nutrition, traditional skills such as Samoan crafts, food preparation, child care, as well as instruction in "grooming" and "relationships" (American Samoa Dept. of Education 1980:14). This assumption of the traditional responsibilities of parents and 'aiga is one consequence of modernization. With the dramatic increase of participation in the wage economy, parents and children have less time for the traditional socialization process of observational and experiential learning. Parents raised in traditional villages may not themselves have the training in such modern technical skills as computer programming to pass on to their children, nor the time to teach the traditional skills of fine-mat making or plantation work.

Adults

Perhaps the most striking aspect of modernization affecting adults in American Samoa is the change in work activities. As Greksa et al. detail in Chapter 13, the dominant work force realignment has been the dramatic change from subsistence agriculturalists to salaried white-collar workers and blue-collar wage laborers; as of 1974, they report "that 71 percent of all adult men and 36 percent of all adult women were employed in the money economy" (Chapter 13:305). Traditionally tautua, or service to family and chiefs, was based on a subsistence agricultural economy described earlier for Western Samoa. In American Samoa as the subsistence base changed to a money economy, the concept of tautua was enlarged to include wage labor and monetary contributions. As Franco (1985a) pointed out, matai still manage the 'aiga labor force, although there is less emphasis on village land management, and more emphasis on the mobilization and reallocation of human resources and money.

Training for positions in the modern work force is provided in a variety of settings. The ease of migration to the United States provides a number of opportunities for higher education and training for business and the professions. The Government of American Samoa sponsors two scholarship programs for higher education. For fiscal year 1980, 170 scholarships were awarded for study in the United States, and 103 for study in American Samoa Community College (American Samoa Dept. of Education 1980). Church-related institutions of higher education have been particularly active in recruiting promising Samoan students and have provided scholarships as well as work-study opportunities. Skills for other job categories have either been learned in high school, through on-the-job training, or in the military.

Older adults

The Samoan cultural agenda includes respect and honor for elders and is transmitted at home and, as we have seen, reinforced in school. Studies on other traditional populations (Cowgill and Holmes 1972) have suggested that modernization has a negative effect on the status of the elderly. Information on the changing roles of older adults in American Samoa is available from the 1976-1977 studies of Rhoads and Holmes in Tau and in the Fagatogo area of Tutuila (Rhoads 1984; Holmes 1978). Rhoads suggests that both in the more traditional Tau and the more modern Fagatogo area roles and status of older Samoans have not been greatly changed by modernization, in spite of the discrepancies in educational attainment, employment status, and income. She proposes that "The idea that prestige based on individual accomplishments accrues not only to the individual but to the entire family seems to temper what could be a more negative effect when the young exceed the accomplishments of their elders" (Rhoads 1984:249). As in Western Samoa, political power at the village level and the national level is still vested in the matais. The government benefits available to the elderly in American Samoa are of economic advantage to the 'aiga and may also explain the findings of these studies. Rhoads (1984) concludes from her study that as long as the 'aiga-related values are maintained in American Samoa, the role of Samoan elders will not change with further modernization.

HAWAII

Data used in interpreting Samoan migrant socialization patterns in Hawaii was collected during the 1976-1977 field sessions on the island of Oahu based on a sample of 265 men and 344 women ages 18-90.

The setting

Hecht et al. have described the geographic and social environments of the Hawaiian studies in some detail in Chapter 3, and pointed to the differences from and similarities to the Samoan environment that migrants to Hawaii encounter. Demographic factors, such as the age of the migrants, their area of origin, and length of residence in Hawaii, and the relative size of the Samoan community in Hawaii also may affect the socialization of Samoans in Hawaii. The implications of these factors for socialization are explored below.

Fifty-seven percent of the 1976-1977 migrant sample studied was between 18 and 39 years of age, 25 percent was between 40 and 54 years of age, and 18 percent was over 55. McGarvey (1980) reported that at the time of the study, the average number of years of residence in Hawaii was 7.5 years, with a standard deviation of 6.1 years for men and 6.7 years for women. Because of the ages of the migrants and the number of years of residence in Hawaii, we can assume therefore that the Samoan adult migrants in our study sample were socialized in American or Western Samoa. We have previously described socialization differences between and within the Samoan islands, from which we can deduce that the area of origin of the migrant sample may have an effect on this process in Hawaii.

McGarvey (1980) classified migrant sample subjects into three groups of areas of origin: the *traditional*, subjects who resided in the villages of Western Samoa before migration, American Samoan villages in the Manu'a group, and the north shore of Tutuila; the *intermediate*, consisting of Samoans who lived in the villages of Tutuila, excluding the north shore and the Pago Pago harbor area; and the *modernized*, who lived in the Apia region and the Pago Pago harbor area of American Samoa. Forty-eight percent of the migrants came from the more modernized areas, 20 percent from intermediate areas, and 32 percent from the more traditional areas of the Samoan islands. There is some variation between these groups in

Table 7.4 Description of the Migrant Sample in Terms of the Area of Origin

	Area of Origin										
	Traditional				Intermediate				Modernized			
	Men		Women		Men		Women		Men		Women	
	N	%	N	%	N	%	N	%	N	%	N	%
Formal Education												
0-3 years	10	15	10	13	5	8	7	8	3	8	0	0
4-9 years	34	52	37	48	22	37	35	40	13	36	23	42
10 years or more	21	33	30	39	33	55	45	52	20	56	32	58
Totals[a]	65		77		60		87		36		55	
Languages Spoken												
Samoan only	14	20	25	30	12	16	21	21	6	14	10	17
Samoan and English	56	80	59	70	61	84	81	79	36	86	48	83
Totals[a]	70		84		73		102		42		58	

Source: Adapted from McGarvey 1980, Table 6.
[a] Data not available for all participants.

pertinent background variables. As shown in Table 7.4, there is a difference in years of education, with the premigrant traditional group showing the least amount of formal education. There are fewer differences between the intermediate and modern groups, with both of these groups containing significant percentages of individuals with 10 years or more of schooling. The traditional group also has the greatest percentage of individuals who speak only Samoan as compared to the other two groups, but the majority of all these groups were bilingual in Samoan and English. While there seem to be some differences between these groups, for the purpose of this discussion, the 68 percent of the migrants coming from the intermediate and modern areas may be considered as one group, since in American Samoa they will have had similar socialization experiences.

One other factor that may be considered separately is the fact that the Samoans in Hawaii constitute only 1.3 percent of the total population of Hawaii (U.S. Bureau of the Census 1983). Minority status may affect socialization in a number of ways, depending on the visibility and size of the community in any one residential area (Macpherson 1978; Kotchek 1977). Franco (1984) reported that in one census tract (Kuhio Park Terrace) Samoans comprised a majority of the total tract population (53.6%), while in two other tracts (Kalihi Valley Homes and Mayor

Wright Housing), they comprised 43.6 and 30.2 percent, respectively. But regardless of the size of the population aggregation, the institutions that control the socialization agenda reflect the views of the state and national majority groups.

Infancy

The physical environment of infants born to migrants living in the rural areas of Oahu does not differ much from that of modernized Tutuila. High-rise, high-density public housing in Honolulu, on the other hand, imposes different constraints on the socialization process in both the physical and social environment. Living space is at a premium and non-Samoan neighbors impose rules on public behavior sometimes at variance from those in Samoa.

As Harbison shows in Chapter 4, fertility has not substantially declined in the Hawaiian migrant community, although there is some variation between the more rural communities (5.6 total fertility) on Oahu and those living in housing developments in Honolulu (4.3 total fertility). A high reproductive rate and large household sizes comparable to those in Western and American Samoa provide the demographic possibilities for a traditional pattern of socialization in infancy in Hawaiian migrant households, subject to the constraints described above. Based on employment status data of the study group, 52 percent of the migrant men and 82 percent of the women were unemployed (McGarvey 1980). The large number of unemployed persons thus increases the at-home population and provides the traditional multiple caretakers described for the islands.

The migrants do not have the same easy access to free prenatal and neonatal care that is available in American Samoa, but in all other respects, our own observations lead us to conclude that Samoan infants in Hawaii are nurtured and indulged in a manner similar to those in the modernized areas of Tutuila. This pattern is at odds with American norms of infant care in small nuclear families.

Childhood

The socialization agenda in modernized American Samoa is a product of both traditional and modern institutions, with individual or family variability in adhering to the agenda accepted in that polity. In Hawaii, on the other hand, the Samoan socialization agenda and the techniques for enforcing it are at variance with the cultural majority. Discipline within the home in the United States has not, until recently, been a matter of public concern. With increased public attention being focused on child abuse, traditional Samoan disciplinary techniques of corporal punishment are being criticized. Franco (1985b) has found that Samoan parents in Hawaii strongly support their traditional child-care practices and consider it good discipline. He concludes that "Perhaps no single cultural adaptation has been more stressful to Samoan parents and children, than their response to American child rearing practices and expectations" (1985b:8).

Most Samoan migrants have had some schooling. The conflicts between traditional socialization and formal education were described previously and may be most stressful for children of migrants from traditional areas of the Samoan islands [cf. other Polynesian groups reported by Graves and Graves (1978)]. For both parents and children who have migrated even from the more modern areas of American Samoa, however, preliminary experiences in schools there may not prepare them for the American school system. Samoan ritual obligations and mobility patterns have disruptive effects on school attendance and achievement in Hawaii, where educational goals and standards are less flexible than those in the islands. The situation is exacerbated by the reported English language difficulties the bilingual children experience (Franco 1984b; Munro 1976). As Ogbu has proposed, minority children "acquire values, attitude and learning styles within their culture that are different from and in conflict with those required for success in public schools" (1978:47). Without the strong political representation in the school system which Samoans have in New Zealand (Macpherson et al. 1978), contemporary Samoan values and behavior may reduce chances for school completion. Franco (1984) finds a relationship between the high rate of unemployment of Samoan youth and their low level of educational attainment. Schools provide the major source of training in the skills necessary for participation in American life; Samoan migrant experiences in the school system have, thus far, not resulted in successful socialization.

Adults

As both Howard (Chapter 8) and Hanna et al. (Chapter 9) point out, in following chapters there is evidence suggesting that Samoan adults experience emotional stress when they are involved in modern society. Although no studies were conducted in Hawaii, we may assume that Samoan migrants to Hawaii respond in similar fashion. In Hawaii several factors may contribute to the stress for Samoan migrants as a group.

Howard has proposed that "migrant populations (who) recreate, where opportunities permit, situations in new environments that replicate those characteristics of their homeland. They do so by founding organizations, performing rituals, arranging living space, initiating activities etc., that make it possible to experience at least a portion of their new lives in terms of prior suppositions" (1982:50). Pitt and Macpherson (1974) discuss this modification of structure and behavior of the 'aiga in Samoan communities in New Zealand, and Kotchek (1978) and Ablon (1971a, 1971b) for Samoan communities on the U.S. mainland. On the other hand, traditional structures may be inappropriate in new environments and may serve to increase stress in individuals who have migrated or who live in areas undergoing rapid modernization (Graves and Graves 1980; Sutter 1980).

As a result of migration, the Samoan 'aiga has become geographically expanded with only modest modifications of the traditional tautua, or service, expected of its members. Matais may reside overseas but do, in most cases, still retain managerial and financial responsibilities for the 'aiga. Malaga, or intergroup visiting and resource sharing, is an essential component of 'aiga relations, and in the past was based on intervillage visiting for ritual occasions. In the modern setting, malaga movements have become wider in scope, and Samoans travel through New Zealand, Samoa, Hawaii, and mainland United States. The modern malaga, which brings migrants home or American Samoans to Hawaii for visits, continually reinforces the pattern of Samoan values, social organization, and behavior. While the financial burden of the fa'alavelave is openly recognized by young people (Ala'ilima and Stover, Chapter 6, this volume; Holmes and Rhoads 1983; Crews and Baker 1982), it is an expected obligation of Samoans everywhere. The ramifications of fulfilling kinship obligations are discussed in Chapter 6.

For this generation of migrants, at least, surveys have shown that they are the largest group of non-English, native language speakers in Hawaii (U.S. Dept. of Health, Education and Welfare 1980), which may not only target them for negative ethnic stereotypes as discussed by Howard in Chapter 16 but also may serve to impede their entrance and successful integration into the Hawaiian labor market (Franco 1984).

SUMMARY

The flexibility of the Samoan sociocultural system has been observed by anthropologists who studied Samoan responses to the first intrusion of Western culture, the introduction of Christianity to the islands (Ryan 1969). Samoan social structure has evolved during 150 years of European contact, and cultural transmission has been successful in maintaining cultural continuity in two different political systems. At present both in Western and American Samoa there is evidence that sociocultural change is modifying the content of the cultural agenda, and that contemporary agents of socialization are attempting to transmit this change. In Western and American Samoa, where Samoans control the agenda, the institutions, and the techniques, the sociocultural change that we have called modernization may be less stressful for participants than in Hawaii, where Samoans are but a small proportion of the sociocultural system, and thus have less control over the process.

Chapter 8

Questions and Answers: Samoans Talk about Happiness, Distress, and Other Life Experiences

ALAN HOWARD

This chapter presents a report of questionnaire data obtained by various participants in, and associates of, the project. As part of the investigation it was deemed desirable to gain some measure of Samoans' self-perceptions, particularly insofar as these were related to health conditions. In the initial phases of the study a heavy reliance was placed on the Cornell Medical Index (CMI), a relatively standardized measure of well-being believed to be sensitive to stress syndromes. The CMI is composed of 195 yes/no questions (190 for women) concerning a variety of symptoms, health-related habits, moods, and feelings. The protocol was administered to three of the research populations, including the 1976 sample of Samoans in Hawaii and American Samoa, the 1979 sample from Salamumu, Western Samoa, and the 1982 sample from Salea'aumua in Western Samoa. The basic data were analyzed by Jay Pearson and are presented in this chapter.

A second body of questionnaire data was obtained during the summer of 1981 from seven villages in American Samoa: Tula and Amouli, located on the eastern tip of Tutuila; Utulei, Faga'alu, and Laulii, located in or adjacent to the Pago Pago harbor area; and Nu'uuli and Vaitogi to the southwest of Pago Pago. Ninety-nine residents, aged 14 to 32 years old, responded to questions concerning household composition; social support; participation in community affairs; residential, migration, and health histories; and occupational goals and attainment. In addition two protocols addressing coronary-prone behavior (Dembroski et al. 1978) and anger arousal and response (Harburg et al. 1979; Harburg et al. 1973) were administered. A variety of physical measurements were also obtained.

Two more specialized studies utilizing survey instruments are also summarized. One, conducted by Scheder in 1982 (Scheder 1983), focused on stressful life events and their correlates among subjects on Manu'a, in American Samoa. The other concerned attitudes toward obesity. It was conducted by Connelly and Hanna in 1977 (Connelly and Hanna 1978) among Samoans residing on the island of Oahu, in Hawaii.

Finally, we consider questionnaire material deriving from a study by Orans in the village of Salamumu on Upolu in Western Samoa. Orans (1978) is concerned with perceptions of happiness and life-satisfaction, and how these are affected by social hierarchy. His findings are of particular importance for what they reveal about the problematic nature of questionnaire research among Samoans.

LIMITATIONS OF THE DATA

For years anthropologists have been warning survey researchers about the pitfalls of relying on questionnaires constructed for use with Western populations in non-Western cultures, and the case for wariness is perhaps as great among Samoans as with anyone. To begin with there is the problem of translation and the standardization of language. Although a significant portion of the Samoan population speaks English, it is with varying degrees of fluency. Most are more comfortable with Samoan, which means that questionnaires have to be translated into Samoan if questions are to be properly understood. Although back-translation techniques provide some measure of assurance that Samoan versions do not go too far astray, nuances of meaning are frequently lost. This means that whereas questions concerning socioeconomic background, genealogical information, and other clear-cut phenomena may be readily understood and responded to, inquiries concerning beliefs and feeling states present serious communication problems (see discussion of Orans below). To make the problem of translation more complex, there are now many Samoans, particularly among those who have emigrated, who are more comfortable with English than with Samoan. For this reason it may be necessary to use two versions of a questionnaire, one in English, the other in Samoan. This, of course, raises questions about their comparability. Other factors may also impede comparability, such as the necessity at times to use translators to explain or administer questionnaires, and the fact that in some instances a subject may be responding only to written versions, in other cases to a reading of the questions. When questions are read, voice intonations and other nonverbal cues may strongly influence whatever affective loadings a question might have. On the other hand, many of the more traditional Samoans have had little experience responding to written questions and find answering them confusing or difficult.

Another problem is the so-called *courtesy bias,* or the tendency of subjects to provide answers that they think will please the interviewer. Related to this is the *social-approval bias*, or the tendency to answer questions in a manner consistent with public norms rather than with private views. As students of Samoa have consistently pointed out, there is a strong emphasis in Samoan culture on the priority of social interests over personal ones (see Shore 1982 for a particularly insightful analysis). Questionnaires, particularly those concerned with attitudes, beliefs, feelings, hypothetical situations, and the like were developed with self-reflective populations in mind, but Samoans, like most other non-Western peoples, are not disposed toward self-reflection. The effect such biases have on response patterns has not been measured among Samoan populations, but there is good reason to presume that they are substantial. One manifestation of the social-approval bias is a reduction in variability among responses, a factor that may have influenced the results found by Scheder in her study of life events (see below).

Despite these limitations, survey research has important contributions to make to studies of the Samoan population. To be of maximum value it needs to be done in conjunction with participant observation, by persons who are fluent in Samoan and familiar with Samoan interactional styles and etiquette. But even under less favorable circumstances it can add important information to broader research programs, such as that reported on in this volume. For one thing it provides important information about variability within study populations, and *despite* the problems involved, the questionnaire data do reveal important aspects of the patterning of variation. The fact is that survey materials complement ethnographic techniques and help to correct what can be misleading impressions that stem from the ethnographic literature. The latter tends to focus on intragroup regularities, on the central tendencies and agreed upon premises for social behavior. Survey materials highlight differences, and therefore raise interesting and important questions, whether or not they provide answers to them. In fact, many of the findings reported on in this chapter are of a provocative nature. Some challenge conventional wisdom and will probably stir debate and stimulate further research. Others make sense in terms of existing theories and appear to add to their consolidation. None of the results should be taken as definitive. Rather they should be looked at as small, often indistinct bits of a large jigsaw puzzle that is continually changing its shape.

THE CORNELL MEDICAL INDEX

The CMI was employed in this study as a method for exploring the ways sociocultural change has affected the overall health status of Samoans. The questionnaire was translated into Samoan by acculturated Samoans in Hawaii and then back-translated in American Samoa by Samoan

Table 8.1 CMI Study: Comparison of Sample Size and Median Age of the Samples

Age	Hawaii		Tutuila		Manu'a		Western Samoa		Age differences[a]
	N	Age	N	Age	N	Age	N	Age	
Men									
18-35	41	27.0	150	25.3	7	30.4	49	22.5	WS < M
36-50	22	43.1	121	43.8	10	47.0	24	44.4	n.s.
50+	23	56.5	139	60.1	34	62.9	24	57.5	n.s.
Women									
18-35	68	24.9	264	26.0	18	25.5	59	24.3	n.s.
36-50	38	41.4	192	42.7	31	42.3	24	45.2	n.s.
50+	20	59.2	184	59.0	50	56.7	22	62.3	M < WS

[a] WS = Western Samoa;   M = Manu'a;   n.s. = not significant at alpha of .05.   Significance levels are based on Kruskal-Wallis tests (alpha = .05) and nonparametric multiple comparisons (alpha = .005).

medical personnel prior to administration. Conceptual equivalence (Kalimo et al. 1970) was used in the translation in order to produce confidence in the meaning of the terms and the intent of the items. In Western Samoa the questionnaires were left with the subjects and collected later. In American Samoa and Hawaii they were filled out either without supervision or during personal interviews. Some individuals failed to provide complete answers and were dropped from the analysis. In the Western Samoan sample 14 percent of the questionnaires were dropped, while 6 percent were dropped from the Hawaiian and 4 percent from the American Samoan samples.

The remaining samples were divided into three age classes: 18-35 years, 36-50 years, and 50+ years. Age composition was generally homogeneous within age categories, although in the 18- to 35-year age class, Western Samoan men were significantly younger than Manu'an men. In the 50+ age class, Manu'an women were significantly younger than Western Samoan women. The sample sizes and median ages for the samples are given in Table 8.1.

Analysis focused on the total number of symptoms reported and the number of symptoms reported in different sections of the CMI. Thirteen sections were distinguished; twelve of these were combined to form a somatic division, while the remaining section was composed of psychosocial symptoms (see Table 8.2). All of the section scores are significantly correlated with the total score for the thirteen sections, although the correlations should be interpreted with caution since the strength of the correlation is significantly associated with the number of questions in the section (Kendall's rank correlations = .53, $p < .01$). The psychosocial, somatic, and nervous system section scores are highly

Table 8.2 Kendall's Rank Correlations of CMI Section Scores with CMI Total and Somatic Scores

Section	Subject	Number of questions	Percent of CMI questions	Total	Somatic
A	Eyes and ears	9	5	.50	.54
B	Respiratory system	18	9	.64	.68
C	Cardiovascular system	13	7	.61	.64
D	Digestive Tract	23	12	.68	.71
E	Musculoskeletal system	8	4	.63	.66
F	Skin	7	4	.57	.58
G	Nervous system	18	9	.71	.72
H	Genitourinary system	11 (6)[a]	4	.50	.49
I	Fatigability	7	4	.62	.62
J	Frequency of Illness	9	5	.56	.56
K	Miscellaneous diseases	15	8	.59	.59
L	Habits	6	3	.57	.56
A-L	Somatic	144 (130)[a]	74	.89	1.00
M-R	Psychosocial	51	26	.72	.60

Note: All correlations were significantly different from zero (*p* < .0001).

[a] Section H has fewer questions for women; therefore, men and women answer different numbers of questions for the somatic section and total questionnaire.

correlated with the total score, and the psychosocial section is moderately correlated with the somatic score.

The median number of symptoms reported by category of subject and section of the CMI is shown in Table 8.3. Contrary to most other studies, which have found women to report significantly more symptoms than men (Al-Issa 1982; Kessler and McRae 1981), Samoan men and women responded to the CMI in similar ways (Table 8.4). In Tutuila men aged 18-35 years had lower standardized total scores than women, but this is balanced by the fact that in Western Samoa women aged 36-50 had lower total scores than men. Men tended to report more detrimental habits and miscellaneous diseases than women. Age category differences were most pronounced in Hawaii and Tutuila, where young Samoans reported fewer symptoms than older Samoans, particularly in the somatic section (Table 8.5). These differences were less important in Manu'a and Western Samoa.

The most surprising finding, however, is that Samoans living in the more modernized settings of Hawaii and Tutuila reported significantly fewer symptoms than Samoans residing in Manu'a and Western Samoa. Almost a third of young Samoans in Hawaii did not report any symptoms on the CMI (Table 8.6). Removing all the individuals who did not report any symptoms, however, did not affect the population differences in CMI scores (Table 8.7).

This finding is inconsistent with the cumulative evidence from

Table 8.3 Median Number of Symptoms Reported on the CMI

Section	Age	Men				Significant comparisons	Women				Significant comparisons
		H	T	M	WS		H	T	M	WS	
Total Score	18-35	6	20	71	65	H,T<M,WS H<T	9	29	51.5	65	H,T<M,WS H<T
	36-50	21.5	33	47	77	H,T<WS	23	28	34	63	H,T<WS
	50+	50	30	49	76.5	H,T,M<WS	48	30.5	42	81	H,T,M<WS
Psycho-social (M-R)	18-35	2	5	17	19	H,T<WS H<M	2	9	17.5	20	H<All T<WS
	36-50	6.5	7	12.5	25	H,T<WS	4.5	7	11	19	H,T<WS
	50+	12	6	9.5	18	H,T<WS	13.5	6	11	25	T<M,WS
Somatic (A-L)	18-35	3	14	49	49	H,T<M,WS H<T	6	19	33	45	H,T<M,WS H<T
	36-50	14	24	34.5	55	H,T<WS	17.5	20	25	42	H,T<WS
	50+	34	24	34	59	H,T<M,WS	34	24.5	27	58	H,T,M<WS

Note: H = Hawaii; T = Tutuila; M = Manu'a; WS = Western Samoa.

Table 8.4 Sex Comparisons of CMI Total and Subsection Scores

Comparison of Standardized Total Scores[a]

Age	Hawaii			Tutuila			Manu'a			Western Samoa		
	M(%)	F(%)	p	M(%)	F(%)	p	M(%)	F(%)	p	M(%)	F(%)	p
18-35	3	5	n.s.	10	15	.002	36	27	n.s.	33	34	n.s.
36-50	11	12	n.s.	17	15	n.s.	24	18	n.s.	39	33	.03
50+	26	25	n.s.	15	16	n.s.	25	22	n.s.	39	42	n.s.

Comparison of Subsection Scores

Age	Hawaii	Tutuila	Manu'a	Western Samoa
18-35	None	Section B F>M (p=.05) Section C F>M (p=.02) Section D F>M (p=.0005) Section G F>M (p=.02) Section J F>M (p=.001)	Section L M>F (p=.01)	Section L M>F (p=.01)
36-50	Section L M>F (p=.05)	Section K M>F (p=.01) Section L M>F (p=.0001)	None	Section A M>F (p=.005) Section B M>F (p=.0005) Section D M>F (p=.02) Section L M>F (p=.0005)
50+	None	Section K M>F (p=.02) Section L M>F (p=.05)	Section K M>F (p=.02)	Section L M>F (p=.01)

Note: Section H was not tested because men and women answer different numbers of questions; n.s. = not significant at an alpha of .05.

[a] Standardized total score = total number of symptoms reported / total number of questions.

Table 8.5 Age Comparisons of CMI Total Scores, Psychosocial Scores, and Somatic Scores

Section	Sex	Hawaii	Tutuila	Manu'a	Western Samoa
Psychosocial	Male	1<3	n.s.	n.s.	n.s.
	Female	1,2<3	.02[a]	n.s.	n.s.
Somatic	Male	1<2,3	1<2,3	n.s.	n.s.
	Female	1<3	1<3	n.s.	.05[a]
Total	Male	1<3	1<2,3	n.s.	n.s.
	Female	1<3	n.s.	n.s.	n.s.

Note: 1 = 18-35-year age class; 2 = 36-50-age class; 3 = 50+ age class; n.s. = not significant.
[a] The Kruskal-Wallis test was significant at the stated *p* value, but no multiple comparisons were significant at alpha of .01.

physiological studies that indicate health risks are markedly higher in Hawaii and Tutuila than in Western Samoa and Manu'a. It therefore raises serious questions about the meaning of self-reported symptoms among Samoans. [1] A number of possibilities exist, but it is not yet possible to choose between them.

McGarvey (1984b) helps to shed some light on the problem. He used the psychosocial section of the CMI data to explore the effects of psychosocial factors on blood pressure. His sample was confined to American Samoa, which he divided into Manu'a (the least modernized), Pago Pago (the most modernized), and villages on Tutuila (intermediate). In order to increase the precision of correlational analysis he constructed a scale of modernization for individuals based on years of education, occupation, travel from Samoa, and language use (see Chapter 15 for a complete description of the study). Among men he found a trend toward individual modernity paralleling community of residence, but it was not statistically significant. Among women the index increases substantially, due to the significantly higher values in the Pago Pago group. Men have higher modernization indices than women in all groups, although the difference is significant only in the intermediate Tutuila group (McGarvey and Schendel, in this book, Chapter 15).

In both sexes there were trends for lower scores on the psychosocial scales from the Manu'a sample when compared to the Tutuila and Pago Pago samples, again suggesting an inverse relationship between modernization and self-reporting of complaints (Kruskall-Wallis One-Way ANOVA: Males, $X^2 = 18.4$, $p = .0001$; Females, $X^2 = 46.0$, $p < .0001$). A positive association between education and CMI scores was found, however, in all the female groups and among the Tutuila males, thus countering the relationship found at the village level. McGarvey points out that "this finding may reflect greater knowledge of psychological states among educated Samaons, or it may also reflect an

Table 8.6 Percent of Sample not Reporting any Symptoms in the Total CMI, Psychosocial Section, and Somatic Section

Section	Age	Men				Significant comparisons	Women				Significant comparisons
		H	T	M	WS		H	T	M	WS	
Psycho-social	18-35	44	27	0	0	H>T,WS	40	20	6	0	H>T>WS
	36-50	23	13	10	0	n.s.	32	24	6	8	.05[a]
	50+	9	16	3	0	.05[a]	5	26	4	0	.05[a]
Somatic	18-35	34	14	0	0	H>T,WS	25	8	0	0	H>T,WS
	36-50	4	4	10	0	n.s.	16	7	3	0	n.s.
	50+	13	3	3	0	n.s.	5	4	0	0	n.s.
Total	18-35	32	12	0	0	H>T,WS	25	8	0	0	H>T>WS
	36-50	4	3	0	0	n.s.	13	6	3	0	n.s.
	50+	9	1	3	0	n.s.	0	4	0	0	n.s.

Note: H = Hawaii; T = Tutuila; M = Manu'a; WS = Western Samoa; n.s. = not significant at alpha of .05.

[a] No multiple comparisons were significant at alpha of .005, but $p < .05$ = significant difference at alpha of .05.

Table 8.7 Median CMI Scores after Removing Samoans not Reporting any Symptoms on the Total CMI

Section	Age	Men				Significant comparisons	Women				Significant comparisons
		H	T	M	WS		H	T	M	WS	
Psycho-social	18-35	4	7	17	19	H,T<WS H<M	6	10	18	20	H,T<WS H<M
	36-50	7	8	13	25	H,T<WS	6	8	11	19	H,T<WS
	50+	12	6	10	18	T<M,WS	14	6	11	25	T<M,WS
Somatic	18-35	8	15	49	49	H,T<M,WS	13	21	22	45	H<All T<WS
	36-50	14	25	34	55	H,T<WS	20	22	25	42	H,T<WS
	50+	39	24	34	59	T,M<WS	34	25	27	58	All<WS
Total	18-35	10	23	71	65	H,T<M,WS	19	32	52	65	H<All T<WS
	36-50	22	35	47	77	H,T<WS	26	32	36	63	H,T<WS
	50+	51	31	50	76	T,M<WS	48	32	42	81	All<WS

Note: H = Hawaii; T = Tutuila; M = Manu'a; WS = Western Samoa.

increased proclivity to identify and express moods, feelings and emotions" (McGarvey 1984b:44). He regards the contrast between these results and those based on residence-based measures of modernization as an example of the ecological fallacy in population research (i.e., the failure to consider the internal distribution of traits in a population).

Closer analysis of the CMI scale with individualized indices of modernization yielded mixed results when combined with physiological measures. McGarvey suggests the possibility that status incongruity between education and occupation may be a key factor among more urbanized Samoans and may explain these results. For example, although there was an overall lack of association between education, blood pressure, and the CMI scores with age and adiposity controlled among Pago Pago men, those with low education and managerial jobs showed a significantly positive association between blood pressure and the CMI scores; among those with some secondary education in managerial positions systolic blood pressure was inversely related to complaints (see Chapter 15 for further discussion).

In a prior analysis of the data McGarvey (1980) found an inverse blood pressure and complaint score association among American Samoan females, but his later work strongly suggests that the inverse association may be due to a positive correlation between fatness and the psychosocial scale of the CMI.

In order to clarify the patterns of psychological complaints among Samoans McGarvey factor analyzed the data, deriving four factors accounting for approximately 80 percent of the variance. The four factors he labeled anger, depression, fear, and anxiety. Among men the anger factor had the strongest loadings, followed by depression, fear, and anxiety. Among women depression loaded most heavily on the first factor, followed by anger, fear, and anxiety. This is an interesting finding in itself, although McGarvey does not follow it up. Anger management is clearly one of the major adjustment problems confronting Samoan men (see Chapter 16 of this book). The significance of depression for Samoan women is less clear, but should be followed up.

It seems clear from these results that self-reported complaints are complex and rather subtle phenomena that cannot be taken at face value. A number of variables appear to be at work in patterning responses to the protocol and these remain to be unraveled. One rather obvious possibility is that the circumstances under which data were elicited in the different communities affected the outcomes. It is also possible that rural, communal environments facilitate talk about symptoms (along with other forms of self-disclosure) that tend to be inhibited in urbanized areas. The whole issue of expressing emotion may in fact be involved here, particularly with that section of the scale utilized by McGarvey. He takes note of studies that document a preoccupation among Samoans with controlling the expression of strong emotions, but this in itself does not seem to shed much light on patterns of response. The fact that the

Samoan scores were intermediate between patients in a general U.S. practice and military populations on the one hand, and samples of neurotic and psychiatric patients on the other (McGarvey 1984b:20), does not make them appear particularly inhibited. Whatever the explanation, the meaning of self-reported conditions will have to be determined before future studies utilizing such techniques to measure health status will be credible.

THE SEVEN VILLAGE STUDY

The sample recruited for the the Seven Village Study was purposefully selected and is limited to youths and young adults.[2] The intent was to obtain a sample of premigrant Samoans. For this reason it was decided to limit the age range of subjects to the 15- to 30-year-old group, the group most likely to migrate in the near future.[3] Originally ten villages were selected with the goal of obtaining a total sample of 200 subjects, approximately 20 from each village. Due to time constraints only seven villages were finally included, and a total of 99 subjects were interviewed. Villages were selected to represent a range from most urban to most rural, and for relatively large population size.

All villages were contacted through the Office of Samoan Affairs (OSA), which handles liaisons with traditional leaders and local governments. The researchers met at the OSA with representatives of the ten villages originally selected, explaining the rationale for the study and some of the mechanics. Five of the ten villages originally contacted participated in the study. Another three were unable to participate due to internal difficulties or conflicting demands on their time. These were replaced by two additional villages whose locations were chosen in order to retain the spatial design of the study. The remaining two of the original ten villages were not included due to time limitations and because there was sufficient representation from villages at similar removes from the Pago Pago area.

In each village the major contact was the village mayor *(pulenu'u)*, who typically discussed the research with the village fono. A list of all residents between the ages of 15 and 30 years of age, who were born in American Samoa, was requested.[4] For three villages, and somewhat less adequately for a fourth, lists were obtained that met specifications. In the first village studies, almost all subjects on the list were interviewed without sampling. Upon obtaining lists for the other three villages, numbers were assigned by age and sex groups and subjects were selected by the use of a random number table. In the remaining three villages, the subjects were preselected by village leaders and, with the addition of a number of walk-ins, constituted the sample.

The actual range of ages in the sample is from 14.4 to 32.9 years, with a mean age of 19.0. Fifty-five of the subjects were male, 44 were female. Only nine subjects were currently married, and no subjects reported prior marriages. All married subjects were age 23 or more, and all reported having children. While the overall sample is not truly representative of the full range of youths and young adults in American Samoa, it can be considered as reasonably representative of those born in the territory.

Health status

The Health Interview comprised a brief personal and family health history and was responded to by 82 subjects. In response to questions

Table 8.8 Seven Village Study: Frequency of Cigarette Smoking and Alcohol Use

Question	Number of respondents	Percent of respondents
Do you now smoke cigarettes:		
Amount/day:		
2 or more packs	1	7
1½ packs	2	13
1 pack	5	33
½ pack	1	7
less than ½ pack	6	40
Total number of current smokers	15	18
Total number of former smokers	10	13
Number who have never smoked	56	69
Do you now drink any alcoholic beverages:		
Number of drinks/week:		
2 or fewer	10	67
3 to 6	2	13
7 to 24	1	7
25 to 40	1	7
more than 40	1	7
no response	1	—
Total number of current drinkers	16	20
Total number of former drinkers	14	18
Non-drinkers	50	62

Table 8.9 Seven Village Study: Age Associations within the Health Interview Data

Variable	Number of responses	Kendall's tau	Probability
History of high blood pressure	99[a]	-.16	.02
Smoker	81	-.30	.002
Alcohol user	80	-.37	.0002
Former smoker	77	-.22	.006

[a] Includes missing cases, coded as zero, so that the probability is likely to be greater than .02.

concerning coronary heart disease, diabetes, and pulmonary diseases, high frequencies of disease states were reported, but misreporting (or miscommunication) is a likely factor.

The frequency of reported cigarette smoking and alcohol use are presented in Table 8.8. Of the 81 respondents who provided information, 18 percent were current smokers and 13 percent were former smokers. Alcohol use was reported by 20 percent of the respondents, while another 22 percent claimed to be former drinkers. Of those currently using alcohol, the mode was for two or fewer drinks per week, with one subject reporting 40 or more drinks weekly.

There were few apparent sex differences in health histories, although males reported more exercise (Kendall's tau = - 0.26, $p < .02$) and more alcohol use ($X^2 = 10.26$, 1 df, $p < .002$). There were, however, several age-related associations and health histories, even within such a constricted age range. Significant negative correlations were found between age and reports of alcohol and cigarette use, and between age and blood pressure (Table 8.9). The negative association of age with blood pressure is unexpected and again raises questions about the validity of the data. It is possible, however, that younger informants felt themselves to be less healthy (see Orans' data, presented below), or that more of them were aware of blood pressure problems. Actual blood pressure measurements, however, contradict the negative relationship with age derived from self-reports. Thus age was positively correlated with systolic and diastolic measurements for the same sample.

Major life experiences

The Major Life Experiences Questionnaire (MLE) was used to obtain information on social background and life histories, with special emphasis on mobility, decision making, and quality of social interaction. Ninety-three subjects completed the MLE. The background data showed this to be a relatively well-educated, bilingual group. Only one subject over the age of 18 had completed less than a high school education, while 13 subjects had some post-high-school training. Likewise, only one subject reported speaking Samoan solely; 88 claimed to speak both Samoan and English; and four listed at least one language in addition to Samoan and English.

Only nine subjects reported current employment, six of them in white collar positions. Half of the total sample, however, had been wage earners at some time in the past. Most informants (73.0%) claimed to receive remittances from relatives abroad, although the majority who did (37 of 65) said they received less than $100 per year.

Mobility

A number of mobility indicators are included in the MLE. They include subject's birthplace in relation to current residence, parents' birthplaces, number of places lived, reasons for moving from one place to the other, length of time on Tutuila, number of places lived outside of Tutuila, and absences from Tutuila during the past five years. On the whole, responses indicate a home-based, but mobile population.

The large majority of subjects, 81.3 percent ($N = 91$), were currently residing in the village of their birth.[5] Since specific villages were often not reported for parents' birthplaces, it was not possible to determine the percentage currently living in parental villages. The information on parental origins is revealing, however. Approximately one-third of subjects' fathers (30 of 91) were born outside of American Samoa, 26 of them in Western Samoa. Twenty-two mothers ($N = 93$) were from Western Samoa, the rest were all from American Samoa.

The mean number of places lived was 2.3, although 50 percent of the sample had never lived in any other village than that of their birth.[6] Twenty-four subjects (27.3%) reported an absence from Tutuila for at least six months during the past five years. Previous residential moves were primarily to Hawaii (35% of all moves), elsewhere on Tutuila (31% of all moves), and to the U.S. mainland (26% of all moves). Thirty-seven percent of the respondents reported that they would move to improve their education or to live with relatives. Almost all other reasons given were based on family concerns, with only one out of 118 respondents suggesting that a job would be a significant reason for moving.

The powerful pull of Hawaii and the mainland United States as future migration destinations is shown by the fact that 70 subjects (76.9%, $N = 91$) said they intended to move away from Tutuila at some point in the future, with Hawaii and California preferred by 60 percent of the respondents.

Social involvement

A number of questions were asked on the MLE regarding the amount and quality of interaction with family and community. In light of popular and ethnographic stereotypes of Samoan village life as being highly communal, the results are extremely revealing. One might expect social pressure on individuals to claim, at least, active participation in such important institutions as the church, the matai system, and other village groups, so the high degree of detachment is surprising. For example, 44.8 percent of those responding ($N = 87$) reported that they did not serve a matai (with females significantly more likely to make such a claim than males, $r = 5.04$, $p < .05$). Interestingly, there was no indication that rural villages differ significantly from urbanized ones in this regard.

Since it was anticipated that everyone would belong to a church subjects were not asked directly about membership, but rather whether they supported their church more, about the same, or less than other people in similar family and financial circumstances. As expected, the most common response was "about the same" (54.4%), but 23.9 percent said "less" and 10.2 percent claimed that they did not support a church at all; only 11.4 percent portrayed themselves as providing more support. While it is possible that canons of modesty are involved here, the results do not seem to confirm the universal depth of involvement in church affairs that is often portrayed in the literature.

With regard to organizational participation (which included village, church, and school groups, among others), the results were even more revealing. Only 28.0 percent of the respondents ($N = 93$) claimed to participate in more than one organization, 36.6 percent were limited to one, and 35.5 percent indicated they did not participate in any.

Correlates of social involvement

The battery of protocols included several sets of questions eliciting information related to social support, in addition to community involvement. These included attitudes toward self-reliance versus other-reliance, attitudes toward accepting or relinquishing responsibility, and feelings about the helpfulness of neighbors and friends.[7] Another set of questions aimed at eliciting self-perceptions about anger arousal and responses to anger. A third set provided information relevant to coronary-prone behavior, and included questions related to time orientation, competitiveness, sense of responsibility, and felt pressure for achievement. The responses were grouped into related sets in order to construct indices, and followed the general procedure for index construction employed by Howard (1974): (1) questions were grouped according to the type of information elicited; (2) frequency distributions of responses to each question were examined and questions not showing any appreciable response variation were dropped; (3) scaled scores were assigned to responses, such that a question score of 0 meant an absence of attribute or no response to the question, whereas a score of 2 or 3 was assigned to strong responses; and (4) component question scores were

equally weighted and summed to yield an index score. All indices are
therefore conservative in that a high index score would only result if an
individual consistently expressed strong responses. Likewise, all indices
are independent in that any given question served as a component of only
one index.

As constructed, the indices show a high degree of internal consistency.
For example, regarding the components of coronary-prone behavior, many
strong intercorrelations were found, suggesting that such a complex exists
within contemporary Samoan society, despite contrary stereotypes. In
general, there is a moderately high level of work orientation and a high
degree of competitiveness in the sample, but there is also a significant
degree of felt time compression, not usually considered characteristic of
Polynesian societies. Among males time orientation correlates highly with
competitiveness (Spearman's $r = .46$, $p < .001$), feelings of pressure for
achievement $(r = .31$, $p < .03)$, and responsiveness to pressure for
achievement $r = .35$, $p < .02$). It also correlates with admissions of experiencing intense anger on occasion $(r = .36$, $p < .01)$. Among females
the syndrome is somewhat less pronounced. The only significant correlate
with time orientation of women is a strong orientation toward work
$(r = .33$, $p < .05)$. Those women who report a positive response to
achievement pressures, however, also report experiencing the most felt
pressure for achievement $(r = .48$, $p < .002)$, intense anger on occasion
$(r = .43$, $p < .006)$, and a tendency to displace their anger $(r = .34$, $p <
.04)$. They also are more likely to invest their earnings in self-
development rather than in social relations $(r = .35$, $p < .03)$. Whether
the coronary-prone syndrome plays a role in the high rates of
hypertension and susceptibility to heart disease among Samoans remains
to be investigated.[8]

Since ethnographic studies have indicated the importance of social
relations and community involvement for Samoans, it is of some interest
to look at the individual correlates of such involvement. As pointed out
above, more variability exists than one might have expected for Samoan
village communities. In urban areas, and in overseas communities, more
options exist and almost certainly allow for greater detachment. Since
evidence demonstrating the importance of social support systems for
mediating the effects of stress has been mounting (Thoits 1983);
Broadhead et al. 1981; Gad and Johnson 1980; Andrews et al. 1978) it
will be important to monitor the effects of social detachment on
individuals as it occurs in the context of change.

The index of community involvement was related to anger patterns for
both men and women. The men who were most involved were less likely
to report expressing felt anger than their less involved peers $(r = .29$, $p
= .05)$, while involved women were more likely to deny having strong
affective responses to angering situations $(r = .43$, $p < .008)$. Thus
community involvement, as one might expect from the ethnographic

literature, appears to require the inhibition of anger, although there are no indications in the questionnaire data that those who are involved experience less anger.

Another indicator of social involvement that showed interesting correlations was attitudes toward neighbors. Among men, the more positive their attitude toward neighbors, that is the more neighbors were perceived as helpful, the less they were expressive of felt anger ($r = .32$, $p < .03$) and the less they reported antisocial responses to anger ($r = .32$, $p < .03$). Positive attitudes toward neighbors, however, was strongly correlated with competitiveness ($r = .44$, $p < .002$). While this may appear incongruous to Westerners, such a finding is consistent with the Samoan pattern of competing for traditional rewards such as titles, for it involves mobilizing social relations to support one's ambitions. Among women friendly attitudes toward neighbors was negatively correlated with the expression of anger through protests ($r = .46$, $p < .003$); when attitudes toward neighbors and friends were combined, a negative correlation also appeared with overall expressiveness of anger ($r = .33$, $p < .05$). Again, the requirement for inhibiting expressions of anger appears to be prerequisite for maintaining social involvement.

Still another measure of social involvement was the degree to which subjects expressed a desire to take responsibility for their own affairs, or reported wanting others, such as matai, to take responsibility for them. Consistent with the above-reported results, those men who were most other-reliant were most likely to inhibit expressions of anger ($r = .42$, $p < .003$), but they reported discussing the reasons for their anger with others more frequently ($r = .39$, $p < .008$). No significant correlations were found for women.

Differences between men and women were slight, and nonsignificant on all measures of social involvement. If anything, women reported being slightly more involved with their immediate families and less involved with the broader community. They also reported more experiences of intense anger and were somewhat less inhibited in expressing it. Women were also less competitive and showed less concern for taking responsibility, but scored somewhat higher on time and work orientations.

Although there were significant differences between villages on some measures, there was no consistent pattern of differences to correspond with an urban/rural continuum. Our conclusion is therefore that as far as our measures are concerned, Tutuila is relatively homogeneous and does not display variations at the village level corresponding to modernizing influences.

Overall, then, data from the Seven Village Study seem to highlight the presence of stressful experiences for Samoans on Tutuila and their patterning according to degrees of social involvement. That there may be important health implications involved is signaled by the finding, using the same sample, that individuals whose urine showed high levels of hormone output in the overnight samples were relatively less involved in community affairs and generally did not view their friends and neighbors as helpful (reported in Chapter 9 of this book).

THE TA'U LIFE EVENTS STUDY

The Ta'u Life Events Study was conducted by Scheder while she was a postdoctoral fellow in medical anthropology at the University of Hawaii.[9] The study site was on the island of Ta'u, in the Manu'a chain, in a village of 500 people, generally regarded as the most traditional village in American Samoa. The sample consisted of 112 subjects (55 males, 57 females). A modified version of the Holmes-Rahe life event scale (Holmes and Rahe 1967), specific to Samoa, was used. Added to the scale were such items as death of the matai; receiving a title; hosting visiting relatives; giving a fiafia (party); moving to another island, Hawaii, or the mainland for work; making the same moves, but to be with relatives; and a change in one's child's church affiliation. Subjects were asked to rate each event according to the strength of their reaction to it. The dates of each event experienced were recorded, and these were later compared to subjects' medical records for association with blood pressure and blood glucose measurements. Data pertaining to cultural stressors and attitudes, social support and social integration, and migration history were also obtained.

Scheder had conducted a similar study among migrant Mexican-American farm workers (Scheder 1981), and her major comparisons were with them. Table 8.10 shows the percentages of Samoans and Mexican-Americans who scored events as causing an "extremely strong reaction" and the comparable rankings of items included on both lists.[10] The most striking difference between the two groups is the lesser degree of variation among Samoans in their responses to the items. Thus the range for Samoans is between 70.9 percent and 45.2 percent, while Mexican-Americans vary between 80.0 percent and 0.0 percent. Whether this indicates that Samoans in fact react more uniformly (and on the whole, more strongly) to events is quite doubtful, but without some comparable contextual anchoring there is no way to tell. The differences in rank ordering are also interesting, although the relative lack of variation among Samoans requires caution with regard to how seriously we can take the rankings. Mexican-American rankings appear to reflect a greater concern for self and immediate family than the Samoan rankings, which reflect a greater concern for maintaining networks of relationship; thus the relatively high ranking among Samoans of argument with relative, serious illness of relative (which Samoans rank higher than serious personal illness), and death of close friend. Mexican-Americans rank serious personal illness just after death of spouse, near the top of the scale. They also rank money and work-related events higher than

Table 8.10 Percent of "Extremely Strong" Reaction and Ranking of Life Events by Samoans and Mexican-Americans

Event description	Samoans		Mexican-Americans		
	Percent	Rank[a]	Percent	Rank[a]	Discrepancy[b]
Death of child	70.9	1	80.0	1	
Death of spouse	69.1	2	74.2	2	
Death of close relative	67.3	3	50.0	4	
Birth of child (for father)[c]	66.0	4	12.5	19	XXX
Birth of child (for mother)			0.00	26	XXX
Miscarriage, stillbirth	65.5	5	33.3	10	
Death of matai	65.5				
Argument with relative	65.0	6	4.8	22	XXX
Child leaves home	64.1	7	25.9	15	
Serious illness of relative	63.6	8	17.9	17	X
Death of close friend	60.0	9	6.9	21	XX
Marriage	59.3	10	16.0	18	X
Son marries with approval	58.8	11	0.0	26	XXX
Argument with sister (1° relative)	58.3	12	26.1	12	
Argument with friend	58.3				
Argument with child	58.3				
Pregnancy	57.9	13	4.5	23	XX
Child married without approval	57.9	13	42.9	6	X
Separation from family	57.7	15	50.0	4	XX
Move from Hawaii to be with relatives	57.5				
Serious personal illness	57.3	16	56.7	3	XX
Move to other island to be with relatives	56.7				
Giving a fiafia (party)	56.2				
Adoption of child	55.9				
Host relatives on malaga (visit)	55.7				
Serious argument with brother	55.3				
Daughter marries with approval	55.3	17	4.5	23	X
Receive a title	53.3				
Marital separation or divorce	52.9	18	38.1	8	XX
Work hours lengthened	52.4				
Argument with parents	52.0				
Argument with spouse	52.0				
Problems with money	52.0	19	42.9	6	XX
Child changes church affiliation	52.0				
Move to Mainland to be with relatives	51.4				
Change jobs	50.0	20	7.4	20	
Menopause	49.5	21	36.8	9	XX
Move to another island for work	49.0				
Work hours shortened	47.1	22	24.1	16	X
Move to Mainland (another state) for work	46.2	23	28.0	11	XX
New boss	46.2	23	4.3	25	
Move to Hawaii for work	45.2				

[a] Only items included on both Samoan and Mexican-American protocols are ranked.
[b] Discrepancy of from 6 to 9 places—X; from 10 to 14 places—XX; from 15 to 19 places—XXX.
[c] Alternate wording included in Mexican-American protocol are enclosed in parentheses.

Samoans. Of interest, too, is the fact that Samoans, unlike Mexican-Americans and most other populations studied (Rahe 1969), but like their Hawaiian cousins, rank marriage higher than divorce. They also rank a son marrying with approval relatively high—higher, in fact, than a child marrying without approval. This reflects, I believe, the great seriousness with which Polynesians enter into new sets of social obligations. To get married is to take on new social burdens, including new in-laws; to get divorced is to remove oneself from them. For a child to get married with permission implies stronger obligations than without permission. Also, since the nuclear family is less emphasized, the trauma of divorce may be less pronounced.

With regard to migration, Scheder offers an interpretation of Samoan responses consistent with this view:

> the largest dichotomy was seen in moving to Hawaii for work, as opposed to the same move to be with relatives. Moving to Hawaii for work was the lowest ranked event. Moving to another island or the mainland for work were also ranked near the bottom. Moving to any of the three places in order to be with relatives was ranked much higher, particularly moving to Hawaii or another Samoan island. It might be interpreted that moving to be with relatives engenders a new set of obligations that are not expected when one migrates for work. At the least, this dichotomy suggests motivations for migration carry different expectations, and are coupled to different emotional responses, but that migration per se does not necessarily provoke a strong reaction.
>
> The differences in subsistence needs of the two populations emerge in the event rankings. Moving for work, loss of property, uncertainty of employment, are givens for migrant farm workers. Samoans, at least in Manu'a, rely on a network of mutual support and effort in providing food and shelter, so that job and monetary changes have less impact. A Samoan who migrates to be with relatives maintains the ties in Samoa, and encounters new obligations among his hosts. The relative ranking of work versus family-based migration by Samoans provides a concise statement of the importance of context. (Scheder 1983:4-7)

To clarify migration expectations Scheder presented her subjects with a protocol eliciting their views of life in the village, Pago Pago, Hawaii, and California. Subjects were asked to state whether a number of conditions would or would not provide difficulties in each of these locations. Results are presented in Table 8.11. The general pattern is for conditions to be perceived as more difficult the further one went from the village. There were two exceptions to the pattern: getting a good job and getting a good education. Getting a good job was viewed as equally difficult regardless of location, while a good education was seen as harder to obtain in Samoa than in California or Hawaii.

The greatest agreement among respondents occurred in relation to

Table 8.11 Views of Life in Samoa, Hawaii, and California: General Patterns

Item Selected by the Greatest Proportion of Respondents

View of	Difficult	Percent	Not difficult	Percent
Village	Education	70.2	Support family	89.5
Pago Pago	Education	74.2	Acting fa'aSamoa	72.4
Hawaii	Good house	80.7	Education	58.6
California	Good house	87.5	Education	49.4

Top Four Items Selected by the Greatest Proportion of Respondents

View of	Difficult	Percent	Not difficult	Percent
Village	Education	70.2	Support family	89.5
	Good job	53.1	Live in good area	86.5
	Medical care	35.1	Acting fa'aSamoa	85.4
	Enough food	29.2	Being safe	85.3
Pago Pago	Education	74.2	Acting fa'aSamoa	72.4
	Job	49.4	Serving a matai	69.7
	Medical care	42.7	Having friends	69.3
	Being safe	39.8	Good things to eat	68.2
Hawaii	Good house	80.7	Education	58.6
	Good area	70.1	Good job	50.0
	Being safe	69.0	Good things to eat	46.0
	Serve a matai	67.8	Enough food	41.4
California	Good house	87.5	Education	49.4
	Good area	85.1	Good job	47.5
	Being safe	74.7	Good things to eat	41.4
	Fa'aSamoa	73.6	Enough friends	37.2

supporting a family in Samoa, with 89.5 percent judging it as not difficult, while in contrast, having a good house in California was seen as difficult by 87.5 percent of the subjects. It therefore appears that migration to Hawaii or the U.S. mainland is seen as a trade-off with some risk attached.

Scheder established an index of life event reactivity by combining responses to six of the items on the list: death of matai, argument with relative, argument with sister, host relatives on malaga, giving a fiafia, and receiving a title. She then performed a discriminant function analysis using subject groups with the highest and lowest responses to these events ($N = 58$), and tested for relationship to health status. She found blood glucose levels to be significantly higher for the high reactor group, and while none of the low reactors had diabetes, 9.5 percent of the high reactors had the disease. Although the numbers are small, her data are suggestive of future research.

PERCEPTIONS OF OBESITY STUDY

The main purpose of this study was to examine the perceived relationship between obesity and health, fertility, age, sex, social status, and role performance. The underlying question motivating the research was: What does obesity signify as a visual social marker? Connelly and Hanna note the widespread traditional Polynesian custom of fattening chiefs, a well-fed chief symbolizing prosperity and the beneficence of the gods. The concern was whether the attitudes of migrants to Hawaii, who show an extremely high incidence of obesity (see Pawson, Chapter 11), reflect such positive associations.

Sixty Samoan migrants, ranging from 8 to 89 years old, were interviewed. They had all participated in a previous survey conducted by Hanna and Baker (see Chapter 1) a year earlier, and were contemporaneously participating in a nutritional survey. Each informant was shown a set of three line drawings depicting several individuals participating in an activity. The activities were typical of life in Samoa, for example, cooking, fishing, and making fine mats. There were separate activities involving children. The figures were Polynesian in appearance, but sketchily drawn so as to emphasize differences in body shape.

Subjects were asked to tell a story about the action in the picture and when finished, were asked a series of standardized questions to elicit information concerning their associations between body size and behavior. They were also asked to sort four male and four female somatotypes, ranging from lean to obese, into several categories. In order to gain some measure of cultural familiarity, two knowledge tests were utilized. One was the Middle Class Conceptual Test (described by Howard 1974), which consists of a set of 20 concepts familiar to most middle-class Americans; the other was a comparable instrument consisting of 24 Samoan terms. The tests were in multiple-choice formats, with four options for each item. The mean score for the MCCT was 3.9, less than chance, indicating a population still unfamiliar with middle American culture. The mean score for the Samoan Conceptual Test was 16.8, indicating a moderate knowledge of traditional Samoan culture.

As expected, for both men and women, the obese figures were perceived as being of high status and the lean figures of low status. When asked to select the most effective *ali'i* (high chief), 80 percent selected the most obese somatotype. The main reason given was that a large-bodied man looks more regal and commands more respect. Several informants conjectured that high chiefs and their wives became obese because they command others to work and do not have to work themselves. Talking chiefs *(tulafale)*, who are required to be more active, were perceived as intermediate between ali'i and untitled men *(taule'ale'a)*. The data are presented in Table 8.12.

When asked whether body size was an influential factor in selecting a

Table 8.12 Reported Perceptual Relationlship between Body Size and Social Status in Somatotype Sorting

Body size	Man's status		
	Untitled (percent)	Talking chief (percent)	High chief (percent)
Husband	(42 respondents)	(42 respondents)	(45 respondents)
Thin	62	57	20
Obese	7	43	80
Wife	(30 respondents)	(35 respondents)	(40 respondents)
Thin	87	43	20
Obese	13	57	80

chief, however, the answers were overwhelmingly negative. Rather informants emphasized intelligence, knowledge of and respect for Samoan custom, thus indicating that obesity is considered a consequence of high status rather than a prerequisite for achieving it. Since age was associated with obesity by 77 percent of the subjects, and since chiefs are generally older than untitled men, perceived age may account for part of the association between chieftainship and large body size. That Samoans do not consider obesity to be a generally positive trait was attested by the fact that 93 percent of the subjects associated it with poor health. They saw it as linked to asthma, difficulty in breathing, tiredness, and high blood pressure. Female beauty, on the other hand, was associated with the thinnest somatotype by 86 percent of the men and 84 percent of the women.

These data suggest that the high rates of obesity among Samoan migrants to Hawaii are not the direct result of values that lead them to try to gain weight.

ORANS' HIERARCHY AND HAPPINESS STUDY

Orans has been studying the relationship between status, self-reports of happiness, and other aspects of well-being in Western Samoa. In conjunction with his field work in the village of Salamumu, on Upolu, he administered a sequence of three survey questionnaires designed to elicit systematic information on the topic. The first was administered in June of 1978, the second about 9 months later, and the third 3 months after that. Orans was interested in examining the relationship between economic variables and happiness/satisfaction, a correlation that has been reported within countries [but not between countries (see Easterlin 1975)]. He found no such correlation in Salamumu, nor did he find significant sex differences, but he did find matai status to be positively associated with such measures.

Orans' experience with the survey instruments is highly instructive and should serve to warn researchers who elicit such data without paying careful attention to their limitations. In the first instance his subjects responded to the global happiness question in an overwhelmingly positive way. The question was: Taking all things together, how would you say things are these days—would you say you're *very happy, pretty happy*, or *not too happy*? Given the fact that 82 percent of his informants chose "very happy," and that the Salamumu mean (2.74, using a 3, 2, 1 scoring system) was far higher than any sample previously studied [inclusive of eight countries (see Easterlin 1975:107)], Orans concluded that "either Salamumu is the happiest place on earth or the survey question didn't work too well" (Orans 1985:28). On the second questionnaire he obtained a more reasonable spread by changing the alternatives to *very happy* (30%), *happy* (32%), and *not so happy* (20%), placing the Salamumu mean (2.23) very close to U.S. scores obtained between 1957 and 1972 (Campbell et al. 1976:26). Using this same wording on the third questionnaire yielded a greater concentration of happy responses (54%) and a slightly lower mean (2.13; Orans 1985).

Orans also learned a good deal about Samoan terms and what they conveyed in the interim between the first and subsequent administrations. In the first questionnaire, for example, he used the term fiafia to denote happiness, but found it to be a stock response to conventional "how are you" questions, and so replaced it in later versions.

The most distressing problem Orans encountered, however, was the lack of cross-questionnaire consistency or reliability. Out of 11 question-and-answer choices that were identical in the second and third questionnaires, only two provided responses that were significantly correlated. Despite this shortcoming, Orans argues for the utility of the instrument on the grounds of reasonable internal consistency, although in the second instance there were some anomalous findings involving the relationship between questions about affect and the happiness scale.

The influence of hierarchy was examined by comparing a group of untitled males with a group of talking chiefs. In general, Orans found support for the proposition that status is positively correlated with various measures of happiness and satisfaction, including the puzzling finding that despite their greater age, the talking chiefs reported better health status (Table 8.13).

Orans is forthright in reporting all the inconsistencies and puzzles presented by his results, which beg for explanation. My own interpretation of the problems he encountered, and this holds for other self-reported data from Samoa, is that responses are patterned far more by the rules of speech behavior and immediate social context than they are reflective of internal states or overall regularities. In addition to a strong courtesy bias and social approval bias, a number of ethnographers

Table 8.13 Reported Health by Social Status

	Very Good		Good		Not So Good	
	N	Percent	N	Percent	N	Percent
Questionnaire II						
Untitled men	9	39	11	48	3	13
Talking chiefs	13	62	7	33	1	5
Questionnaire III						
Untitled men	6	19	18	58	7	23
Talking chiefs	9	50	6	33	3	18

have pointed to the importance of immediate context in the patterning of Samoan experience and behavior. Thus the lack of consistency between Orans' second and third sets of responses may well have been due to individuals responding to questions in terms of their most recent set of experiences, rather than in terms of some longer term trend (even though the questions are structured to elicit the latter).

The influence of the courtesy bias and social desirability shows up, I believe, in the relatively high degree of acceptance of statements in their strong forms, which occurs in both Orans' and Scheder's data. Social desirability may also be what lies behind the anomaly of the older chiefs reporting better health than the younger untitled men. In Polynesia, as in many other parts of the world, the virility of a chief is metaphorically synonymous with the prosperity of his group, including an abundance of crops, the fecundity of women, and other blessings of nature. A chief's health is indicative of his mana, of his support from the gods, or in the modern context, from God. Thus if a chief were to admit to poor health, he would be admitting to ineffectiveness in a most basic way. Since tulafale, as talking chiefs, are particularly expected to demonstrate virility and action, in contrast to the more passive role of ali'i, it would not be surprising if they de-emphasized health problems in their talk to interviewers.

The tantalizing complexities of such data are underscored by James's analysis, using Orans' data in conjunction with physiological measures of the same population by the Penn State research team. James (n.d.) divided the sexes into two age groups, 18-39 years and 40 or more. Among the males he found significant negative associations between measures of fatness and Orans' well-being indices for those under 40, whereas significant positive associations occurred between these indices and measures of fatness in the older group. No such relationship was found among women. One might speculate that obesity in older men results from inactivity associated with high status, whereas among young men it is a hindrance to high-level performance and symptomatic of ineptitude.

SUMMARY

Although the data presented in this chapter must be treated with caution, a number of interesting questions have been raised that beg additional research. I conclude by citing those that I see as most important for understanding Samoan adaptation to the processes of "modernization," or as I would prefer to call them, the processes of cultural diversification (see Chapter 16).

Perhaps most pertinent for health-related research are the questions raised about Samoan self-reports of health status. The CMI data, the health interviews from the Seven Village Study on Tutuila, and Orans' questionnaire material all raise questions about the meaning of the information. McGarvey raises the right question when considering the significance of CMI responses. That is, he sees them as a form of complaint behavior, and poses the question of what factors influence the disposition to make complaints. This is a complicated issue and needs to be studied if health researchers are to make headway using self-reported information. As it stands, two anomalies arise from self-reports concerning health. One is the inverse correlation between indices of modernization and CMI complaint scores, the other is the fact that younger men report poorer health than older men. Both findings are contradicted by physiological measures of health status. Also intriguing is the fact that no significant differences appear in the amount of complaint behavior between men and women, since research elsewhere has consistently shown women to report symptoms more readily than men. The finding that anger has the strongest factor loading for men's psychological complaints, while depression has a stronger loading for women, makes ethnographic sense, but until we have a clearer understanding of the meaning of symptoms to Samoans, and of complaints, we will not be in a position to clearly interpret such data.

McGarvey's finding relating complaint behavior to status incongruities (low education, managerial responsibilities) among Pago Pago males, while it must be viewed cautiously in the light of the problem of interpreting CMI data and the relatively small sample size, presents an interesting and potentially important hypothesis for further study. A substantial segment of the Samoan population in urban areas and in enclaves abroad are in socioeconomic positions that involve such incongruities. It would be worthwhile to follow up McGarvey's finding in other settings with larger samples.

Among women McGarvey's data suggest eating and activity patterns may implicate complaint behavior, although why that should be the case is not clear. Whether or not these particular correlations continue to hold, what is needed most at this juncture in Samoan research are hypotheses, like McGarvey's, that go beyond correlations between gross measures of modernization and health status indicators.

Another important set of issues revolves around the effects of social involvement on Samoan health patterns. The fact that questionnaire data reveal a good deal of variability with regard to community involvement is itself an interesting finding, but it suggests a line of research of special relevance to the Samoan case. While stress studies have placed increasing emphasis in recent years on social support as a mitigating variable, the Samoan data are ambiguous on this issue. The results so far seem to suggest a U-shaped relationship, with stress occurring at both the uninvolved and strongly involved ends of the scale (see Chapter 16 for a more extensive discussion of this issue). Given the overwhelming importance of family obligations and engaging in systems of reciprocal exchange within the Samoan value system this variable would appear to be a crucial one for understanding that population's vulnerability to stress. More refined measures, aimed not only at degree of social involvement but its forms and qualities, are needed if further headway is to be made.

The finding that the coronary-prone syndrome occurs in Samoa also is suggestive of lines of inquiry. On the one hand research aimed at understanding the genesis of type A patterns among Samoans would be enlightening; on the other hand, the health consequences of such a disposition ought to be studied. Even without further elaboration the questionnaire data indicate that the pressures of modern life have seeped down to the village level, at least in American Samoa.

Scheder's use of the life event reactivity index provides yet another research lead. Although her sample is too small to be convincing, the notion that differential vulnerability to life crises implicates health outcomes should be followed up with additional research. Life event protocols, carefully administered and checked against other sources of information, may turn out to be among the more useful data obtainable through questionnaires. Properly adapted to the Samoan context, life event protocols may have the advantage of circumventing some of the more delicate problems involved in eliciting information about beliefs, attitudes, and feeling states.

Finally, there are the more general problems associated with the issues of questionnaire validity and reliability raised by Orans' research. Of particular importance, in my view, is the need to understand the relationship between context and response. As Shore (1982) and others have made clear, context is all important for interpreting social action in Samoa. I believe it is equally important for interpreting questionnaire data. What is needed, perhaps as a top priority research project, are studies of how Samoans interpret questions and construct answers under a variety of interviewing conditions. We might then be in a better position to relate Samoans' talk about happiness and distress to both their subjective experience and the objective conditions of their physical well-being.

NOTES

1. In other studies the CMI has correlated in the expected way with objectively measured symptoms; see Dutt and Baker 1978, for example.

2. The description of the sample villages was prepared by Hecht. Dr. Hecht also did a preliminary analysis of the health and major life experiences data.

3. The plan is to do a follow-up study comparing migrants with nonmigrants in order to assess the impact of migration on various health measures.

4. The ideal sample was to include only subjects who had not previously resided outside of American Samoa. Since we were told that many residents of American Samoa in the targeted age range were from Western Samoa or Tonga, we restricted our request to those born in American Samoa. A news brief in *Pacific Magazine* (March-April 1982:9) supports this assertion. Citing the Government *News Bulletin*, Pago Pago, it indicates that at least 37 percent, or 12100, resident aliens were included in the territory's population of 32000. Legal aliens numbered 10500 Western Samoans and 1100 Tongans, as well as 500 others. While no age breakdown is presented, the numbers suggest that aliens would probably be well represented in this age group.

5. Despite our request to include only individuals born in American Samoa on lists of eligible subjects, five were born in Hawaii, one on the U.S. mainland, and one in Melanesia. In accordance with our request, however, none were born in Western Samoa.

6. Number of places lived correlates weakly with age, Kendall's tau = .1514, $p < .035$.

7. The material presented in this section was analyzed by Martz. Portions of the findings were presented by Martz et al. 1984.

8. As part of their study of Pacific island immigrants to New Zealand, Graves and Graves (1985) included measure of type A attributes. They found these to be present in Samoans, although to a lesser extent than among European New Zealanders. Among all groups, including Samoans, type A attributes correlated significantly with symptoms of poor health. In the Seven Village sample reported on here, however, no correlation was found between hormonal indices of arousal and type A attributes.

9. The results described in this section were originally presented in Scheder 1983.

10. There were some minor differences in the wording of certain items. The event descriptions in Table 8.10 are those used with the Samoan sample, with phrases included in the Mexican-American format in brackets.

Chapter 9

Hormonal Measures of Stress

JOEL M. HANNA
GARY D. JAMES
JOANN M. MARTZ

Since its inception, the Samoan research has emphasized the concept of stress as a focus and as a unifying theme. The earliest surveys provided a description of a population that was undergoing morphological, physiological, and demographic changes reflecting responses to the stresses of migration from Samoa to Hawaii. As background studies from Western Samoa were added, it became evident that the stresses associated with modernization were at least equivalent to those of migration and that much of the stress-inducing exposure to modern society was actually taking place in Samoa. Thus when it became time to isolate specific stressful events and situations that have an impact on the health and well-being of Samoans, the research shifted to American and Western Samoa. As a result we are beginning to understand complex factors that in the long run may have influences upon health. While these factors have not been systematically studied in migrant communities, there is evidence that they may operate in these as well.

Stress research has followed two general lines. One line has investigated the stress of modernization and migration through survey and epidemiological techniques. These studies have used modernization and migration as class variables in order to contrast Samoan subjects from several communities. Dependent variables have been individual risk factors or aggregations of intercorrelated risk factors—blood pressure, obesity, blood glucose and lipids, as well as changes in mortality. This approach and its findings are illustrated throughout this volume.

The second approach involved examining the association between potentially stressful specific events, situations, and life-styles and psychological and physiological responses that have been linked to arousal and stress. Research techniques have included questionnaires, interviews, and collection of urinary hormones. In this context, migration and modernization are included as variables, but are viewed as more pervasive processes, influencing specific situations by providing a perspective for interpretation of events. This chapter discusses several such studies carried out independently in Western and American Samoa and considers their broader implications.

DEFINITIONS OF STRESS

Of the numerous attempts to conceptualize stress, we have selected a modification of that given by Rabkin and Struening (1976:1014). We define *stress* as an individual's response to some external stressor that is patterned and may be immediate, delayed, or continued for a period of time. From an anthropological perspective this definition is especially useful. Stress ultimately has an external origin, hence is amenable to systematic investigation using anthropological techniques. There is also a recognized patterning of responses that follow a stressful experience. While the details may change from situation to situation and vary between individuals, there is a patterning in the arousal that includes elevated hormone excretion, increased blood pressure, and in some cases a change (usually negative) in health status (Brown 1981). The delayed aspect of response is also important, because it means that the stress may continue after the event has passed and may persist for extended periods of time. This perspective provides an explanation of the delayed and negative health outcomes that follow stress and allows stress to be studied in the field, outside of the constraints found in controlled laboratory investigations.

We have added one further modification to this concept, that of *mediating factors*. We propose that certain individual and situational factors—mediating factors—may modify perception of and responses to stressors. These provide the background for interpretation of stressors and explain to some extent why there is variation in response between individuals exposed to the same stressor. Mediating factors may be environmental conditions such as the Three Mile Island disaster (Fleming et al. 1982); individual behavioral characters such as type A behavior (Dembroski et al. 1978) or social phenomena such as family support (Graves and Graves 1980). Mediating factors are a constant reference for interpretation of individual perceptions and responses to stressors. Modernization and migration are seen as mediating factors that pervade much of contemporary Samoan life.

MEASURES OF STRESS

Without well-defined, measurable dependent variables, the study of stress becomes difficult. Fortunately, recent investigations have defined several empirical measures that may be used to establish the existence of stress. Negative health outcomes have been widely used, since stress is frequently associated with dramatic changes in physical and mental health. Life events research has convincingly demonstrated that negative life events and crises may precede mental and physical breakdown. Some of the major consequences are heart disease, diabetes, stroke, and family disintegration. Much current epidemiological investigation into risk factors has attempted to define the relative importance of life events.

A more direct and equally useful measure has been the analysis of urinary hormone excretions, particularly the excretion of epinephrine and norepinephrine and of cortisol. Extensive laboratory and field research has documented these hormones as excellent indicators of states of arousal and they have been tied to specific stressful situations (Mason et al. 1979;   Frankenhauser 1975). These hormones are particularly useful because they are more than simply correlates of arousal. The catecholamines (epinephrine and norepinephrine) and cortisol are the effectors of the central nervous system by which perceived physical and social stressors are translated into physiological response. By monitoring the excretion of these hormones we can directly observe a part of the response itself. Upon release, catecholamines cause an immediate mobilization of the cardiovascular and metabolic systems for physically dealing with a perceived stress. Cortisol plays a complementary role in providing large quantities of glucose for energy and in mobilizing metabolic resources toward its increased production. One result of this mobilization is a suppression of the immune system and catabolism of protein. Hormonal variation also reflects more long term responses as well as immediate arousal. Catecholamine levels have remained elevated for months after perceived stressful events such as the Three Mile Island disaster (Fleming et al. 1982). Long-term, stress-induced hormone release may also have profound effects upon health due to extensive catabolic effects on body tissue.

STUDIES OF STRESS ON SAMOANS

Western Samoan adults

James (1984) and James et al. (1985) examined whether life-style differences that result from the sociocultural alterations occurring in Western Samoa influence urinary hormone and blood pressure levels. Some 123 young Samoan men living in Western Samoa provided urine samples which were assayed for epinephrine and norepinephrine. The men were divided among 4 groups based on place and type of occupational activity. These groups were rural village agriculturalists, urban manual laborers, urban sedentary workers, and urban university students. Some of the biological characteristics of the groups are shown in Table 9:1. The villagers are older and less fat than the urban groups, while the students are on the average younger.

Despite a common Samoan heritage, there were significant between-group sociocultural differences. The individuals that comprised the three urban groups were either employed or attending school in Apia. Consequently, they were continually exposed to or involved in a wage economy. The rural villagers were from Salamumu, a village whose residents adhere to more traditional Samoan ways of life. Although sociocultural differences occur among the groups, none of the men in the study were completely isolated from either traditional or modern influences since there is daily bus service between Apia and the rural villages (James 1984).

Two timed urine samples were collected from each individual using techniques described by Jenner et al. (1980). The first was an overnight sample collected early in the morning, and the second, a midmorning sample, was collected after 2 to 4 hours of typical morning activity. Aliquots of 15 to 10 ml were taken from these specimens and sent to the University of Oxford in the United Kingdom for analysis using high performance liquid chromatography (Jenner et al. 1981).

Psychological and social data were also collected as was information on activity, diet, and stimulant intake (James 1984). These data were used to define a number of life-style variables. The psychological and social data were summarized as 6 indices of several questions each. These indices described the degree of life satisfaction, emotional stability, agreement with Samoan customs (denoted attitudes), familial responsibility, coronary-prone behavior, and integration into more modern life-styles (denoted Western profile). Type of activity during the morning was partitioned into three categories (mental and physical effort, and non-taxing activity). The activity of an individual was described by the percentage of time spent performing tasks in these categories over the study period. Dietary intake was described by recalled number of calories and carbohydrates consumed over the morning, the stimulant intake as the number of cigarettes smoked and caffeine drinks consumed over the same period.

Mean hormone excretion values are given in Table 9.2, where group differences in excretion are observed. There is a uniform increase in hormone excretion of catecholamines in the daytime as compared to overnight values, which probably reflects a general increase in activity (Jenner et al. 1980; Matsuda et al. 1972; Patkai 1971). There are similar overnight values among the villagers and laborers and among the sedentary workers and students, with the latter exhibiting higher rates. At midmorning, the villagers maintain lower levels while the other three groups are higher, particularly the students.

To facilitate comparisons, the excretion values were adjusted for volume of urine collected, volume average of the sample group, age, and

Table 9.1 Biological Characteristics of the Sample Groups

Trait	Villagers (N = 31)		Laborers (N = 28)		Sedentary workers (N = 33)		Students (N = 31)	
	$\bar{X}$	SD	$\bar{X}$	SD	$\bar{X}$	SD	$\bar{X}$	SD
Age	31.4	9.4	26.9	6.2	26.4	6.3	21.6	2.8
Height	168.0	5.5	171.4	4.6	170.9	5.4	172.7	5.8
Weight	69.1	10.4	75.1	10.3	75.6	11.4	75.8	10.4
Triceps skinfold	8.5	4.7	11.4	7.2	11.6	6.9	12.4	6.2
Midaxillary skinfold	8.1	3.6	10.9	6.2	11.1	5.8	10.7	5.3
Subscapular skinfold	15.0	6.1	18.6	10.8	17.6	9.6	18.1	9.8
Waist skinfold	10.5	6.5	16.6	14.5	15.9	12.2	14.3	8.6

Source: Adapted from James 1984.

Table 9.2 Unadjusted Catecholamine Excretion Rates Overnight and Midmorning by Sample Group (ng/min)

Trait	Villagers (N = 31)		Laborers (N = 28)		Sedentary workers (N = 33)		Students (N = 31)	
	$\bar{X}$	SD	$\bar{X}$	SD	$\bar{X}$	SD	$\bar{X}$	SD
Overnight								
Epinephrine	2.85	2.06	2.56	1.87	4.05	2.92	4.21	2.29
Norepinephrine	9.59	6.77	9.12	6.18	12.66	8.48	13.25	6.19
Midmorning								
Epinephrine	4.62	3.53	6.66	4.14	6.33	3.49	8.82	4.93
Norepinephrine	13.62	12.40	24.46	14.39	22.20	14.20	24.32	9.93

Source: Adapted from James 1984.

waist skinfold (James 1984). Because the distributions of the rates were found to be positively skewed, square root transforms were used in the analysis. No appreciable differences in the pattern of group contrasts are evident after this adjustment (James et al. 1985).

The relationships between the adjusted hormone values and blood pressure were also explored through correlational analysis (James et al. 1985). There are significant associations between norepinephrine excretion and systolic pressure in the pooled data, which was diminished after partialling the effects of the mean differences among the groups. This pattern of association suggests that the life-style differences among the groups may influence casual blood pressure (James et al. 1985).

The mean values of the life-style variables are presented in Table 9.3. Villagers reported being more satisfied with life, like Samoan customs, seem more emotionally stable, assume more familial responsibility, and have less of a modern orientation as compared to the other groups. In most respects the other three groups are similar, except that the laborers were intermediate between the villagers and the other Apia groups in Western profile and the percentage of time spent in mental effort activity.

The correlations of the life-style variables and the hormone excretion rates are presented in Table 9.4. Higher overnight epinephrine excretion is positively associated with the percentage of time spent in mental effort activity over the morning and modernization, while overnight norepinephrine is similarly associated with length of mental work but also with decreased life satisfaction. Increased hormone excretion during daytime activity is associated with less familial responsibility (epinephrine) and greater modernization (norepinephrine).

To examine whether the individual life-style characteristics are facets of other complex behavioral phenomena, James performed a principal components factor analysis. Three factors were retained from this analysis (see Table 9.5). The first of these is dietary, as indicated by the high loadings of calories, carbohydrates, and caffeine drinks consumed. The second was a modernization work-level complex, which can be described as indicating that individuals with more modern exposure tend to expend more mental effort. The third describes a psychosocial complex of increased life dissatisfaction, dissatisfaction with Samoan customs, and emotional instability among the urban groups. These factors were significantly associated with overnight epinephrine ($p < .03$), but not norepinephrine. As a final analysis, day of the week was related to level of daytime hormone excretion. Rates of both catecholamines were found to be higher among individuals whose samples were collected on Friday or Saturday (James 1984).

American Samoan adults

A second study of urinary hormone excretion and stress has been described by Martz et al. (1984). This study investigated specific events

Table 9.3 Comparisons of the Life-style Characteristics among the Sample Groups

Characteristic	N	Range		Villagers		Laborers		Sedentary workers		Students		p < F
		Low	High	X̄	SE	X̄	SE	X̄	SE	X̄	SE	
Life satisfaction	123	28	9[a]	13.50	.66	16.80	.70	17.90	.64	17.70	.66	.0001
Emotional stability	121	16	48	37.10	.85	35.90	.90	35.60	.85	33.40	.85	.05
Attitudes (to tradition)	123	13	4[a]	5.40	.44	8.10	.47	7.60	.43	9.10	.44	.0001
Coronary-prone behavior	121	28	14[a]	19.10	.36	19.60	.37	19.10	.35	20.00	.36	n.s.
Responsibility	123	11	4[a]	7.70	.33	8.50	.35	8.20	.32	9.50	.33	.0001
Western profile	123	1	8	3.20	.23	4.10	.24	6.00	.22	5.80	.23	.0001
Cigarettes smoked[b]	123	0	6	1.40	.28	1.30	.29	0.63	.27	0.32	.28	.05
Caffeine drinks[b]	123	0	3	0.52	.13	0.93	.14	0.75	.13	1.13	.13	.01
Mental effort[c]	123	0	100	3.20	5.10	0.00	5.40	54.10	5.00	60.00	5.10	.0001
Physical effort[c]	123	0	100	54.30	3.80	66.90	4.00	0.20	3.70	1.90	3.80	.0001
Nontaxing activity[c]	123	0	100	42.50	6.00	33.10	6.30	45.70	5.90	38.30	6.00	n.s.
Calories[b]	123	0	2060	481.70	81.70	340.90	85.90	384.30	79.20	585.10	81.70	n.s.
Carbohydrates (gm)[b]	123	0	546.4	56.80	14.40	56.50	15.20	65.30	14.00	92.10	14.40	n.s.

[a] Higher scores on these indices indicate decreasing life satisfaction, coronary-prone behavior, familial responsibility, or attitudes to tradition, respectively.
[b] Amount consumed over study period.
[c] Percentage of time spent in activity over study period.

Table 9.4 Correlations Between the Life-style Characteristics and Catecholamine Excretion Rates

Characteristic	Overnight epinephrine	Overnight norepinephrine	Activity epinephrine	Activity norepinephrine	Average epinephrine	Average norepinephrine
Life satisfaction[d]	.04	.25[b]	.09	.07	.06	.25[b]
Emotional stability	.05	-.09	-.10	.00	.03	-.06
Attitudes (to tradition)	.06	.04	.10	.12	.12	.12
Coronary-prone behavior[d]	.03	-.09	.01	.02	.09	-.01
Responsibility[d]	-.03	.02	.19[a]	.11	.08	.11
Western profile	.17[a]	.04	.13	.19[a]	.23[b]	.18[a]
Cigarettes smoked[e]	-.14	-.03	.05	.15	-.13	.02
Caffeine drinks[f]	.05	.11	.00	.04	.05	.14
Mental effort[f]	.23[a]	.22[a]	.06	.02	.21[a]	.19[a]
Physical effort[f]	-.15	-.10	.00	.07	-.15	-.09
Nontaxing activity[f]	-.12	-.14	-.09	-.09	-.08	-.13
Calories[e]	.03	.10	-.19[a]	-.28[c]	-.08	-.09
Carbohydrates[e]	.08	.13	-.16	-.23[b]	-.02	-.03

[a] $p < .05$.
[b] $p < .01$.
[c] $p < .005$.
[d] Higher scores on these indices indicate decreasing life satisfaction, coronary-prone behavior, familial responsibility, or attitudes to tradition, respectively.
[e] Amount consumed over study period.
[f] Percentage of time spent in activity over study period.

Table 9.5 Rotated (Promax) Factor Pattern of the Life-style Characteristics

Characteristic	Factor 1	Factor 2	Factor 3
Calories	.87768[a]	-.01929	-.04419
Carbohydrates	.85661[a]	-.01021	.06961
Caffeine drinks	.39367[a]	.02838	.00873
Mental effort	.03767	.64094[a]	.03044
Physical effort	-.05784	-.68057[a]	-.01860
Western profile[b]	-.09456	.56755[a]	.03475
Cigarettes smoked	.14326	-.22821	.24669
Urban-rural residence	-.03126	.21853	.36984[a]
Life satisfaction[c]	.00736	-.02018	.59388[a]
Attitudes[d]	-.06251	.08405	.42981[a]
Emotional stability[e]	-.08131	.02127	-.51338[a]
Nontaxing activity	.03206	-.00199	-.02203
Responsibility[f]	.15628	-.05268	.08907
Coronary-prone behavior[g]	-.08816	-.04047	-.07505

[a] Important factor loadings.
[b] Higher scores indicate increasing Western contact.
[c] Higher scores indicate decreasing life satisfaction.
[d] Higher scores indicate decreasing agreement with Samoan customs.
[e] Higher scores indicate lower emotional reactivity.
[f] Lower scores indicate higher familial responsibility.
[g] Lower scores indicate type A behavior.

and mediating factors within a single modernizing community in American Samoa. From a survey population of 99, 30 men and 24 women provided overnight urine, and 21 men and 20 women provided an additional urine sample during the day. The techniques were patterned after those of Jenner et al. (1980), which were also used by James. The hormones collected included epinephrine, norepinepherine, and cortisol. In addition, each individual provided extensive questionnaire data as to demographic, ethnographic, health, and social status. Anthropometric and blood pressure data were also collected. These are summarized in Table 9.6.

From the questionnaire data a number of indices were constructed to measure social support, anger arousal and response, and components of coronary-prone behavior (Table 9.7). The general procedure described by Howard (1974) involved the following 4 steps: (1) questions from the various protocols were grouped on the basis of type of information elicited; (2) the frequency distribution of answers for each question were examined and those without variation were dropped; (3) responses were coded such that a good range was expected (thus *no* and *never* answers were given a value of zero while extremely positive responses were given the highest score); (4) questions were equally weighted and responses summed to produce an index. The resulting indices were independent in that any given question was part of only one index and they were consecutive

Table 9.6 Anthropometric Characteristics of the Sample

	Males (N = 30)		Females (N = 24)	
	Mean	SD	Mean	SD
Age (yrs)	18.4	3.9	17.9	4.0
Height (cm)	171.7	6.1	160.1	7.5
Weight (kg)	75.3	15.5	68.1	13.0
Triceps skinfold (mm)	10.8	5.2	19.8	6.0
Body mass index (wt/ht^2)	25.5	4.7	26.3	5.0
Systolic pressure (mmHg)	115.6	10.1	114.0	12.8
Diastolic pressure (mmHg)	71.6	9.3	72.7	10.6

Table 9.7 Social and Psychological Indices Employed in Analysis

		Index Score		
			Actual	
Area Index	Number of questions	Theoretical	Males	Females
Social interaction				
Self-reliant behavior	8	0-16	0-14	2-16
Attitude toward relin-				
quishing responsibility	3	0-3	0-3	0-3
Community involvement	6	0-12	0-10	0-9
Helpfulness of kin	2	0-4	0-4	0-4
Helpfulness of friends				
and neighbors	4	0-8	1-8	2-8
Anger arousal and response				
Intensity of anger	11	0-33	0-31	9-27
Behavioral response to anger				
Inhibition	12	0-12	0-12	0-11
Displacement	12	0-12	0-6	0-4
Protest	12	0-12	0-5	0-9
Discussion	12	0-12	0-9	0-6
Affective response to anger				
Expressive	12	0-12	0-8	0-8
Inhibitive	12	0-12	0-10	0-12
Denial	12	0-12	0-9	0-6
Coronary-prone behavior				
Feels pressure for				
achievement	4	0-8	0-7	0-7
Responds to pressure				
for achievement	4	0-8	0-6	0-4
Sense of responsibility	3	0-6	0-4	0-3
Work orientation	5	0-10	0-8	0-8
Competitiveness	4	0-8	0-8	0-8
Time orientation	15	0-30	0-25	8-27

because several extreme scores were required to produce a high or low index score.

For purposes of analysis hormones were divided into overnight and daytime samples. The rationale was that overnight levels would represent "background" hormone excretion and be less influenced by daily events, while samples collected during the day would reflect the results of specific events superimposed upon the background excretion (Matsuda et al. 1972; Patkai 1971).

A 3-level classification system based upon epinephrine and cortisol proved to be a useful discriminator for several of the social variable indices. Since epinephrine and cortisol covaried, their values were averaged into a single number. Classification was by quartile with a high, low, and intermediate class, consisting of the mid-two quartiles (Table 9.8).

Table 9.8 Mean Hormone Values of the Sample by Overnight Hormone Classes[a] and Gender

	Overnight hormone class							
	High		Intermediate		Low			
	Males (N = 7) Mean	Females (N = 6) Mean	Males (N = 16) Mean	Females (N = 12) Mean	Males (N = 7) Mean	Females (N = 6) Mean	F	$p > F$
Epinephrine: creatinine (ng/ml)	3.64	2.48	1.72	1.45	0.78	0.53	13.73	.0001
Cortisol: creatinine (g/ml)	4.38	4.63	2.48	3.19	1.76	2.25	43.30	.0001
Norepinephrine: creatinine (ng/ml)	13.61	11.42	7.51	7.79	7.81	5.32	3.85	.028
Epinephrine (ng/ml)	6.17	3.20	3.83	3.24	1.69	1.38	3.95	.026
Cortisol (mg/l)	7.17	6.51	5.14	5.31	3.62	6.01	3.19	.050
Norepinephrine (ng/ml)	22.86	15.40	15.66	14.13	16.54	13.85	.063	n.s.[b]
Creatinine (mg/ml)	1.64	1.42	2.09	1.78	2.13	2.67	3.99	.025

[a] Created on the basis of overnight epinephrine and cortisol levels, as discussed in the text.
[b] n.s. = not statistically significant.

Men and women showing the highest overnight levels of hormone excretion reported relying the least upon family members when faced with decisions about personal problems, financial matters, and household matters (Table 9.9). For each of these cases, they are more self-reliant and less family-reliant. Also, among those indices describing the quality of relationships with friends and with family, individuals reporting the highest quality social relations also were in the lowest hormone excretion category. There is thus a social factor related to overnight hormone excretion such that family reliance and high-quality social relationships yield lower levels of hormone release.

Martz and her colleagues next investigated hormone excretion and more widely used measures of social activity including: the number of people in the household, number of people the subject "felt close to," and the frequency of interactions with the latter. While household membership varied from 3 to 19 and was not associated with hormone excretion, the number of people listed as "close" was strongly related to levels of excretion in a negative way. The greater number of close individuals, the lower the hormone excretion.

Participation in community activities also bore a significant relationship to hormone excretion; however, it was not so clear as in other cases (Table 9.7). Those who were most involved excreted less hormone than those with intermediate involvement. The significance of this relationship is not clear. The low involvement, low excretion levels may be from peripheral members of the community who neither expect nor get much in the way of interaction, while the high involvement, high excretion group may be those who have been more intensively involved. Numerous other interpretations are possible, so further work is required to determine the significance of the relationship.

Despite an appreciable variation in the various components of coronary-prone behavior and anger arousal and response, this variation was not significantly associated with overnight hormone excretion, as shown in Table 9.10. The only exception is found in the index of time orientation, where class differences approach statistical significance. Among men, reported time orientation increases from the high to the low hormone classes. Among women, only the high-hormone-class individuals differ in having lower scores. For both sexes then there is an association between relatively high levels of overnight hormone excretion and less preoccupation with being on time, doing things rapidly, or generally feeling a sense of time urgency.

A subsample of 21 men and 20 women provided one or more additional urine samples at the same time during the day as well as a detailed account of concomitant activities. Several kinds of additional information were found from the analysis of those materials. (1) The three hormones—epinephrine, norepinephrine, and cortisol—increased during the daytime, suggesting that overnight levels represented a base line and other activities are superimposed upon them during the course of the day.

Table 9.9 Mean Scores of Social Interaction Indices by Overnight Hormone Classes and Gender

| | Overnight Hormone Class | | | | | | | |
| | High | | Intermediate | | Low | | | |
	Males (N = 7) Mean	Females (N = 6) Mean	Males (N = 16) Mean	Females (N = 12) Mean	Males (N = 7) Mean	Females (N = 6) Mean	F	$p > F$
Self-reliant behavior	10.7	12.7	8.8	9.0	7.0	9.2	4.28	.020
Attitude toward relinquishing responsibility	0.6	1.6	1.0	1.8	1.4	1.7	0.64	n.s.[b]
Community invovlement	3.4	2.5	5.7	5.3	3.3	2.3	6.86	.002
Helpfulness of kin	2.6	1.3	2.2	2.6	1.9	2.3	0.69	n.s.
Helpfulness of friends and neighbors	4.7	3.0	3.5	4.3	5.7	5.3	3.65	.034

[a] Created on the basis of overnight epinephrine and cortisol levels, as discussed in the text.
[b] n.s. = not statistically significant.

Table 9.10 Mean Index Scores of Coronary-prone Behavior Components, by Overnight Hormone Classes[a] and Gender

| | Overnight Hormone Class | | | | | | | |
| | High | | Intermediate | | Low | | | |
	Males (N = 7) Mean	Females (N = 6) Mean	Males (N = 16) Mean	Females (N = 12) Mean	Males (N = 7) Mean	Females (N = 6) Mean	F	$p > F$
Feel pressure for achievement	1.4	2.7	3.3	2.3	2.0	3.2	0.58	n.s.[b]
Respond to pressure for achievement	0.9	2.2	2.1	1.4	1.9	3.3	1.41	n.s.
Sense of responsibility	1.1	1.0	1.8	1.0	0.7	1.2	0.85	n.s.
Work orientation	4.6	5.7	4.6	5.4	4.1	5.3	0.08	n.s.
Competitiveness	3.1	3.3	4.1	3.7	4.0	2.7	0.64	n.s.
Time orientation	14.6	15.2	19.0	18.9	13.0	20.3	3.03	.057

[a] Created on the basis of overnight epinephrine and cortisol levels, as discussed in text.
[b] n.s. = not significant.

Although there is a daily circadian pattern (Patkai 1971), daily activities seem to be superimposed upon this as well (James et al. 1985; Matsuda et al. 1972). (2) There is no clear, simple relationship between norepinephrine excretion and level of physical activity as has been described in laboratory situations (Dimsdale and Moss 1980). Although there were several examples of vigorous physical activity—lugging sand, playing volleyball, working on the plantation —none of these were clearly associated with increases in norepinephrine. It would thus seem that either the level of physical exercise was inadequate to cause increases in norepinephrine excretion or that norepinephrine does not relate to physical activity in any simple manner. (3) Extreme increases in cortisol and epinephrine were associated with activities and events carrying negative affect. A continuum of proportional increases (daytime level/overnight level) in hormone was constructed and the most extreme values for each hormone identified. The activity in which the individual had been engaged prior to the urine collection was then examined. Four of the 5 most extreme increases in cortisol and each of the 4 extreme increases in epinephrine were associated with disliked activities. These included a disliked white collar job, babysitting, and waiting. In each case, the daily increase was extreme. It should also be noted that there were also cases of disliked activities that did not yield hormonal increases, but the extreme increases were always associated with negative affect.

Western Samoan school children

In a third study, which took place in Western Samoa, Sutter (1980) employed urinary norepinephrine analysis to document the stressor of beginning school on 5- to 7-year-old children resident in a rural and an urban village. Within each village he also contrasted students with children who did not attend school. His hormonal data for the most part confirmed expectations for rural to urban differences. Urban children consistently showed higher levels of excretion than rural children, regardless of their attendance at school. There were other interesting patterns, however. The urban school children manifested a dramatic increase in norepinephrine excretion at the beginning of school but adapted, and gradually declined in output during the next 10 weeks of the study. The rural children showed relatively low excretions at the beginning of school, but sustained a continual increase over the 10-week period. In Sutter's interpretation, the urban children were stressed at the beginning of school and adapted to the new environment, but the rural children had difficulty in doing so. This was because they generally relied upon group-oriented behavior, characteristic of traditional Samoan culture in solving problems, but in school this was inappropriate. Each student was required to work independently and was forced into a self-reliant pattern resulting in conflict. As the school term continued the conflict became more intense, leading to greater levels of hormone production as children were forced to assume greater individual responsibility in school. The urban children had no such problem because urban behavior is normally individual-reliant.

When students and nonstudents were contrasted in both villages, the students showed greater fluctuation in daily excretions related to school stresses and a decline in excretion rates from weekday to weekend. Nonstudents were more regular during the week, but showed an increase at the weekend. Again it would seem that some aspect of the weekend, or the preparation for traditional feasting and socializing, is stressful for more traditional villagers. It is also noteworthy that one individual who had a tearful confrontation with an older brother showed a doubling of norepinephrine excretion that continued throughout that day. Martz et al. (1984) reported a similar phenomenon.

Sutter's study suggests that the usual rural-urban pattern of stressors is evident early in life and that some aspect of formal schooling is also a stressor, perhaps foreshadowing the elevated hormones among college students observed by James.

EVALUATION

Our studies of hormone excretion in Western and American Samoa have shown that this approach holds great promise in furthering our understanding of the origins of stress and the factors that modify its final expression. To a degree we can now begin to answer Mechanic's (1974) famous query: "What stress, under what conditions?" We have not been able to identify all the major stressors in contemporary Samoan society, but we have isolated some of those associated with modernization. The mediating influences of social support and integration into Samoan society have been documented and several classes of stressful events have been observed. We can also make some methodological suggestions as to simplification and improvement of data collection and analysis.

As previously noted, Samoans seem influenced by the stresses associated with modernization and migration. The shift in disease patterns and risk factors documented elsewhere in this volume seem closely associated with rapid cultural change and are frequently called stress diseases. Hormone analysis has suggested that the stresses of modernization are associated with daytime activities and probably include mental effort, a nontraditonal occupation, or formal education. This is apparent from Table 9.2. Salamumu villagers have the lowest daytime and overnight excretion rates, while urban sedentary workers and students have the highest rates. The manual workers are intermediate, excretion rates are high during the day, but fall to village levels overnight. The manual laborers worked in the same urban conditions as the other

Apia groups, thus it would appear that some aspect of daily activity, not simply exposure to the urban environment, caused a continued elevated level of hormone excretion. Further, those aspects of the daily activity seem tied to sedentary work or to education, for as part of a factor analysis on his pooled hormone data, James (1984) found three of the factors associated with elevated hormone excretion were mental effort, lack of physical effort, and a modern orientation. In Western Samoa such a complex is most likely a manifestation of modernization. Since we do not have hormone data on the migrant communities we cannot speculate as to what additional factors might be added by migration. Since much of the disease pattern associated with the modernization is already evident in American Samoa, however, there may be very little additional stress associated with migration to Hawaii or California.

Some support for the view that mental effort combined with low physical activity is stressful is gained from two other studies. In their study of Oxfordshire Jenner et al. (1980) found professionals and managers excreted hormones at significantly higher rates than other groups of workers. These high rates of epinephrine excretion continued throughout the week and into weekends including both daytime and overnight collections. In that population the critical factors were also in the mental sphere—a nonmanual job, being rarely bored during day, and ending the day mentally tired (Reynolds et al. 1981). These characteristics may also describe aspects of our Samoan urban sample.

Sutter's work (1980) provides additional support for the importance of such mental stressors of non-Samoan origin. Although he studied 5- to 7-year-old children and employed only norepinephrine analysis, a similar pattern emerged. In both the rural and urban villages, those children attending school excreted hormones at a substantially higher rate than children from the same villages who did not attend school. He attributed these increased rates to an individual-achievement-oriented behavior that was instilled by schools, as opposed to a group-oriented behavior that is typical of life in a traditional village. This early exposure to mental stressors and the resulting cultural conflict may have long-term effects and may, in part, be the origin of the higher rates of hypertension and other stress-related diseases that are observed in young Samoans (McGarvey and Baker 1979; Hanna and Baker 1979). Clearly this is an area of work requiring further investigation.

Hormone excretion studies have also established the mediating effects of social factors in ways that are not apparent using other kinds of data. Martz et al. (1984) contrasted the high to intermediate to low overnight excretions in several villages in American Samoa. This was a more modern sample in that 98 percent were bilingual in English and 75 percent were still in school, hence, sociologically they correspond to the sedentary workers and students studied in Western Samoa. Observed variation within this relatively homogenous population corresponds to the previous observation: Those who were most self-reliant, made their own decisions, and did what they (as opposed to the group) thought, were in the high excretion category, while those who were more group oriented in reliance and decision making excreted at a lower rate. The modern profile of self-reliance—making one's own decisions, choosing one's own friends, and having high educational and occupational aspiration—is very clear. Those in the lower hormone category also reported more social activity, more people they could seek for assistance, and more individuals to whom they felt close. The greater amount of social support available also correlated with lower rates of hormone excretion. Even within this modernized Samoan community, the mediation of stress by adherence to aspects of Samoan culture is still evident. A similar role for the social mediation of stress has been described at Three Mile Island. Those individuals with the most social support continued to excrete hormones at significantly lower rates than those with little support for several months after the disaster (Fleming et al. 1982). Since the manner by which support modifies reactions to stress is not clear, the mediation could be in the perception of the stressors or through reduction of reactivity.

It is difficult to generalize as to how orientation and the effects of support could operate in other situations, for in migrant communities, self-reliant behavior might prove more adaptive than Samoan-oriented behavior. For example, in Hawaii and California access to jobs and education might be less stressful than retaining a Samoan identification. This is in agreement with observations of Pawson and Janes (n.d.), who have suggested that in California social obligations to other Samoans may outweigh the advantages of accrued social support and may actually increase the level of stress (see Howard, Chapters 8 and 16). Hanna and Baker (1979) also believe that this may serve as a possible explanation for the higher blood pressures observed in tradition-oriented communities as compared to those more fully integrated into urban Honolulu.

In addition to modernization, several other aspects of life in contemporary Samoa have been identified as stressful. James (1984) reported significant increases in hormone excretion toward the weekends, a time when preparations are underway for feasting and socializing. Sutter also observed elevations in non-school children, but it did not occur in children attending school. There may then be an increase in stress associated with the traditional weekend activities in Samoa. The social demands and duties associated with weekend feasting may be somewhat stressful, at least for our young sample.

Several stressors of a less pervasive nature have also been identified. Sutter (1980) and Martz et al. (1984) have reported extremely high levels of daytime hormone excretion associated with interpersonal conflict. In each case after fights with family members, the individual in the inferior position excreted hormones at an elevated rate. Martz et al. have also reported that those who excreted at the highest levels during the daytime were engaged in disliked activity. Further study of such unrelated events might enable us to identify and classify some apparently "individual" causes of stress.

Coronary-prone behavior (CPB) or type A behavior has also been investigated through questionnaire and hormone analysis. Studies in the United States have suggested that CPB is associated with higher reactivity to stress and greater catecholamine release (Herd 1978). James (1984) could find no relationship in his sample between a composite CPB index and hormone excretion rate. Similarly, Martz et al. focused upon the three components of CPB—work orientation, competitiveness, and time orientation—and found only a weak, negative relationship with time orientation. As in some other studies (Frankenhauser 1983), CPB does not appear to bear any major relationship to hormonal measures of stress in Samoa.

Understanding of hormonal-related stress requires some selection of hormones and selection of the proper times for collection. Physiologically epinephrine and cortisol seem more significant for stress analysis than does norepinephrine (Summers et al. 1983), because these hormones are more exclusively associated with an intense degree of arousal than is norepinephrine. Norepinephrine is a sympathetic neurotransmitter that is released as part of normal sympathetic maintenance of internal homeostasis, hence its appearance is not necessarily associated with external states of stress. While norepinephrine is released as part of the arousal response, it may also be released in conjunction with nonstressful, trivial internal events reducing its utility in identification of stressors (Axelrod and Reisine 1984). Epinephrine and cortisol are adrenal hormones and their release is more closely tied to sympathetic stimulation. These are major, long term effector hormones and may be expected to more closely correlate with important stressful events and response to them. For the most part our field work (James 1984, Martz et al. 1984) and that of Jenner et al. (1980) has corroborated controlled studies (Foresman 1982, 1981; Dimsdale and Moss 1980; Frankenhaeuser 1975) to demonstrate that epinephrine is more sensitive to mental stressors than norepinephrine. Similarly, Martz et al. (1983) have corroborated Mason et al. (1979) and Dimsdale and Moss (1980) in showing epinephrine and cortisol are more stress sensitive than is norepinephrine.

A second step in understanding hormone/stress relation is the timing of collections. Most field studies that have collected hormones overnight and during the day have shown a significant increase during the day, presumably associated with the stresses of daily activities. (Jenner et al. 1980; Matsuda et al. 1972; Patkai 1971). This diurnal variation could obscure relationships by adding hormonal responses to more transient and less consequential daily stressors to responses to the more meaningful long-term ones. Differentiating between the two classes is difficult, but overnight samples are largely free from contamination by more trivial or unique events and can be taken as background levels of excretion (Matsuda et al. 1972). Night-time levels do correlate with significant daytime stresses such as occupation (Jenner et al. 1980) and disasters (Fleming et al. 1982). Thus we suggest that overnight hormone excretions are more representative of significant long-term stresses unencumbered by variation due to day-to-day, unrelated events.

Chapter 10

Growth and Body Composition

JAMES R. BINDON
SHELLEY ZANSKY

The study of growth among Samoans offers important insights into several aspects of human biology. Whereas many studies have concentrated on describing differences between migrant and nonmigrant children, this research has had limited success in relating differing growth patterns to specific environmental conditions. It is difficult to isolate the mechanisms related to changing patterns of growth when the environment to which a group migrates contrasts greatly with the home region. In order to identify which components of the new environment are affecting the growth of the children, it becomes necessary to categorize the environment into discrete units. In contrast to a research design comparing migrants with nonmigrants, a model that distinguishes a population according to varying degrees of exposure to modernization may provide a base upon which more detailed analysis of environmental influences on growth may be examined.

The comparisons between different Samoan populations afford such an opportunity to examine how specific environmental factors affect growth. The contrast between samples from Western and American Samoa permits an examination of the influences of *in situ* modernization on growth. By virtue of the separate political and economic histories of the two Samoas, a natural experiment is available for study whereby a relatively homogeneous group has been subjected to very different processes of sociocultural change. Similarly, comparisons between the growth of children in American Samoa and Hawaii or California allows an investigation of the influence of migration while at least partially controlling for modernization.

Studying the growth of Samoan children also provides an opportunity to look for precursors of the extreme fatness found among Samoan adults (see also Chapter 11, this volume; Bindon and Baker 1985; Pawson and Janes 1981). Several studies suggest that there is a definite association between infant, childhood, adolescent, and adult fatness or relative weight status (Charney et al. 1976; Abraham et al. 1971; Eid 1970). Recently, Rolland-Cachera and coworkers have suggested that adult obesity can be predicted by measurements taken during childhood (Rolland-Cachera et al. 1984). By using the comparisons between different groups of Samoan children outlined above, a more detailed appraisal of childhood obesity precursors may be attained.

This chapter presents the results of several studies of growth among various groups of Samoan children (see Table 10.1 for sample sizes, locations, and data source). The first aspect of growth considered is birth weight, followed by growth during infancy. Finally, growth and body composition during childhood and adolescence are examined.

Table 10.1 Sample Characteristics of Studies of Growth and Body Composition

Location	Data set	Dates collected	N
	Birth weight studies		
American Samoa	Well Baby Clinic- LBJ Tropical Medical Center	1/76-7/82	5007 infants
	OB/GYN Clinic- LBJ Tropical Medical Center	1/75-7/75	238 mothers/ offspring
	Infant growth studies		
Western Samoa	Salea'aumua, Upolu	2/77-5/82	55 infants
American Samoa	Well Baby Clinic- LBJ Tropical Medical Center	10/74-9/76	1186 infants
	Well Baby Clinic- LBJ Tropical Medical Center	1/76-7/82	6587 children
	Childhood and adolescent growth studies		
Western Samoa	Salamumu, Sa'anapu, Salea'aumua, Upolu	1979, 1982	497 children
American Samoa	Tutuila (40 villages), Fitiuta, Ta'u	1978, 1982	674 children
Hawaii	Island of Oahu (11 communities)	1975, 1977	556 children
California	San Francisco Bay Area	1979-1983	144 children

BIRTH WEIGHT

Variability in birth weight, both between and within populations, can be attributed to a variety of genetic and environmental influences on both the mother and fetus (Morton 1955). Some of the maternal attributes that have been found to influence birth weight include maternal height, prepregnant weight, weight gain during pregnancy, maternal age, and parity. Neonatal attributes that influence birth weight include gestational age and sex. In this chapter, we report on two studies based on records of American Samoan births with an analysis of the effects of some maternal and neonatal characteristics on birth weight.

The first set of birth weight records is from the Well Baby Clinic (WBC) of the Lyndon Baines Johnson (LBJ) Tropical Medical Center on the island of Tutuila, American Samoa. Records on 6587 children, representing approximately 95 percent of the children born between January 1976 and July 1982, form the basis of this study. Of the total, 6267 children were classified as Samoan in the records. Birth weight was available on 5007 Samoan infants, or about 80 percent of the Samoans in the records.

The average birth weights of children in American Samoa are presented by year of birth and sex in Table 10.2. The overall mean birth weight was 3.52 kg (± 0.56 kg). The averages by year range from 3.51 to 3.56 kg from 1976 to 1981. Throughout this time, boys are slightly heavier than girls. The birth weight of the American Samoan children is slightly above the very high median birth weight found in the United States (Hamill et al. 1977). By comparison to other populations, only North American Indians have a higher average birth weight (3.60 kg, according to Meredith 1970). The data of Rosa (1974) yield an average birth weight for Western Samoan infants of 3.36 kg, 160 g lower than the average for American Samoa, and falling in the middle of the range of birth weights of world populations noted by Meredith (1970).

Table 10.2 Birth Weight (gm) of Infants in American Samoa (1976-1982) by Year of Birth and Sex

	Boys			Girls		
Year	Mean	SD	N	Mean	SD	N
1976	3557	622	366	3467	523	306
1977	3596	514	387	3468	533	327
1978	3540	579	346	3503	511	306
1979	3578	582	362	3544	569	361
1980	3550	560	419	3484	532	444
1981	3532	564	505	3484	578	415
1982	3600	538	227	3319	561	236

In addition, there is an unpublished study of birth weight among American Samoan children based on records from the Obstetrics and Gynecology Clinic at LBJ. Records were available on 500 Samoan women who gave birth from January to July of 1975. After exclusion of unusual pregnancies (still births, premature births, twins, or neonatal death) and newborns weighing less than 2500 g, a sample of 469 births remained. We attempted to assemble a sample with complete data on birth weight, sex of infant, maternal age, parity, height, weight at 3 months of pregnancy, and weight gain from 3 to 9 months of pregnancy. A subsample of 75 women had such a complete data set. In addition, a subsample of 238 women with data on the more frequently reported maternal age, parity, height, and infant sex was assembled. Information on these samples is presented in Table 10.3.

The mean birth weight in the subsample with complete data ($N = 75$) was significantly lower than the birth weight of the total sample ($N = 469$) and the larger subsample ($N = 238$). In addition, the women in the subsample with complete data were more likely to be primiparous than either of the other samples, and they were also younger than the total sample. Average maternal weight in the complete sample was 71.3 kg, and average height was 159.4 cm. Thirty-three of the women (44%) were 20 percent or more above ideal weight (ideal weight taken from the 1959 Build and Blood Pressure Study as noted in Guthrie 1979). This is a lower frequency of obesity than for women in the general population of American Samoa (Bindon and Baker 1985). Average maternal age was 25.7 and weight gain was 11.0 kg in the sample with complete data.

The most statistically significant determinants of birth weight among the 75 women with complete data were maternal age, parity, and sex of the infant. Maternal height, weight, and weight gain were not found to be significant influences on birth weight when other significant effects were controlled. In the larger subsample of 238 women, maternal height was found to be significantly associated with birth weight, but this was not the case for the smaller subsample.

The lack of association between maternal weight or weight gain during pregnancy and birth weight differs from the findings of other studies. Most investigations indicate that women with small pregnancy weight gains have smaller neonates than do women with larger weight gains (Niswander and Jackson 1974; Hytten and Leitch 1971; Wiehl and Thomkins 1954). One possible explanation for the difference between this and other studies is that there may be a threshold effect of maternal weight on birth weight increases with maternal weight to about 72.7 kg (160 lb), but no additional increments in birth weight occur above this value. Simpson et al. (1975) reported such an association in their work. Since the average weight of the Samoan women was 71.3 kg, most of the mothers would be above the postulated weight threshold. There are, however, no significant maternal weight influences even among the mothers with weights below 72.7 kg.

Table 10.3 Sample Characteristics for the American Samoa Birth Weight Study Based on OB/GYN Clinic Records, January - July 1975

	Total sample with birth weight >2500 gm			Sample with maternal height, age, and parity (N = 238)			Sample with complete maternal data set (N = 75)		
	Mean	SD	N	Mean	SD	N	Mean	SD	N
Maternal characteristics									
Height (cm)	159.6	5.7	314	159.7	5.8		159.4	5.2	
Weight (kg)	71.6	13.5	96	—	—		71.3	13.2	
Weight gain (kg)	11.2	5.3	80	—	—		11.0	5.3	
Age (yr)	26.7	5.6	468	25.9	5.4		25.7	4.7	
Parity number of mothers									
1			70			34			20
2			99			62			17
3			79			44			17
4+			219			98			21
Neonate characteristics									
Birth weight (gm)	3557	449	469	3557	442		3462	427	
Number of neonates									
Male			224			114			36
Female			232			124			39

In summary, birth weights in American Samoa are among the highest in the world, while birth weights in Western Samoa are in the middle of world values. In American Samoa, birth weight appears to be influenced by maternal age and parity, as well as by the sex of the infant, but birth weight appears to be independent of maternal weight or weight gain. This lack of association may be due to the high average weight of women in American Samoa.

INFANT GROWTH

The period of infancy is a time of rapid hyperplastic growth (Tanner 1962). The pattern of growth during the first years of life is important for the subsequent morphology and body composition of the individual (Fomon 1980; Mack and Johnston 1976). There are many determinants that affect infant growth including genetics, disease, and nutrition. Three studies of infant growth, one in Western Samoa and two in American Samoa, have been conducted, using Well Baby records from the health services in the Samoas.

Records of infant growth in weight for the village of Salea'aumua on the southeast coast of the island of Upolu, Western Samoa, have been analyzed by Bindon and Pelletier (1986). This village is described in Chapter 3. The sample for this study consists of 55 children born between February 1977 and May 1982. Babies are seen by the nurse on about a monthly basis, at which time weight is measured. Length, head circumference, and chest circumference are measured at the first visit only.

Age and anthropometric characteristics of the sample at the first nurse's visit are presented in Table 10.4. Boys were 29 days and girls 20 days old, on average. At this visit, boys averaged 4.5 kg and girls averaged 4.2 kg. These values approximate the 75th percentile of U.S. weight-for-age [National Child Health Survey (NCHS) standards from Hamill, et al. 1977]. Lengths averaged 54.7 cm for boys and 53.6 cm for girls, between the NCHS 50th and 75th percentiles. Head circumferences and chest circumferences averaged 37.4 cm and 36.6 cm, respectively, for boys, and 35.8 cm and 35.0 cm, respectively, for girls. The head circumference values are near the U.S. median values.

Semilongitudinal curves of weight for the boys and girls are presented in Fig. 10.1. Average weights fall between the NCHS 50th and 75th percentile to 9 months, and then fall on or below the U.S. median values. Differences between the Western Samoan and U.S. children increase up to 5 months for girls and 6 months for boys, and then differences decrease until the Samoan averages fall below U.S. medians. This same pattern of growth in weight was noted for infants in American Samoa by Malcolm (1954) before bottle feeding and prepared infant foods were widely available.

Table 10.4 Characteristics of Rural Western Samoan Infants Noted at First Nurse's Visit

	Boys		Girls	
	Mean	SD	Mean	SD
Age (days)	29	—	20	—
Weight (kg)	4.5	1.1	4.2	1.2
Length (cm)	54.7	5.3	53.6	3.7
Head circumference (cm)	37.4	3.3	35.0	1.9
	N	Percent	*N*	Percent
Birth order				
Firstborn	3	11	2	8
2-5	11	39	9	36
6-14	14	50	14	56
Birthplace				
Home	27	93	24	96
Hospital	2	7	1	4

Weight change in grams per day, plotted against U.S. standards (Fomon 1974) is presented in Fig. 10.2. Mean increments in weight are high relative to U.S. standards up to 6 months, after which time weight gain is generally below that found in the United States. Such a falloff in rate of weight gain in the second semester of life is taken to represent undernutrition in many developing countries (Underwood et al. 1981; Chavez et al. 1973; Venkatachalam et al. 1967). The children in most developing countries, however, are substantially smaller and lighter than the Samoan children. Whether the second semester decrease in weight gain represented a significant health threat to the Samoan children is not known.

In addition to the above, two studies based on records from the Well Baby Clinic (WBC) of the LBJ Medical Center in American Samoa have been conducted. Pelletier and Bindon (1986) analyzed the records of 1186 children born between October 1974 and September 1976, representing approximately 60 percent of the births in American Samoa during this period. This study described growth in weight and length on the basis of semilongitudinal data up to age one, and analyzed infant feeding influences on rate of weight gain on a subsample of 458 infants. Bindon (1986) surveyed the records of 6587 children born between January 1976 and July 1982, representing approximately 95 percent of the children born in American Samoa during this interval. In this study, growth in weight and length was described from birth to 6 years, and the influences of infant feeding and health patterns on subsequent growth were analyzed.

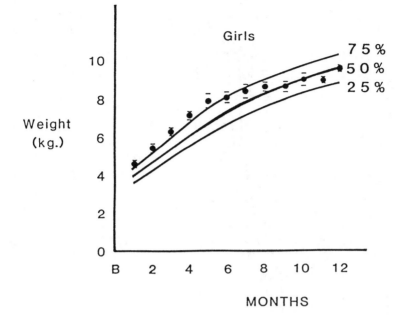

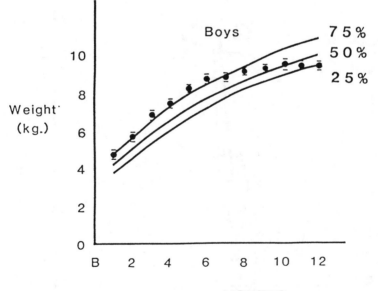

Fig. 10.1 Semilongitudinal curves of weight for age of Samoan
infants from the village of Sale'a'aumua, Western Samoa. Mean
+ s.e. (NCHS median from Hamill et al. 1977)

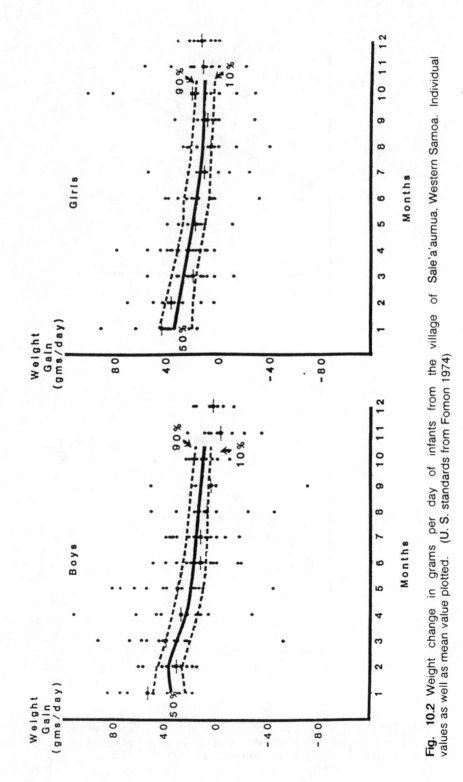

Fig. 10.2 Weight change in grams per day of infants from the village of Sale'a'aumua, Western Samoa. Individual values as well as mean value plotted. (U. S. standards from Fomon 1974)

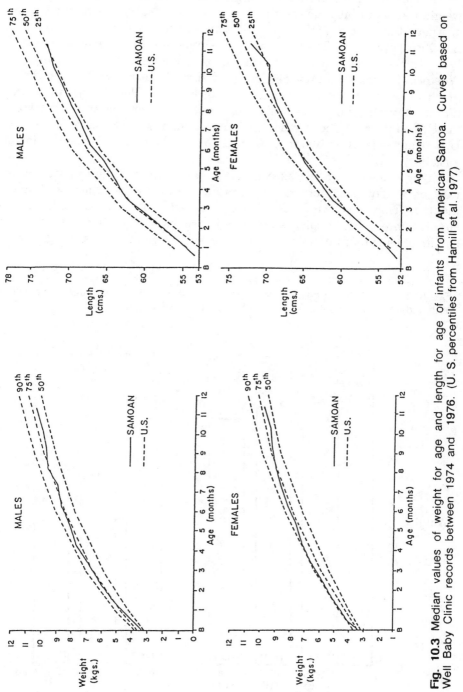

Fig. 10.3 Median values of weight for age and length for age of infants from American Samoa. Curves based on Well Baby Clinic records between 1974 and 1976. (U. S. percentiles from Hamill et al. 1977)

In Fig. 10.3, median values of weight and length of the American Samoan children have been plotted against NCHS percentiles from birth to 12 months. Both the boys and girls in American Samoa have median weights substantially above the U.S. values. For the boys the medians are near the NCHS 75th percentile, and for the girls the median is near the 90th percentile. These values are similar to those reported above for Western Samoan infants up to the age of 6 months. After 6 months, the Western Samoan infants decline in weight relative to U.S. standards much faster than the American Samoan infants. The American Samoan infants have median values of length that are at or below median values for U.S. children. The American Samoan lengths decrease relative to the U.S. standards after about 4 months for the boys and 6 months for the girls.

Figure 10.4 shows the association between infant feeding patterns and rate of weight gain for Samoan infants between the ages of 1 and 4 months. The 458 infants were divided into breast-fed, bottle-supplemented, and solely bottle-fed groups. The boys showed no significant feeding influences on rate of weight gain. The average weight

Fig. 10.4 Rate of weight gain in grams per day of American Samoan infants between birth and age 1 to 4 months, according to reported source of milk.

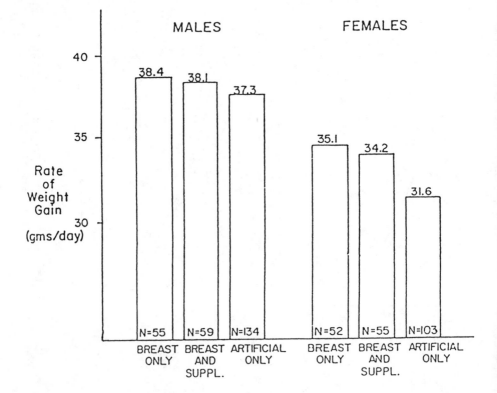

Table 10.5 Characteristics of Children from Well Baby Clinic Records, American Samoa, 1976-1982

	Boys	Girls
Ethnic classification		
Samoan	3296	2971
Other Polynesian	46	35
White	28	30
Oriental	23	19
Unclassified	75	64
Characteristics of the Samoan children		
Place of Birth		
LBJ Tropical Medical Center	2752	2454
Home (in village)	109	110
Western Samoa	185	169
U. S. mainland	124	93
Hawaii	72	96
Other (New Zealand, etc.)	40	35
Missing	40	35
Year of Birth		
1976	512	441
1977	527	437
1978	472	407
1979	485	477
1980	514	520
1981	553	449
1982	235	247

gain for the boys was just below the 75th percentile of U.S. breast-fed boys (Fomon 1974). For the girls, those who were exclusively bottle-fed had a significantly lower rate of weight gain than those in the other two feeding groups. Even the bottle-fed girls had weight gains that are near the 75th percentile of breast-fed children in the U.S.

The second survey of Well Baby Clinic records involved children born between 1976 and 1982. Table 10.5 presents several demographic characteristics of this sample. There were 3296 Samoan boys and 2971 Samoan girls, out of the total of 6587 children. Most of the children (84 percent) were born at the LBJ Tropical Medical Center, about 5 percent were born in Western Samoa, 3 to 4 percent were born in their homes and in the United States, with about another 3 percent born in Hawaii. The birth dates of the children were evenly distributed over the 6.5 years of records. Seventy-four villages were represented by the sample.

The pattern of visits to the Well Baby Clinic was as follows: Normally the first visit occurred during the first month of life, a second visit took place between 2 and 4 months, a third between 6 and 8 months, and a fourth visit at about one year. After the first year, visits became much

Table 10.6 Percent of American Samoa Infants Deriving Milk from Three Sources, 1976-1982

Trimester	Breast only		Mixed		Bottle		Total	
	N	Percent	N	Percent	N	Percent	N	Percent
First	2122	49	763	18	1410	33	4295	100
Second	1248	42	368	12	1366	46	2982	100
Third	726	35	179	9	1173	56	2078	100
Fourth	404	31	84	7	806	62	1294	100

more infrequent and irregular in spacing. The average age at a first trimester visit was 54 days for boys and 53 days for girls, or about a week under 2 months of age. The average age at a third trimester visit was 232 days or about 7.5 months, and during the fourth trimester the average age was 322 days or 10.5 months.

Table 10.6 depicts the source of milk for American Samoan infants by trimester. During the first 3 months, nearly 50 percent of the sample reported breast milk as the only source of milk. This figure dropped to about 30 percent by the fourth trimester. Combined breast and bottle milk accounted for 18 percent during the first trimester, dropping to 7 percent during the fourth trimester. Bottle milk alone was used by 33 percent of the sample during the first trimester, increasing to 62 percent during the fourth trimester. By comparison, in 1976, about 20 percent of mothers in the northeastern, north central, and southeastern United States were breast-feeding their 6-month-old infants, while the percentages on the Pacific Coast and in the western United States was 37 percent (Filer 1978). In Scandinavia during the 1970s, the percentage of breast-feeding at 6 months ranged from a low of 1 percent to a high of 27 percent, and in England and Wales, 9 percent were breast-feeding at 6 months (Underwood et al. 1981). In Western Samoa, the percentage of rural children being breast-fed was reported as 89 percent from 0 to 5 months, 73 percent from 6 to 11 months, and 52 percent from 12 to months (King 1975). Thus, the percentage of breast-fed infants in American Samoa is comparable to the number on the West Coast of the United States, about half that found in Western Samoa, and much higher than is found in most of the developed nations.

Analyses of variance were calculated to examine the association between source of milk and weight and weight gain by trimester. Age, within the boundaries of the trimester, was used as a covariant, and was found to have a significant effect on weight, with the older children being heavier. In addition, boys were found to be heavier than girls during all four trimesters, and the boys had higher weight gains during the first two trimesters.

Table 10.7 presents the average weight by source of milk during the

Table 10.7 Average Weight (kg) among Infants in American Samoa by Postnatal Trimester and Source of Milk

Trimester		Boys							Girls					
	N	Breast	N	Mixed	N	Bottle		N	Breast	N	Mixed	N	Bottle
First[a]	1107	5.8	400	5.7	737	5.6		1007	5.4	362	5.3	667	5.2
Second[b]	650	8.0	194	7.9	712	7.9		597	7.4	173	7.4	651	7.3
Third	414	9.3	95	9.0	676	9.3		375	8.5	112	8.7	562	8.7
Fourth[a]	274	10.1	66	9.9	513	10.3		254	9.3	56	9.6	413	9.7

Note: Significance of differences between infants with contrasting sources of milk was assessed by Analysis of Covariance, adjusting for age and sex.
[a] $p < .001$.
[b] $p < .05$.

Table 10.8 Average Weight Gain (gm/day) among Infants in American Samoa by Postnatal Trimester and Source of Milk

Trimester		Boys							Girls					
	N	Breast	N	Mixed	N	Bottle		N	Breast	N	Mixed	N	Bottle
First[a]	997	40.5	368	39.3	666	37.6		918	35.0	349	33.5	604	31.2
Second[b]	613	25.7	187	25.7	649	27.0		560	23.7	166	23.1	591	24.4
Third[a]	345	13.1	75	14.9	571	15.3		301	12.1	90	15.5	472	15.1
Fourth[a]	190	8.0	43	10.4	398	11.0		168	8.5	40	10.4	314	11.5

Note: Significance of differences between infants with contrasting sources of milk was assessed by Analysis of Covariance, adjusting for age and sex.
[a] $p < .001$.
[b] $p < .05$.

first four trimesters of life. Breast-fed infants are significantly heavier than bottle-fed infants through the first two trimesters, and bottle-fed infants were heaviest during the fourth trimester. Weight gain was calculated by subtracting the weight at the previous trimester (or the birth weight in the case of first trimester weight gain), and dividing by the number of days between visits (or from birth to the first trimester visit), and it is presented in grams per day (g/day) in Table 10.8. During the first trimester breast-fed infants have higher rates of weight gain than the bottle-fed infants do. From the second trimester on this pattern is reversed, with the bottle-fed infants having higher average weight gains than the breast-fed infants.

There are several factors that might account for the patterns of weight demonstrated for this sample. The mothers with the lightest infants during the first trimester are likely to be encouraged by doctors to supplement with bottle feedings. It should be noted that the bottle-fed infants have the lowest rate of weight gain during the first trimester. The low first trimester weight of the bottle-fed infants might be an artifact of following doctor's orders among infants who are not gaining weight as rapidly as others. In addition, bottle-fed infants are more likely to be getting solid food as well as milk or formula. No significant associations could be demonstrated, however, between solid foods and weight or weight gain throughout the first year. Beyond the first semester, bottle-fed infants outgain their breast-fed counterparts, and end up as heavier infants.

Moli Pa'au, Chief Public Health Nurse, surveyed infant feeding practices and health in American Samoa in 1979 (unpublished data). Her conclusions point to a complex set of interactions between socioeconomic, growth, and health factors. She suggested that mothers wean children early to return to work. This is apparently a recent phenomenon based on the marketing of milk and infant formula in Samoan stores and the availability of jobs for women in American Samoa. She reported that the bottle-fed babies of the working mothers are far more likely to contract gastroenteritis than breast-fed babies. In spite of the relationship between bottle-feeding and gastrointestinal problems, catch-up growth seems to favor the bottle-fed children as they are likely to have higher weights and rates of weight gain (as noted above) than their breast-fed counterparts.

In summary, Samoan infants gain weight during the first 6 months of life at a very rapid rate whether they live in Western or American Samoa. After 6 months of age, infants from a rural village in Western Samoa dramatically slowed in rate of weight gain, while infants in American Samoa maintained rapid weight gain. These results suggest that the predisposition for obesity among Samoans may start at birth with a tendency for rapid weight accumulation. Recent socioeconomic changes in American Samoa may be causing increased bottle-feeding, which also appears to be associated with even more rapid weight gain.

CHILDHOOD AND ADOLESCENT GROWTH AND BODY COMPOSITION

There are two studies that have analyzed associations between infancy and growth during early to middle childhood, both of which are based on the Well Baby Clinic records from American Samoa. The first study used the 1974-1976 records to choose a sample based on source of milk (breast or bottle during the first year of life) and weight-for-length status (< 25th percentile versus > 75th percentile) (Bindon 1984a.) Seventy-four children with an average age of 6.6 years were measured in 1982 (Bindon 1984a). Children who had been high weight-for-length infants had larger skeletal measurements (height, sitting height, tibial height, bicondylar femoral breadth, bistyloid wrist breadth, total arm length, upper-arm length, biacromial diameter, and biiliocristal diameter). Children who had been bottle fed as infants had greater adiposity measurements (skinfolds from the triceps, biceps, subscapular, forearm, midaxillary, suprailiac, anterior thigh, and medial calf sites). In addition, both the high weight-for-length infants and the bottle-fed infants were heavier with larger arm and calf circumferences as children.

The finding that high weight status infants become heavy and tall children is consistent with the results of many other studies (Mack and Johnston 1976; Eid 1970). It is likely that the initial high weight status was due in part to large muscle and skeletal components. The relationship between infant source of milk and childhood fatness is more complex. There are confounding socioeconomic factors associated with bottle feeding (such as having working mothers and higher incomes) that could play a role in childhood adiposity. Among these children in American Samoa, however, there was no significant association between income and any measure of adiposity.

The second analysis of Well Baby Clinic (WBC) records involves children born between 1976 and 1982. The characteristics of these children were presented in Table 10.5. In addition to visits during the first year, there were irregularly spaced visits up to the age of 6.

Weight and length (or height) are measured by the WBC nurses at each visit, using spring scales and linen tapes. While the measurements are not necessarily taken according to standard procedures, it is felt that the measurements are sufficiently reliable to serve as an indicator of general growth patterns. Average weight and length/height by month of age is presented for the American Samoan children up to 72 months in Figs. 10.5 and 10.6, respectively. The monthly age groups include all children who had a visit recorded within 2 weeks of a monthly anniversary. The sample sizes for these growth curves are presented in Table 10.9. Samples are greatest during the first several months, and sample sizes become quite small as age increases beyond 2 years. The weights for both the boys and girls fall above the NCHS 50th percentile for age from 2 months on, consistent with the analysis of WBC records

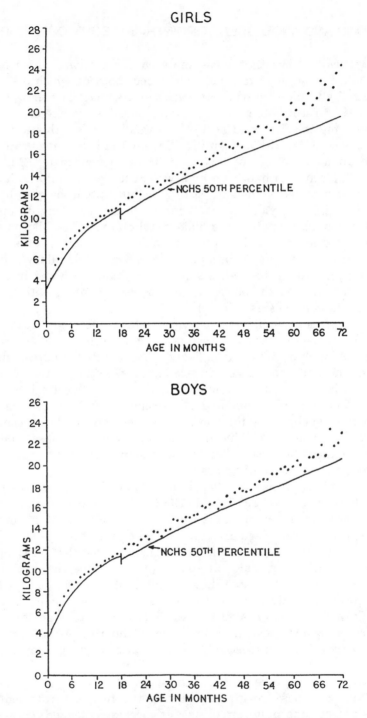

Fig. 10.5 Semilongitudinal curves of weight for age from birth to 72 months of American Samoan children. Values plotted represent monthly means. (NCHS median from Hamill et al. 1977)

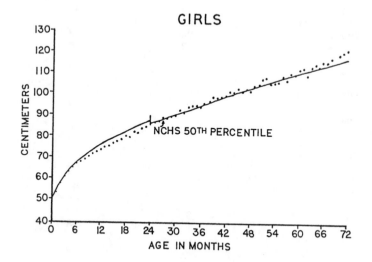

GIRLS

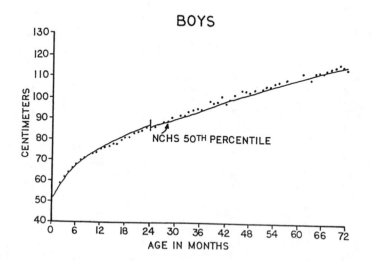

BOYS

Fig. 10.6 Semilongitudinal curves of length/height for age from birth to 72 months of American Samoan children. Values plotted respresent monthly means. (NCHS median from Hamill et al. 1977)

Table 10.9 Sample Sizes for Measurements of Weight and Length/Height from the American Samoan Well Baby Clinic, 1976-1982

Weight (gm)

Age (mo)	Boys	Girls	Age (mo)	Boys	Girls	Age (mo)	Boys	Girls
0	2613	2394	25	119	91	50	56	52
1	2254	2056	26	115	96	51	56	34
2	1769	1540	27	93	94	52	40	35
3	471	432	28	102	97	53	40	38
4	1208	1100	29	106	94	54	43	31
5	539	477	30	99	83	55	35	36
6	837	734	31	85	89	56	38	51
7	503	449	32	95	81	57	32	30
8	628	564	33	98	76	58	43	38
9	467	395	34	81	72	59	19	34
10	457	423	35	74	87	60	37	23
11	394	301	36	87	72	61	25	23
12	369	344	37	68	49	62	25	30
13	318	240	38	74	63	63	23	21
14	250	225	39	71	65	64	31	20
15	310	240	40	61	61	65	26	17
16	221	194	41	70	67	66	24	18
17	200	184	42	81	55	67	23	19
18	171	141	43	57	56	68	19	5
19	178	150	44	86	51	69	15	15
20	162	130	45	47	41	70	17	18
21	150	121	46	75	47	71	10	7
22	142	118	47	65	44	72	12	2
23	145	115	48	59	41			
24	100	105	49	53	45	Total	17 566	15 466

Length/Height (cm)

Age (mo)	Boys	Girls	Age (mo)	Boys	Girls	Age (mo)	Boys	Girls
0	120	95	25	112	83	50	57	50
1	2248	2051	26	108	90	51	54	34
2	1767	1538	27	87	90	52	39	35
3	468	429	28	99	91	53	41	37
4	1202	1094	29	93	88	54	44	31
5	535	475	30	93	75	55	37	36
6	834	731	31	80	84	56	38	51
7	496	447	32	89	77	57	32	30
8	622	563	33	93	72	58	43	28
9	460	393	34	80	65	59	19	35
10	450	419	35	67	81	60	38	23
11	394	297	36	82	67	61	26	23
12	365	342	37	62	47	62	25	20
13	309	235	38	66	59	63	23	22
14	246	219	39	71	61	64	31	20
15	301	241	40	51	56	65	26	17
16	220	186	41	57	63	66	23	20
17	195	180	42	73	50	67	23	20
18	166	134	43	49	52	68	19	5
19	174	142	44	78	45	69	15	15
20	154	125	45	42	31	70	17	18
21	140	115	46	66	41	71	10	7
22	135	109	47	60	42	72	11	2
23	136	113	48	55	40			
24	97	97	49	50	43	Total	14 788	12 942

from 1974-1976 presented above. The American Samoan children have average weights during the first few months that are near the NCHS 50th percentile. The distance between the average Samoan weight and the NCHS median increases with age, however, until the Samoan boys are near the 75th percentile and the girls are near the 90th percentile by the time they are 6 years old.

For length/height, the American Samoans have mean values near the NCHS 50th percentile up to 6 months of age, and then the Samoan averages decline to near the 25th percentile at 24 months, after which time the NCHS standards are based on height. The apparent lag in growth in length of the Samoan children between 6 and 24 months may simply reflect differences in measurement techniques. After 2 years, the monthly averages for the children in American Samoa are near the NCHS 50th percentile. It should be noted that while average length or height of the Samoan children is below or near the 50th percentile of U.S. standards, the average weight of the Samoan children is above the 50th percentile, indicating that the children in American Samoa are heavy for length.

The influence of infant feeding pattern on growth was also investigated for this sample (Bindon 1986). Analyses of covariance for the influence of source of milk on weight, length or height, and body mass index (BMI) [weight (kg) ÷ length (m)^2] are presented in Tables 10.10 and 10.11. For this analysis, infant feeding has been categorized as breast-fed if there was no other source of milk reported for the first year of life; bottle-fed if at any time during the first year the infant received milk or formula bottles, but not breast milk; and mixed-fed if both breast and bottle sources of milk were reported simultaneously.

Source of milk influences were assessed at yearly intervals from 1 to 4 years. Infants from 11 to 13 months, inclusive, were grouped into the 1-year sample; the 2-year age group included those from 22 to 26 months; the 3-year sample was from 33 to 39 months; and the 4-year sample included children from 45 to 51 months of age. The variation in age within each yearly age group proved to be a significant covariant for weight and height up to age 4. Significant sex differences are found through age 3 for all measurements, and at age 4 for weight, with the boys being consistently larger than the girls.

There are significant associations between source of milk and weight up to age 4 (Table 10.10), height up to age 3 (Table 10.11), and BMI up to age 2. Breast-fed babies are shorter and lighter, with lower BMIs. Sometimes the bottle-fed babies are larger, and sometimes the mixed-fed babies are larger, but the breast-fed babies were consistently smaller on all measurements.

The influence of age at the introduction of solid foods was also investigated with a small sample of children. The timing of the introduction of solids was found to have a significant relationship with length through age 2. The regression coefficient was negative, indicating

Table 10.10 Average Weight (kg) by Primary Source of Milk for American Samoan Children

Age range (months)	Boys						Girls					
	N	Breast	N	Mixed	N	Bottle	N	Breast	N	Mixed	N	Bottle
11-13[a]	275	10.3	88	10.5	551	10.6	272	9.6	88	9.8	551	10.0
22-26[a]	149	13.0	55	13.5	251	13.5	130	12.1	35	12.7	202	12.7
33-39[b]	128	15.1	40	15.5	218	15.6	135	14.4	22	14.9	178	14.9
45-51[b]	91	17.2	41	17.7	161	18.0	70	16.6	21	17.1	116	17.4

Note: Significance of differences between infants with contrasting sources of milk was assessed by Analysis of Covariance, adjusting for age and sex.
[a] $p < .001$.
[b] $p < .01$.

Table 10.11 Average Length/Height (cm) by Primary Source of Milk for American Samoan Children

Age range (months)	Boys						Girls					
	N	Breast	N	Mixed	N	Bottle	N	Breast	N	Mixed	N	Bottle
11-13[a]	275	74.8	88	75.4	551	75.6	272	73.3	88	74.0	551	74.0
22-26[a]	149	86.0	55	86.8	251	87.1	130	84.7	35	85.5	202	85.8
33-39[b]	128	95.5	40	96.3	218	96.4	135	94.7	22	95.6	178	95.7
45-51	91	103.4	41	103.8	161	104.5	70	102.6	21	103.0	116	103.8

Note: Significance of differences between infants with contrasting sources of milk was assessed by Analysis of Covariance, adjusting for age and sex.
[a] $p < .001$.
[b] $p < .05$.

Table 10.12 Associations between Infant Health and Growth among American Samoan Children

	Age: 1 year		Age: 2 years	
	Boys	Girls	Boys	Girls
Weight (kg)				
Gastrointestinal Symptoms				
Present	10.3^a	9.6^a	13.0	12.4
Absent	10.6^a	9.9^a	13.2	12.5
Body Mass Index (kg/m^2)				
Gastrointestinal Symptoms				
Present	18.2^a	17.4^a	17.5	16.2
Absent	18.5^a	18.1^a	17.8	17.0
Respiratory Symptoms				
Present	18.4^a	17.8^a	17.8	16.9
Absent	18.6^a	18.2^a	17.7	17.0

Note: Significance of differences between groups with and without symptoms was assessed by Analysis of Covariance, adjusting for age and sex.
[a] $p < .05$.

that the later solids were introduced, the shorter the child was at one year, and vice versa.

The influence of infant health on growth was analyzed by noting the presence or absence of symptoms of respiratory or gastrointestinal illness under the age of 12 months. Respiratory infections were significantly associated with a lower BMI at age 1, and gastrointestinal problems were significantly associated with lower weights and BMIs at the same age (Table 10.12). All significant differences disappeared by age 2, however, suggesting that whatever illness-induced growth retardation had occurred during the first year, was overcome through catch-up growth during the second year.

Since 1975, four surveys of growth have been conducted among various segments of the Samoan population: in Western Samoa, American Samoa, Hawaii, and California. Results from these surveys on children between the ages of 4 and 20 years are presented here. The age and sex distribution of the samples from the four areas are presented in Table 10.13.

The samples from Western Samoa come from three rural villages on the island of Upolu, described in Chapter 3. Children from the villages of Salamumu and Sa'anapu were surveyed in 1979 by a team from The Pennsylvania State University, and children from Salea'aumua were measured in 1982 (Salea'aumua data courtesy of Pelletier). Nearly all of the children in Salamumu and Salea'aumua were measured, but only a small fraction of the population of Sa'anapu was surveyed. There were

Table 10.13 Age and Sex Distribution of Four Groups of Samoan Children

Age (yr)	Western Samoa Boys (N)	Western Samoa Girls (N)	American Samoa Boys (N)	American Samoa Girls (N)	Hawaii Boys (N)	Hawaii Girls (N)	California Boys (N)	California Girls (N)
4	11	9	21	20	14	14	1	—
5	15	13	20	16	19	23	3	—
6	30	14	34	40	21	21	4	1
7	27	13	29	25	19	20	2	4
8	12	22	29	26	22	19	5	6
9	18	15	29	21	13	19	2	7
10	21	14	34	20	16	14	7	4
11	19	26	16	26	26	17	5	5
12	15	20	22	28	20	24	4	2
13	18	14	17	14	12	20	4	9
14	12	15	22	16	15	13	5	9
15	20	10	18	11	13	16	6	9
16	9	20	16	20	13	16	2	7
17	6	8	21	16	25	17	5	6
18	14	7	7	15	15	16	2	2
19	15	11	5	17	9	11	1	5
20	3	1	2	1	—	4	3	3
Totals	265	232	342	332	272	284	61	83

265 boys and 232 girls surveyed, representing most of the population of these villages, but less than 1 percent of the children in Western Samoa.

The sample from American Samoa was surveyed between 1978 and 1982 by researchers from The Pennsylvania State University and the University of Alabama. The subjects for this survey come from over 40 villages on the island of Tutuila and from the village of Fitiuta on the island of Ta'u. Three surveys have contributed to this sample, each with slightly different research goals and sampling strategies. Standard measurement techniques, however, were used throughout the studies. There were 342 boys and 332 girls surveyed, representing about 5.5 percent of the population in this age group based on the 1980 census.

The sample of Samoans living in Hawaii was surveyed during the summer of 1975 and again during the winter of 1977 by two field teams from The Pennsylvania State University with assistance from personnel from the University of Hawaii. Most of the research was conducted at the household of the subjects. There were also several church-related clinics held that attracted participants. Eleven neighborhoods and communities representing both urban and suburban groups were surveyed on the island of Oahu, Hawaii. For the purpose of cross-sectional analyses, only the first set of measurements (i.e., the 1975 measurements) have been used for those subjects seen more than once. There were 272 males and 284 females surveyed, representing about 9.8

percent of the Samoan population in this age group when the 1980 census figures are used as a base. The socioeconomic characteristics of some of the parents in these households have been found to be representative of the migrants in general (Bindon 1981).

The sample from California was surveyed between 1979 and 1983 by a team from the University of California School of Medicine, San Francisco, under the direction of Dr. Pawson (Chapter 11). Church congregations and night school classes in the San Francisco Bay Area were contacted in an attempt to locate Samoan migrants as potential participants in a survey of health risk factors among the migrants. There were 61 boys and 83 girls representing less than 5 percent of the Samoan children in San Francisco and San Mateo counties (population as of 1980 from the California State Census Data Center). Only height, weight, and triceps skinfold measurements are available on the California sample.

Standard anthropometric techniques, described in Weiner and Lourie (1969), were used throughout all of these surveys. Height, weight, upper-arm circumference (AC), triceps (TSF), subscapular, midaxillary, bi-iliocristal, forearm, and calf skinfolds were measured. In addition, two indices of body composition were calculated: body mass index (BMI) and upper-arm muscle circumference. The calculation of BMI has been noted above. Arm muscle circumference (AMC) is calculated as: AMC = $\{[(AC \div \pi) - TSF] \times \pi\}$. The sum of trunk skinfolds has been calculated as the sum of the subscapular, midaxillary, and bi-iliocristal skinfolds. The sum of 6 skinfolds is the additive combination of the trunk skinfolds plus 3 extremity skinfolds: triceps, forearm, and calf.

Distance curves of growth in height and weight are presented in Fig. 10.7. Among the boys, the sample from rural Upolu has the smallest averages for all age groups. The rural Upolu sample is below the NCHS 50th percentile for both height and weight throughout the age ranges represented in the figure. Both the height and the weight measurements for the sample from Upolu are similar to values obtained in a survey conducted on Upolu in 1965 (Wilkins, 1965). The earlier survey included children from Apia and from a rural area of Upolu. Both of Wilkins's samples are close to the averages for the three villages represented in the present study, although the boys were slightly shorter in the 1965 survey. The American Samoan and Hawaiian samples have mean values for height that are near the NCHS 50th percentile, while their average weights are around the NCHS 75th percentile. The California sample has average heights that are near the 50th percentile to about 12 years, and then from 12 to 17 their heights greatly exceed the NCHS norms. In this same 12- to 17- year age span the Samoan boys in California exceed the NCHS 90th percentile for weight. Differences between the group from Upolu and the other samples are moderate during childhood, and reach a maximum during adolescence. Adolescent boys in California average about 20 kg heavier and 10 cm taller than their counterparts in Western Samoa.

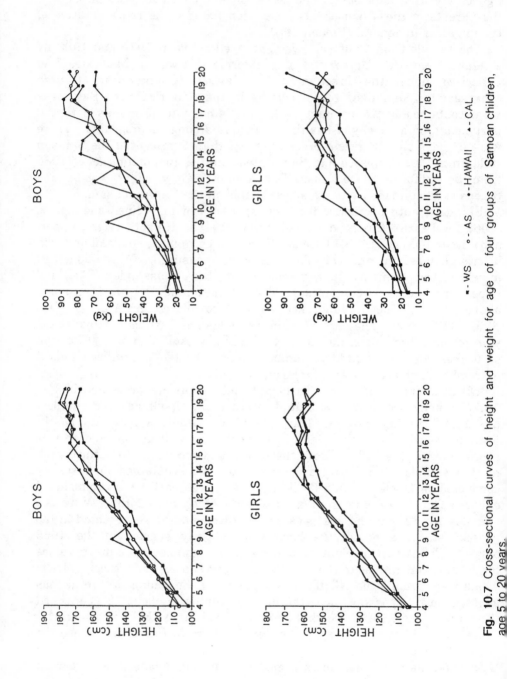

Fig. 10.7 Cross-sectional curves of height and weight for age of four groups of Samoan children, age 5 to 20 years.

BOYS

GIRLS

BOYS

GIRLS

■- WS   ○- AS   ●-HAWAII   ▲- CAL

Among the girls, the sample from Western Samoa is consistently shorter and lighter than the other groups, with the American Samoan sample next smallest (except weight in 14- and 15-year-old girls). The young (under 8) and older (13 to 17) girls from California are tallest and heaviest. Height for all but the Western Samoans is at or above the NCHS 50th percentile. Weights from age 10 on are between the NCHS 75th and 90th percentile for the American Samoan, Hawaiian, and Californian samples. The girls from Western Samoa are similar in height and weight to the girls surveyed in 1965 by Wilkins (1965). The differences between the four current samples is at a maximum at age 13, when the California group averages 25 kg heavier than the Western Samoan group, and 12 kg heavier than the sample from American Samoa. In the same age group the sample from California is 12 cm taller than the girls from Western Samoa, but only 3 cm taller than the girls in American Samoa.

A small sample of Samoan children in Hawaii was surveyed a second time after an average interval of 1.7 years. There were 55 males and 53 females aged 5 to 20 years in this sample. This small semilongitudinal sample has been used to produce the velocity curves for growth in height and weight that are presented in Fig. 10.8. The girls reached their peak height and weight velocity between 12 and 13 years of age, while the boys reached a peak height velocity around 13 to 15 years and peak weight velocity from 15 to 17 years. The average age at menarche was found to occur among the migrant girls between 13 and 14 years, according to self-reporting by a sample of women aged 15 to 20 at the time of the 1975 survey (Lieberman and Baker 1976). This timing of

Fig. 10.8 Semilongitudinal velocity curves of height and weight for Samoan children living in Hawaii, remeasured after approximately 1.7 years (1975-1977).

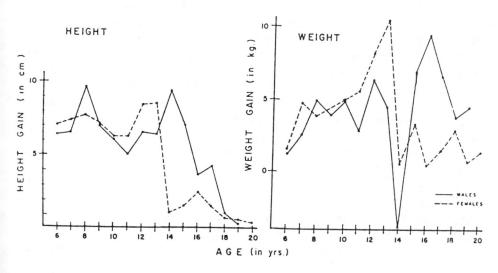

menarche relative to peak height velocity agrees well with the standards for the sequence of events during adolescence established by Tanner (1962).

Plots of mean values for body mass index and triceps skinfold measurements are presented in Fig. 10.9. The boys from Western Samoa have smaller average BMIs than the other three groups at all ages. The discrepancy between the boys from Western Samoa and the other groups increases with age, throughout adolescence. The boys from California have the highest BMIs from age 10 to 19. The values for the samples from Hawaii and American Samoa are intermediate. For triceps skinfolds, the boys from Western Samoa also have the lowest averages, showing little change in average measurements from age 4 through 19 (8 to 10 mm). There is overlap between the average values for the other three samples; however, the sample from California shows slightly greater average values, corresponding to their higher BMIs.

Among the girls, lower average BMIs occur in Western Samoa for all but the oldest age group (Fig. 10.9). The highest averages occur in California for all but the oldest girls. Triceps skinfolds were lowest in Western Samoa, except for the 18- and 19-year-old girls. Through childhood and early adolescence, the girls in Hawaii have the largest skinfolds, but from age 14 on, the girls in American Samoa have the highest average values. The girls from California have skinfold measurements that are intermediate to the American Samoan and Hawaiian averages.

Additional measurements of body composition are not available for the children from California, so the remaining figures are based on the samples from Western and American Samoa and Hawaii. Upper-arm circumferences and upper-arm muscle circumferences are presented for the three samples in Fig. 10.10. The groups from American Samoa and Hawaii have higher average measurements than the sample from Western Samoa, for both sexes at all ages, except for the older girls. The values for muscle circumference are highest among the children in Hawaii. The muscle circumferences of the boys in Western and American Samoa are comparable. Muscle circumference among the girls from both groups is also close, with the girls from Western Samoa having higher average measurements from age 15 on. The differences between the samples are much greater for arm circumference than for muscle circumference, suggesting that most of the difference between the groups is based on subcutaneous adiposity. The Western Samoans have small adipose stores, and muscle circumferences that are close to those for the other groups.

The sum of 6 skinfolds is presented in Fig. 10.11 for the three groups of Samoan children. For the boys, the Western Samoan sample has substantially lower average sums at all ages. The boys from Hawaii generally have the highest averages, except the oldest age group from American Samoa. Among the girls the same pattern prevails, with the

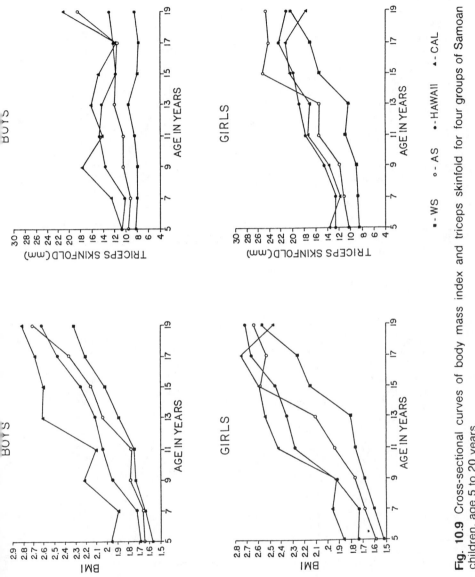

Fig. 10.9 Cross-sectional curves of body mass index and triceps skinfold for four groups of Samoan children, age 5 to 20 years.

●-WS   ○-AS   ●-HAWAII   ▲-CAL

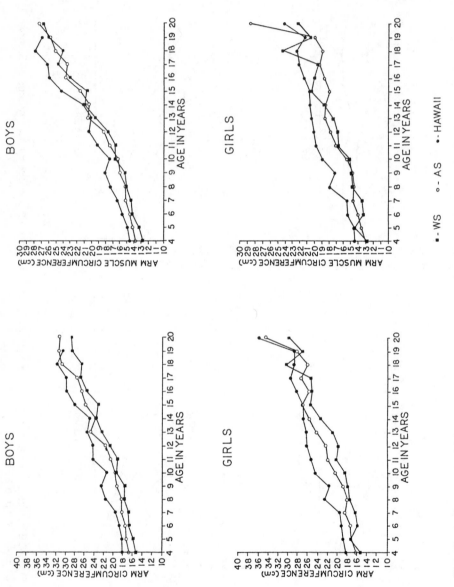

Fig. 10.10 Cross-sectional curves of upper arm circumference and upper are muscle circumference for three groups of Samoan children, age 5 to 20 years.

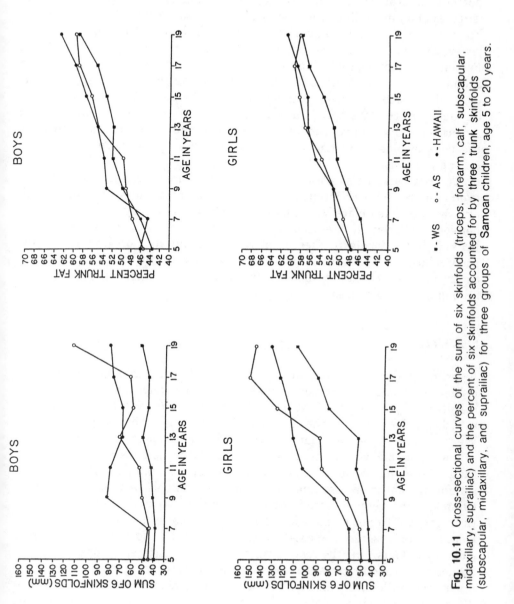

Fig. 10.11 Cross-sectional curves of the sum of six skinfolds (triceps, forearm, calf, subscapular, midaxillary, suprailiac) and the percent of six skinfolds accounted for by three trunk skinfolds (subscapular, midaxillary, and suprailiac) for three groups of Samoan children, age 5 to 20 years.

■ - WS   ○ - AS   ● - HAWAII

251

Western Samoan girls having much smaller sums than the girls from the other areas. The girls in Hawaii have the highest averages to about age 14, and then the girls from American Samoa have higher averages. The percent of total (sum of 6) skinfolds accounted for by 3 trunk skinfolds (subscapular, midaxillary, and suprailiac) is also presented in Fig. 10.11. For the boys, higher trunk depositions occur in Hawaii from age 9 on, and lower relative fat on the trunk in Western Samoa from age 13 on. Percentages increase with age from about 45 percent at ages 4 and 5 to around 60 percent at ages 18 and 19, indicating the increasing trend for trunk deposition with age. Among the girls, the percentages are lowest at all ages in the Western Samoan sample. The girls in Hawaii and American Samoa have similar values. The same pattern as shown for the boys of increasing trunk fat over extremity fat with age also occurs among the girls.

The comparisons of growth and body composition between the different groups of Samoan children indicate that the major alteration in growth and fatness takes place with modernization (as shown by the comparisons between children in Western and American Samoa), and migration to Hawaii or California further accelerates the patterns of growth and fat accumulation. In quantitative terms, however, migration appears to contribute far less than modernization to the changes in growth seen among Samoan children.

SUMMARY

A primary finding of the growth studies of Samoan populations is a tendency among the Samoan children to gain weight rapidly. This is manifest only during the first 6 months of life in Western Samoa. After 6 months, differences in the pattern of weight gain between infants in Western and American Samoa occur. From the age of 6 months on, infants and children in American Samoa tend to be substantially heavier than those in Western Samoa. In American Samoa, Hawaii, and California, children continue to increase in height, weight, and triceps skinfold measurements much more rapidly than Western Samoan children throughout childhood and adolescence. Differences in all measurements were greatest between Western Samoan children and the other Samoan children. Thus, in the comparison between different populations of Samoans to evaluate environmental influences on growth, the most important influence appears to be socioeconomic modernization (as illustrated by the comparison between Western Samoa and American Samoa samples) rather than migration (as illustrated by the American Samoan and Hawaiian or Californian samples).

In examining the development of obesity among Samoans, the difference in weight between children in Western Samoa versus those in American Samoa and Hawaii was shown to be largely due to excess fat among the children in more modernized areas, as indicated by skinfold measurements. While there may be some genetic differences between the various groups of Samoans (Parsons 1982), it appears that the environmental contrast between Western Samoa and the other areas is responsible for the major difference. In previous work, Bindon (1981) had demonstrated that there is a high degree of familial resemblance for weight and skinfolds among Samoan adults living in American Samoa and Hawaii. In fact, parent-offspring correlations for triceps and subscapular skinfolds are higher than for stature. It can be argued that the relatively high correlations for adiposity were likely to be due to shared environmental conditions, rather than strong genetic control. Since one of the important environmental conditions associated with obesity is occupation (Bindon and Baker 1985; Bindon 1981), and since the shift from traditional to modern jobs appears to be proceeding unabated for Samoans, it seems likely that children in American Samoa, Hawaii, and California will continue to exhibit overweightness and obesity as they mature.

Chapter 11

The Morphological Characteristics of Samoan Adults

IVAN G. PAWSON

This chapter presents an overview of the morphological characteristics of adult Samoan populations studied and compares them to other Pacific and U.S. populations.

A recurrent theme that runs through this volume is the large physical size of Samoans, especially those who have moved to the United States. As Baker pointed out in Chapter 1, the second phase of the Samoan research project was greatly influenced by variations in body morphology among different Samoan subpopulations, and the high prevalence of obesity in a broad spectrum of the migrant population in Hawaii.

The increased political visibility of the U.S. Samoan population, helped in part by its inclusion as a separately identified entity in the U.S. 1980 National Census (U.S. Bureau of the Census 1982), and the emergence of legislative programs geared toward the special needs of Pacific island populations in this country, has helped to focus attention on the various economic problems that beset these minorities. The present volume should contribute additional information that bears on specific health needs, in particular risks associated with high levels of obesity.

One of the problems that affects attempts to draw parallels between migrant and nonmigrant representatives of a population is the biocultural homogeneity of the samples selected for study. Although the ability to draw generalizations from comparisons of the Samoan populations of, for example, Hawaii and American Samoa has been simplified by the homogeneous nature of these two subgroups, other comparisons, such as those that involve the California subpopulation are more problematic, because of the special structure of this large and widely distributed community. Pawson and Janes (1981) have pointed out that the diversity of the California Samoan population may severely limit the extent of comparisons that can be made with Samoan subpopulations elsewhere. Additional problems in presenting an overview of the morphological characteristics of these populations concern the relatively large numbers of investigators who collected the data, an unavoidable consequence of conducting numerous relatively self-contained studies under different conditions.

Because large physical size is such a noticeable and common feature of Polynesian peoples in general, and of Samoans in particular, it becomes necessary to consider to what extent conventional criteria of obesity and overweight are appropriate for describing the distribution of body morphology in these groups. For example, the standard criterion of 20 percent over a standard weight for individuals in the U.S. population would likely include the majority of a sample of Samoans drawn at random from any of the locations described in this volume, except perhaps Western Samoa. Even though epidemiological studies have adopted such criteria because they match observed patterns of increased risk for developing weight-related degenerative disease, the uncertainty of how these risks operate among Samoans renders these criteria of little use.

The Samoan migrant populations described here are generally young, having a median age about half that of the general population; in this respect they resemble other migrant groups. Since only a small part of the population is exposed to one of the primary risk factors for obesity-related disease, namely age, analyses that seek to stratify risk by levels of fatness will of necessity be imprecise because of the relatively low rates of mortality. It is not the purpose of the present chapter to discuss health-related outcomes in relation to excess weight among Samoan subpopulations; rather, the material presented here will set the stage for the detailed descriptions of biochemical, behavioral, and other correlates of obesity that follow in subsequent chapters.

BACKGROUND

Peoples of the Pacific islands have long been known for their large physical size, although the development of the type of obesity among migrants to the United States that is described in this volume is probably a recent phenomenon. Virtually all the research that seeks to trace links between modernization and health outcomes points to increased weight as one of the primary results of rapid adoption of "Western" behavior patterns and diet. Pacific island peoples, however, seem to be especially prone to becoming obese when they migrate, more so than other migrants, such as those coming to the United States from Japan or Latin America.

There have been several previous attempts to describe the effects of migration and modernization on Pacific island peoples. Additional information has come from a series of studies on the impact of migration on persons of Japanese descent, in particular, those that have moved to Hawaii and California. The significance of body morphology in general, and of adipose tissue in particular, is a consistent theme that runs through most of this literature.

Much of the data concerning weight gain among Japanese migrants to the United States derives from the Honolulu Heart Study, which showed a gradient of increasing body mass and skinfold thicknesses among Japanese men living in Japan, Hawaii, and California (Kagan et al. 1974). Virtually all the differences were ascribed to behavioral or environmental variables, such as diet. Marmot and Syme (1976) found that retention of traditional Japanese behavior patterns contributed to a lower prevalence of most risk factors for coronary heart disease, including weight. While results of these studies are not directly comparable to the study of weight gain among Samoans moving to the United States, they provide an indication of how migration affects peoples of different ethnic backgrounds.

Indigenous peoples of the Pacific comprise three main groups, Polynesians, Melanesians, and Micronesians. Samoans, who are Polynesians, represent the largest component of this geographically widespread group, which includes the Maoris of New Zealand, the native inhabitants of Tonga, the Cook Islands, Tahiti and the Marquesas, as well as those of many other smaller island groups. Native Hawaiians are also considered to be Polynesians.

In 1966 a typhoon destroyed most of the coconut plantations on the Tokelau islands, a series of atolls 480 mi north of Samoa. Being deprived of their principal livelihood, many Tokelauans moved to New Zealand, providing the stimulus for an extensive series of studies, collectively known as the Tokelau Island Migrant Study, of the impact of migration on health (Prior et al. 1981, 1974). This study had the advantage of providing an opportunity to study the "migrants" before they had migrated. Compared with individuals who were to remain on the island, males in this group were taller and heavier, and among younger individuals, had somewhat high blood pressure (Prior 1977b; Stanhope and Prior 1976). Following migration, individuals exhibited several indicators of increased cardiovascular risk, including increased low density lipoprotein cholesterol and serum triglycerides, decreased high density lipoprotein cholesterol, increased weight gain, and adiposity (Stanhope et al. 1981).

A series of health surveys among rural and urban Maori populations of New Zealand and the Cook Islands showed that rural Maori males in New Zealand tend to be more obese than their urban counterparts (Prior and Davidson 1966). Maori populations from Rarotonga and Pukapuka in the Cook Island group have significantly lower levels of obesity, about 12 percent and 8 percent, respectively, of the adult male population. Among Maori females, approximately 45 percent of an island sample from Rarotonga exhibited obesity, similar to the prevalence among rural Maori females in New Zealand. Pukapuka females have a comparatively low rate of obesity, about 22 percent. While there is little doubt that Maoris share the Polynesian propensity to become obese when they live in an urban setting, the significant European admixture this group has undergone over the past few decades makes comparisons with other Polynesian populations such as the Samoans difficult.

Similar responses to modernization have been noted among the inhabitants of Palau, (Labarthe et al. 1973), and among the Chamorro, a Micronesian population who have migrated over the past two decades to Hawaii and the mainland United States (Reed et al. 1970).

The links between modernization and the development of obesity are not confined to peoples of the Pacific. Indian tribes of the southwestern United States are also known to develop considerable obesity. One such group, the Pimas, have levels of obesity that match some of those reported here for Samoan migrants to Hawaii and California. Pima males have an average weight 22 percent above that considered desirable for their height; Pima women exhibit mean weights 49 percent above the norm (Bennett et al. 1976). Among individuals above 150 percent of their ideal weight, the risk of diabetes was 6.4 times that of individuals under 125 percent of their ideal weight.

While most of these studies cite dietary modification as one of the principal antecedents of obesity, the link is not a simple one. Bindon (1984b) points out that Pacific island women tend to be heavier as nonmigrants than as migrants, a trend documented among the population of Rarotonga (Prior and Davidson 1966), and among the the Chamorro (Reed et al. 1970). In addition, the finding that "premigrant" Tokelauan males were taller and heavier than their nonmigrant counterparts suggests a degree of bias in the composition of migrant populations.

MATERIALS AND METHODS

The data base on which the material presented in the present chapter is based represents the cumulative results of almost 10 years of field studies conducted by contributors to this volume and others. It is not possible to address concerns over comparability of methods across the various investigations. Each investigator underwent the appropriate training in anthropometric methods, however, and detailed descriptions of the various data collection procedures used appear in the appropriate literature.

POPULATIONS SAMPLED

Five principal Samoan subpopulations have been examined during the course of this project, those of Western Samoa, two in American Samoa, Hawaii, and California. In Western Samoa, the principal research location was the village of Salamumu. As Bindon and Baker (1985) report, approximately 90 percent of the adult population of this village was examined, including 142 men and 199 women over the age of 18. Compared to other villages on the island, Salamumu is relatively affluent; the extent to which this has introduced bias into the data cannot be ascertained because of the lack of comparative data on the morphology of residents of other communities in the region. However, Zimmet's survey (Zimmet et al. 1981) of two rural Samoan communities, Tuasivi, on the island of Savai'i, and Poutasi, not far from Salamumu on the main island of Upolu, suggests little difference in the overall distribution of weight among Salamumu adults reported here.

The Manu'a group in American Samoa exhibits many contrasts with its comparatively urbanized neighbors. Other contributors to this volume point out that Manu'a is considerably less developed and, by most criteria, less acculturated than Tutuila; as the material contained in this and other chapters illustrates, the Manu'a population differs in several socio-economic and biological respects.

The sample of Tutuila adults was drawn from 38 villages across the island. The 1460 individuals over the age of 18 who comprise this sample represent approximately 12 percent of the adult population of the island. American Samoa is something of an economic anomaly in the South Pacific because of the large amount of economic support it receives from the U.S. Government. The population has numerous ties with Samoans in Hawaii and on the U.S. mainland; differences between this population and the neighboring inhabitants of Western Samoa represent some of the most marked biocultural contrasts within a geographically circumscribed group of peoples that exist anywhere.

The Hawaiian sample of Samoans described here was drawn from eleven locations on the main island of Oahu, including urban dwellers from Honolulu, and communities located on the northern and western ends of the island. The total sample of individuals, 585 adults, represents somewhat less than 10 percent of all adult Samoans believed to be resident in Hawaii at the time (Bindon 1984b). The structure of the Hawaiian subpopulation is discussed in the present volume, as well as by Hanna and Baker (1979).

The California sample of adult Samoans represents the smallest of the groups surveyed during the course of this project. The 228 individuals examined all lived in San Francisco or the adjacent peninsula. They were drawn from several church congregations after pilot studies among this geographically diverse population had shown this to be the only practicable manner of reaching sufficient numbers of individuals to provide adequate information on body size and other characteristics. Janes and Pawson (in press) have recently described historical, economic, demographic, and other features of this population. Because of the relatively small size of our sample, the extent to which the material presented here is representative of the larger California Samoan

Table 11.1 Age and Sex Distribution of Samoan Subpopulations

| Age (yr) | Western Samoa | | American Samoa | | | | | | Hawaii | | California | |
| | | | Manu'a | | Tutuila | | | | | | | |
	Male	Female	Male	Female	Male	Female	Male	Female	Male	Female		
18-24	37	46	2	17	121	149	53	104	12	22		
25-34	26	49	16	16	124	195	57	82	22	31		
35-44	20	32	15	39	100	183	50	59	28	25		
45-54	29	27	24	250	124	157	44	47	20	17		
55-64	17	22	25	34	95	102	29	31	21	12		
65-74	10	19	10	9	47	44	11	6	5	4		
75+	0	4	4	4	6	9	5	2	2	3		
Total	142	199	96	169	620	840	250	335	112	116		
Both	341		265		1460		585		228			

population is uncertain. Additional limitations are placed on comparisons with other Samoan groups described here by structural and other contrasts present within the California population. Table 11.1 summarizes the age and sex distribution of each of the five subpopulations described above.

Comparability of samples

One of the principles that has governed the design of this project from its inception has been that the various Samoan subpopulations are sufficiently similar, in both biological and cultural attributes, to warrant comparisons between them. As Baker points out (Chapter 1), the rationale for making such comparisons has been the likelihood that they will produce generalizations concerning the impact of modernization on human populations. The problems inherent in such an undertaking have been addressed here, as well as by other investigators who have sought to link health-related outcomes to the process of migration among Pacific island peoples (Prior et al. 1974; Reed et al. 1970). Foremost of the concerns that surround these comparisons is representativeness of the samples selected for study. With the exception of the California sample, the numbers of individuals surveyed during this project are adequate to allow the necessary stratification by age, sex, and other characteristics, for detailed analysis of morphological features. As indicated, the representativeness of the sample of individuals from Salamumu, Western Samoa, may be in doubt because of the comparative affluence of this village. This and other concerns over sampling strategy within the five subpopulations should not, however, affect the validity of the comparisons presented here, which are made for the purpose of describing the overall pattern of morphological variation within migrant and native Samoan populations.

Methods of analysis

A complete set of Samoan adult anthropometry was assembled and organized within the structure of an Statistical Analysis System (SAS) data set. The author is grateful to Dr. Bindon for his assistance in assembling information for the Western Samoan, American Samoan, and Hawaiian populations. For the purposes of data presentation, templates were constructed from weight/height standards for the U.S. population (Abraham et al. 1979). The standards were subjected to a certain degree of smoothing, using a cubic spline nonlinear interpolation procedure, in order to improve their visual appearance. Individuals were grouped, by age, into 10-year cohorts, as illustrated in Table 11.1, and median values for anthropometric measurements calculated for each cohort (Table 11.2). Before calculating the distribution of weight/height relationships in the various samples, heights were rounded to the nearest 5 cm to facilitate the formation of cohorts grouped by height. Where appropriate, statistical comparisons between the various samples were undertaken using analysis of variance treating age or other variables, as covariates.

As Bindon and Baker (1985) and others have pointed out, skinfold data for the very obese individual are prone to be imprecise. Among the California sample, we found that several individuals had triceps skinfolds that exceeded the maximum capacity of the instrument we used to collect these data, about 55 mm. Lohman (1981) has suggested that 30 mm may be the maximum skinfold thickness that can be reliably measured with skinfold callipers. We have also suggested (Pawson and Janes 1981) that differences in fat patterning may exist among Samoan populations according to their degree of adiposity. Because of these and other uncertainties, detailed analyses of skinfolds data for the purpose of the present overview of Samoan adult morphology was deemed inappropriate.

RESULTS

Anthropometric data collected among the various Samoan subpopulations studied during the course of this project are summarized in Table 11.2. For the purposes of comparisons with U.S. standards, measurements are presented according to the age grouping used by the National Center for Health Statistics (Abraham et al. 1979). More detailed descriptions of these data are contained in reports prepared by individual investigators; the material presented here is intended as a synthesis of their principal findings.

Table 11.2 Morphological Characteristics of Samoan Adults

Sex Age	Western Samoa X̄	S.D.	N	Manu'a X̄	S.D.	N	Tutuila X̄	S.D.	N	Hawaii X̄	S.D.	N	California X̄	S.D.	N
Males 18 - 74															
Age (yr)	39.8	16.0	139	49.4	12.7	92	41.6	15.5	611	38.8	14.7	244	42.4	13.7	108
Height (cm)	169.8	5.7	138	171.0	5.5	91	170.4	6.2	607	170.4	5.7	243	172.0	6.6	107
Weight (kg)	76.0	14.9	138	83.0	15.6	91	86.3	18.5	611	89.8	17.8	241	98.8	16.9	108
Skinfolds (mm)															
Triceps	10.6	6.0	139	15.9	8.3	91	17.1	10.1	607	17.8	8.5	242	16.4	7.6	100
Midaxillary	10.6	7.3	138	19.6	10.9	91	23.4	13.5	603	21.3	10.7	241	—	—	—
Subscapular	16.3	9.2	139	24.7	12.3	91	24.6	13.2	606	25.0	11.5	242	—	—	—
Arm circumference (mm)	30.7	3.7	139	31.7	3.5	91	32.4	4.0	607	33.8	4.1	242	—	—	—
Males 18 - 24															
Age (yr)	21.2	1.9	37	22.3	2.8	2	21.1	1.9	121	20.6	2.0	53	21.5	21.6	12
Height (cm)	169.7	6.0	37	171.4	1.3	2	172.3	5.9	121	173.7	5.7	53	172.6	6.4	12
Weight (kg)	70.1	10.5	36	79.8	15.0	2	77.7	11.7	121	84.5	14.7	52	84.0	13.4	12
Skinfolds (mm)															
Triceps	7.2	3.2	37	19.5	9.2	2	11.8	5.7	121	14.4	7.5	53	14.2	8.9	11
Midaxillary	7.1	2.9	37	26.0	9.9	2	16.1	9.8	121	16.5	10.0	53	—	—	—
Subscapular	11.3	4.7	37	27.5	10.6	2	15.5	8.8	121	18.3	10.6	53	—	—	—
Arm circumference (mm)	28.9	3.1	37	29.5	3.5	2	30.5	3.0	121	32.2	3.8	53	—	—	—

Continued next page

Table 11.2 (continued) Morphological Characteristics of Samoan Adults

Sex Age	Western Samoa X̄	S.D.	N	Manu'a X̄	S.D.	N	Tutuila X̄	S.D.	N	Hawaii X̄	S.D.	N	California X̄	S.D.	N
Males 25 - 34															
Age (yr)	29.8	2.9	26	31.4	2.9	16	30.4	2.8	124	29.5	2.9	57	30.1	3.0	22
Height (cm)	171.4	5.5	26	174.1	5.8	16	171.5	6.5	124	171.4	5.0	57	176.3	8.0	21
Weight (kg)	75.9	15.1	26	91.4	14.1	16	89.5	17.5	124	90.4	17.0	57	101.5	19.2	22
Skinfolds (mm)															
Triceps	8.8	4.8	26	15.4	5.7	16	16.6	10.0	124	17.8	9.3	57	16.1	8.4	22
Midaxillary	9.0	6.3	26	21.3	8.8	16	24.7	13.2	124	21.0	10.7	57	—	—	—
Subscapular	14.3	7.6	26	28.6	11.5	16	25.2	12.6	124	23.9	11.2	57	—	—	—
Arm circumference (mm)	30.7	3.3	26	33.6	2.7	16	33.1	3.6	124	34.1	3.7	57	—	—	—
Males 35 - 44															
Age (yr)	40.0	2.9	20	40.3	2.8	15	39.9	2.9	100	40.0	2.8	50	38.6	2.8	28
Height (cm)	170.4	7.2	20	171.9	6.4	15	169.8	6.3	99	169.4	6.0	50	170.6	5.6	28
Weight (kg)	77.9	17.7	20	88.8	16.7	15	88.8	18.2	100	93.5	17.9	49	100.4	15.4	28
Skinfolds (mm)															
Triceps	11.0	5.4	20	18.1	9.8	15	17.4	8.9	99	18.8	7.1	49	17.9	7.2	25
Midaxillary	11.7	7.5	20	23.3	12.3	15	26.3	12.7	98	23.5	8.7	49	—	—	—
Subscapular	18.8	9.9	20	30.9	16.2	15	27.4	13.0	98	29.0	10.3	49	—	—	—
Arm circumference (mm)	31.4	3.8	20	33.1	3.6	15	33.2	4.0	99	35.1	3.6	48	—	—	—

Continued next page

Table 11.2 (continued) Morphological Characteristics of Samoan Adults

Sex Age	Western Samoa			Manu'a			Tutuila			Hawaii			California		
	$\bar{X}$	S.D.	N	$\bar{X}$	S.D.	N	$\bar{X}$	S.D.	N	$\bar{X}$	S.D.	N	$\bar{X}$	S.D.	N
Males 45 - 54															
Age (yr)	49.8	2.8	29	50.7	3.1	24	49.8	3.0	124	50.1	3.3	44	49.5	3.1	20
Height (cm)	169.2	4.5	29	171.4	5.0	23	170.0	6.2	123	170.4	5.4	43	171.5	5.4	20
Weight (kg)	79.4	12.1	29	79.2	14.1	23	93.3	21.5	124	90.6	19.1	43	101.6	16.2	20
Skinfolds (mm)															
Triceps	12.4	8.0	29	13.8	6.1	23	21.6	11.9	123	18.4	9.0	43	14.8	7.7	18
Midaxillary	11.9	7.5	29	16.6	9.1	23	29.2	14.9	123	21.3	8.6	43	—	—	—
Subscapular	18.9	10.8	29	19.1	8.4	23	30.7	14.0	123	26.4	9.5	43	—	—	—
Arm circumference (mm)	32.0	3.4	29	31.1	3.5	23	33.9	4.0	123	34.2	4.8	44	—	—	—
Males 55 - 64															
Age (yr)	61.8	2.3	17	60.1	2.8	25	59.6	2.5	95	59.3	3.2	29	59.1	2.8	21
Height (cm)	169.1	5.8	17	170.0	5.0	25	169.2	5.2	95	168.7	4.9	29	170.7	6.3	21
Weight (kg)	81.1	21.8	17	78.6	15.9	25	83.5	16.8	95	89.8	15.4	29	100.4	16.1	21
Skinfolds (mm)															
Triceps	12.9	6.6	17	16.1	10.5	25	18.0	10.9	95	20.2	7.3	29	17.0	5.9	19
Midaxillary	15.6	11.2	16	19.0	13.5	25	21.5	13.4	94	25.3	11.8	28	—	—	—
Subscapular	21.3	11.9	17	23.4	12.0	25	23.9	11.4	95	28.2	9.8	29	—	—	—
Arm circumference (mm)	31.0	4.9	17	30.4	3.8	25	31.7	3.9	95	33.5	3.2	29	—	—	—

Continued next page

Table 11.2 (continued) Morphological Characteristics of Samoan Adults

Sex Age	Western Samoa			Manu'a			Tutuila			Hawaii			California		
	$\bar{X}$	S.D.	N	$\bar{X}$	S.D.	N	$\bar{X}$	S.D.	N	$\bar{X}$	S.D.	N	$\bar{X}$	S.D.	N
Males 65 - 74															
Age (yr)	68.1	1.4	10	67.6	2.2	10	69.3	2.8	47	69.4	3.0	11	69.2	1.8	5
Height (cm)	167.3	5.2	9	167.4	4.2	10	168.0	6.4	45	169.0	5.9	11	168.0	4.3	5
Weight (kg)	75.9	10.4	10	81.1	15.1	10	82.2	21.6	47	92.8	31.1	11	95.6	18.1	5
Skinfolds (mm)															
Triceps	15.3	5.2	10	17.4	8.4	10	17.4	9.2	45	21.1	12.5	11	17.8	10.0	5
Midaxillary	14.2	6.6	10	18.7	7.9	10	21.0	11.9	43	24.9	18.1	11	—	—	—
Subscapular	19.3	6.0	10	24.3	11.6	10	25.7	13.8	45	29.8	18.9	11	—	—	—
Arm circumference (mm)	31.5	2.1	10	31.4	2.6	10	30.9	4.8	45	33.1	6.1	11	—	—	—
Females 18 - 74															
Age (yr)	39.3	15.5	195	46.2	13.1	165	40.0	14.2	830	35.1	13.6	329	37.7	14.0	111
Height (cm)	158.5	5.3	193	160.1	5.8	165	159.8	5.5	830	160.1	5.8	327	162.6	5.0	110
Weight (kg)	71.3	15.5	194	81.4	16.9	161	84.0	18.8	808	83.8	18.8	320	90.4	19.7	110
Skinfolds (mm)															
Triceps	20.4	7.5	193	35.2	12.9	165	35.4	14.9	830	31.3	11.5	326	28.9	9.9	103
Midaxillary	18.3	8.4	186	39.9	14.8	163	34.2	15.5	820	33.6	14.9	297	—	—	—
Subscapular	23.8	9.5	193	43.3	14.1	165	37.6	14.1	825	35.4	13.7	318	—	—	—
Arm circumference (mm)	29.7	4.7	193	33.4	4.9	165	34.0	5.6	830	33.5	5.8	328	—	—	—

Continued next page

Table 11.2 (continued) Morphological Characteristics of Samoan Adults

Sex Age	Western Samoa			Manu'a			Tutuila			Hawaii			California		
	$\bar{X}$	S.D.	N	$\bar{X}$	S.D.	N	$\bar{X}$	S.D.	N	$\bar{X}$	S.D.	N	$\bar{X}$	S.D.	N
Females 18 - 24															
Age (yr)	21.4	1.6	46	22.0	1.7	17	21.4	1.9	149	21.2	2.0	104	21.1	1.6	22
Height (cm)	157.7	5.8	46	164.4	5.4	17	160.8	5.8	149	160.8	5.7	103	166.0	3.5	22
Weight (kg)	63.9	10.0	46	78.2	17.5	16	72.3	13.5	142	74.4	14.7	100	80.6	19.7	22
Skinfolds (mm)															
Triceps	17.3	6.2	46	26.5	9.8	17	25.1	11.1	149	25.0	9.6	104	22.7	8.9	22
Midaxillary	15.2	6.7	45	29.8	11.0	17	24.1	11.6	146	25.6	11.0	96	—	—	—
Subscapular	19.1	7.4	46	32.2	12.6	17	27.3	11.3	148	28.2	11.6	104	—	—	—
Arm circumference (mm)	26.8	2.9	46	29.7	4.3	17	29.8	4.8	104	32.2	3.8	53	—	—	—
Females 25 - 34															
Age (yr)	30.2	2.4	49	29.5	2.4	16	29.8	2.9	195	29.5	2.8	82	28.9	2.9	31
Height (cm)	159.2	5.6	49	161.6	5.8	16	160.3	5.7	195	161.7	6.1	82	162.7	4.9	31
Weight (kg)	70.4	12.5	49	80.5	16.2	16	83.5	19.0	185	89.4	20.4	79	88.1	17.2	31
Skinfolds (mm)															
Triceps	19.1	6.9	49	30.8	11.8	16	33.9	14.4	195	32.9	11.6	81	27.7	8.7	29
Midaxillary	17.3	7.3	49	37.8	13.6	16	33.1	14.8	194	34.8	15.0	71	—	—	—
Subscapular	22.8	7.1	49	39.1	14.6	16	35.9	13.7	193	37.3	14.7	78	—	—	—
Arm circumference (mm)	28.6	3.7	49	31.8	3.7	16	33.4	5.2	195	33.9	5.4	82	—	—	—

Continued next page

Table 11.2 (continued) Morphological Characteristics of Samoan Adults

Sex Age	Western Samoa			Manu'a			Tutuila			Hawaii			California		
	X̄	S.D.	N	X̄	S.D.	N	X̄	S.D.	N	X̄	S.D.	N	X̄	S.D.	N
Females 35 - 44															
Age (yr)	40.0	2.8	32	41.2	2.8	39	39.8	2.7	183	40.3	2.9	59	39.8	3.3	25
Height (cm)	158.0	4.2	32	162.0	5.2	39	159.9	5.4	183	159.5	5.0	59	161.9	5.7	25
Weight (kg)	74.0	15.2	32	88.7	17.6	36	88.3	18.6	178	90.8	19.3	58	97.9	22.1	25
Skinfolds (mm)															
Triceps	23.2	7.6	32	38.7	12.5	39	38.5	14.2	183	35.6	10.5	59	34.7	9.4	22
Midaxillary	20.9	8.6	31	44.1	14.1	38	37.7	14.6	182	39.6	15.6	55	—	—	—
Subscapular	27.5	10.1	32	47.4	12.7	39	41.6	12.7	182	39.8	13.3	56	—	—	—
Arm circumference (mm)	30.9	4.0	32	34.1	5.0	39	35.0	4.9	183	36.5	5.6	59	—	—	—
Females 45 - 54															
Age (yr)	49.3	2.3	27	50.7	2.9	50	49.9	2.8	157	49.0	2.8	47	48.8	2.8	17
Height (cm)	159.7	4.9	27	158.9	5.8	50	159.9	4.8	157	158.8	4.6	46	161.3	4.5	17
Weight (kg)	77.3	18.4	27	82.6	15.8	50	90.9	18.3	157	84.9	15.2	46	95.9	17.7	17
Skinfolds (mm)															
Triceps	25.2	6.5	27	37.9	11.7	50	42.0	14.8	157	35.5	10.4	46	32.7	8.8	15
Midaxillary	22.0	7.5	25	44.0	14.5	50	40.1	16.0	155	37.5	14.0	42	—	—	—
Subscapular	28.1	11.2	27	47.1	12.4	50	43.4	14.0	156	40.5	11.7	46	—	—	—
Arm circumference (mm)	33.5	4.7	27	34.7	4.8	50	36.2	5.8	157	35.4	4.7	42	—	—	—

Continued next page

266

Table 11.2 (continued) Morphological Characteristics of Samoan Adults

Sex Age	Western Samoa			Manu'a			Tutuila			Hawaii			California		
	X̄	S.D.	N	X̄	S.D.	N	X̄	S.D.	N	X̄	S.D.	N	X̄	S.D.	N
Females 55 - 64															
Age (yr)	59.3	3.0	22	59.2	2.8	34	59.6	2.7	102	59.5	3.0	31	60.0	3.0	12
Height (cm)	158.4	5.4	21	157.8	4.7	34	158.2	5.0	102	157.1	6.4	31	158.9	4.3	11
Weight (kg)	78.2	20.8	22	77.0	15.5	34	84.3	19.2	102	86.2	19.4	31	90.9	16.0	11
Skinfolds (mm)															
Triceps	21.5	9.6	21	34.6	13.9	34	37.9	14.5	102	33.3	10.6	30	30.3	7.9	12
Midaxillary	21.4	12.2	21	38.1	15.2	34	36.8	15.4	99	38.0	14.8	29	—	—	—
Subscapular	26.3	12.7	21	43.3	13.9	34	39.6	13.1	102	39.2	13.0	29	—	—	—
Arm circumference (mm)	31.4	6.4	21	33.8	4.7	34	35.6	5.5	102	36.0	5.0	30	—	—	—
Females 65 - 74															
Age (yr)	67.9	2.9	19	69.7	3.1	9	68.5	2.9	44	68.9	3.9	6	69.5	3.3	4
Height (cm)	157.3	4.6	18	156.2	5.6	9	157.9	5.4	44	157.1	6.5	6	162.7	5.3	4
Weight (kg)	70.3	15.9	18	69.2	16.3	9	81.1	17.2	44	78.2	14.4	6	90.7	23.7	4
Skinfolds (mm)															
Triceps	18.6	5.1	18	30.8	15.3	9	35.0	13.7	44	32.0	17.2	6	20.0	6.6	3
Midaxillary	15.6	6.6	15	27.0	10.3	8	32.3	15.1	44	49.0	13.3	4	—	—	—
Subscapular	22.7	6.7	18	33.1	15.7	9	37.9	13.4	44	36.6	10.0	5	—	—	—
Arm circumference (mm)	30.4	4.4	18	32.8	5.0	9	35.0	5.4	44	33.8	6.0	6	—	—	—

Height and weight among Samoan adults

Figs. 11.1 and 11.2 illustrate height/weight relationships among Samoan adults in the five study populations, compared with selected percentiles for the U.S. population (Abraham et al. 1979). The data clearly indicate that Samoans, whether migrant or nonmigrant, are significantly heavier for a given height than the U.S. norm. Even the sample of individuals from Western Samoa, where excessive obesity is uncommon, fell above the U.S. median; the difference was particularly noticeable among Western Samoan women. Among men, each group, with the exception of the California sample, fell between the 50th and 90th percentile of height for weight in the U.S. population; the latter group exceeded the 90th percentile except among shorter individuals (160-170 cm). Women who are consistently heavier for a given height than men, show less intergroup variation in the height/weight relationship, generally falling along the 90th percentile.

Analysis of intergroup differences indicated that there is significant overall variation in height between these five subpopulations. As other summary anthropometric data in Table 11.2 show, adult men aged 18-74 are shortest in Western Samoa (169.8 cm), and tallest in California (172.0 cm). Comparisons of individual sample pairs, controlling for the effect of age, showed that California men were significantly taller than each of the other groups, with the exception of Manu'an men.

Among women, individual variation in the pattern of height change with age prevented comparisons of adjusted mean weights. It was apparent, however, that California women increased in weight more sharply than did women of any of the other groups. Compared to this pattern of intergroup variation in height, men and women belonging to the same subpopulation are relatively homogeneous. Although men are, as expected, taller than women, there is relatively little variation between the sexes in the relationship of height with age, as judged from the similarity of regression slopes. Except for the absolute difference in height, the overall pattern of variation in these populations suggests a considerable degree of morphological homogeneity within individual groups.

As the data in Fig. 11.1 suggest, there was significant overall variation in weight among the five subpopulations. Among men, the California sample had the highest mean weight (98.8 kg) and the Western Samoan group the lowest (76.0 kg). There was a strong gradient in weight among adult men of Manu'a (83.0 kg), Tutuila (86.3 kg,) and Hawaii (89.8 kg). Women exhibited a similar pattern of variation. As before, California women were the heaviest (90.4 kg) and Western Samoan women the lightest (71.3 kg). Women from Manu'a had an average weight of 81.4 kg, compared to 84.0 kg among the Tutuila sample and 83.8 kg among Hawaiian Samoans. Comparisons between the various groups indicated similarity between the male samples in the

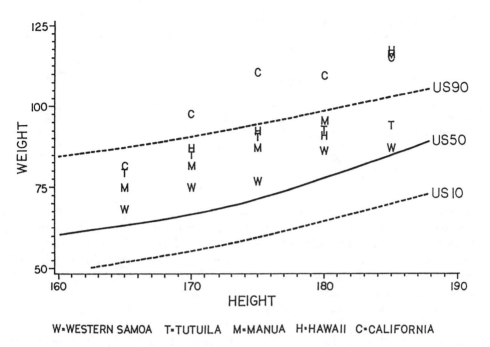

W=WESTERN SAMOA T=TUTUILA M=MANUA H=HAWAII C=CALIFORNIA

Fig. 11.1 Weight for height among Samoan men.

Fig. 11.2 Weight for height among Samoan women.

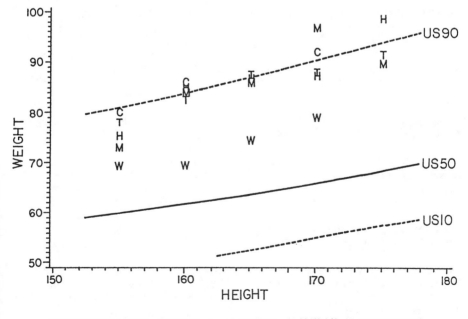

W=WESTERN SAMOA T=TUTUILA M=MANUA H=HAWAII C=CALIFORNIA

Table 11.3 Weight/Height Ratio (Body Mass Index) for Adults over 20 Years

	Men			Women		
	N	Mean	SD	N	Mean	SD
Western Samoa	125	2.65	0.46	187	2.84	0.57
Manu'a	95	2.80	0.49	162	3.16	0.61
Tutuila	570	3.00	0.58	777	3.30	0.70
Hawaii	219	3.12	0.56	285	3.33	0.70
California	103	3.53	0.50	105	3.45	0.69

overall relationship of weight with age. Judging from individual regression analyses, men in Manu'a and Western Samoa tend to lose weight with age; other samples show a positive trend of weight with age; however, the overall pattern of intergroup variation was not significant. As expected, statistical comparisons of the various subpopulations indicated a highly significant pattern of variation in weight.

Table 11.3 illustrates mean values of the weight/height ratio (weight/height^2) for the five Samoan populations described here. With the exception of the California sample, women in each of the populations surveyed have significantly larger weight/height ratios than men. As indicated, values for the Western Samoan sample closely resemble those

Table 11.4 Weight of Migrant and Nonmigrant Samoan and Tokelau Islanders (kg)

	Age			
Group	25-34	35-44	45-54	55-64
A. Nonmigrants				
Tokelau Males[a]	76.1	77.9	78.9	74.1
Western Samoan Males	75.9	77.9	79.4	81.1
Tokelau Females[a]	71.8	77.7	78.1	74.2
Western Samoan Females	70.4	74.0	77.3	78.2
B. Migrants				
Tokelau Males (New Zealand)[b]	83.0	87.0	81.0	82.0
Samoan Males (Hawaii)	90.4	93.5	90.6	89.8
Tokelau Females (New Zealand)[b]	77.0	84.0	85.0	80.0
Samoan Females (Hawaii)	89.4	90.8	84.9	86.1

[a] Data from Prior et al. 1984.

[b] Data values estimated from graph (Prior 1981, Figs. 3 and 4).

reported by Zimmet et al.(1981) for rural populations of Savai'i and Upolu although unpublished data collected in other rural areas of Western Samoa do suggest a moderate degree of intrapopulation variability in body size (Schendel et al. 1983; Schendel personal communication). The extent of this local variability, however, is considerably less than that which exists between the data reported here and those for the other Samoan populations surveyed. Compared to Tokelauans, the only other Polynesian population for which data are available, Western Samoan men are of similar weight, except among individuals over 55. Western Samoan women tend to be lighter than their Tokelauan counterparts, except for those over 55. Table 11.4 summarizes differences in weight between migrant and nonmigrant Tokelauans and Samoans (Hawaian sample). The Tokelau islands sample is that classified as "nonmigrant," compared to "premigrant" (Prior et al. 1977). Data for Tokelauan migrants to New Zealand were estimated from a graph illustrating mean weights for males and females (Prior 1981, Figs. 3 and 4).

While there are some differences between the Tokelauans and Western Samoan nonmigrant samples, they are far exceeded by weight differences between the migrant samples. As indicated above, the tendency of Samoans to become obese when they migrate is illustrated even more by the data from California.

Body fat

Although skinfold thickness measurements have been traditionally used as indicators of overall adiposity, there is some doubt, as pointed out earlier, about their value as accurate indicators of fatness in the very obese person. Fig. 11.3 illustrates selected percentiles of the triceps skinfold, the most commonly used indicator of fatness, among Samoan adults. Although the Western Samoan samples have low levels of fat, differences between the two samples from American Samoa (Tutuila and Manu'a) and those from Hawaii and California do not correspond to the pattern of weight variation illustrated in Fig. 11.1. The inconsistency is particularly pronounced among the California sample, which exhibits similar or lower median triceps skinfolds than the lighter (and presumably less obese) samples from Hawaii, and American Samoa. Among women from American Samoa, Hawaii and California, there appears to be an inverse relationship between the triceps skinfold and weight. This suggests either a methodological problem, such as greater fat compressibility in the extremely obese person, or a difference in fat patterning among these groups. In any case, the imprecise relationship of the triceps skinfold with weight in these samples renders further analysis of intergroup variation of doubtful value. Similarly, indices based on this measurement, such as the Upper-Arm Muscle Circumference Index (upper arm-circumference $\div \pi -$ triceps skinfold), are likely to be inaccurate.

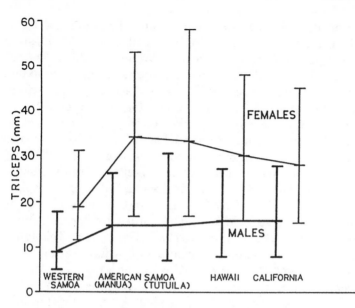

Fig. 11.3 Triceps skinfolds in Samoan subpopulations. The 10th, 50th, and 90th percentiles are represented.

Additional measurements of skinfold thicknesses were obtained from Western and American Samoan samples, and the Hawaiian sample. As the data in Table 11.2 indicate, one is left with a general impression of relatively low levels of fatness measures among Western Samoans, and a relatively homogeneous distribution among the samples from American Samoa and Hawaii. Unfortunately, only triceps skinfolds are available for the California population.

Influences on body morphology

Bindon (1984b) and Bindon and Baker (1985) have recently discussed the influence of several socioeconomic factors on body morphology in Samoans from Western Samoa, American Samoa, and Hawaii. Occupational status was assessed from reports of habitual daily activity and grouped into broad categories, including *traditional,* comprising subsistence farmers or fishers, housewives, and unemployed persons, *unskilled,* including laborers, fish cannery, agricultural, and other workers, and *skilled,* including professional persons, police, and electricians. Comparison of these different job categories suggested that individuals with the most highly skilled jobs also tended to be the most obese. Educational status (number of years of school attendance) suggested that taller individuals also tended to be better educated. Although number of years spent in Hawaii might by expected to be associated with the development of obesity among Samoan migrants, comparisons of recent migrants with those who had spent longer away from Samoa showed a complete lack of association of time spent in the United States with any of the anthropometric measures described here. Bindon (1984b) points out that the limited data available in this study cannot meet the demands of a rigorous analysis of the underlying causes of obesity in these populations.

SUMMARY

Despite certain inconsistencies in the very large amount of morphological data collected during the course of these investigations, the information summarized here effectively portrays these populations as being some of the largest and heaviest known anywhere, even compared to other migrant Polynesian populations. That such a high prevalence of excessive weight should exist in a relatively small population should be a matter of major concern to local health authorities in areas that support large Samoan migrant communities, especially in view of the clearly demonstrated links between obesity and serious metabolic disorders among other Pacific island peoples. Most comprehensive reviews of the effects on health of obesity (e.g., Mann 1974, Parts 1 and 2) identify a long list of harmful health effects of excessive weight gain, including hypertension, degenerative heart disease, and risks during surgery and pregnancy. That populations surveyed during the course of this project, at least those in American Samoa, Hawaii, and California, exhibit this trend to a pronounced degree, ought to confirm the findings of numerous studies that have sought to link obesity with these and other outcomes. While there have been several reports that link the large physical size of Samoans to specific health effects (e.g., Hornick and Hanna 1982; Hanna and Baker 1979) some puzzling inconsistencies remain. For example, in a review of 10 years of mortality data for Samoans, identified from the California State Death Registry, we found that mortality from heart disease and stroke were far below what might be expected on the basis of projections from other published mortality data (Pawson and Janes 1981). In the absence of evidence for any degree of significant underreporting of deaths among this, the largest Samoan subpopulation, we concluded that either the relatively low median age of this population (19.4 years) accounted for the absence of significant mortality, or that Samoans were uniquely protected against the risks of excessive obesity by the maintenance of strong sociocultural bonds within the community that helped to reduce stress. We were unable to identify specific mechanisms whereby obesity-related risks might be reduced in the California population, however, so this latter hypothesis must remain tenuous at best.

Despite the extensive data that this project has made available on the distribution of obesity among Samoan migrant populations, the antecedents of this condition are still poorly understood. Since fat represents stored energy, it might be supposed that a genetic predisposition toward obesity (if any) among Polynesians might have developed in response to the need to withstand periods of starvation or long ocean voyages. As Zimmet (1981) has pointed out, the prevalence of diabetes, which might be indicative of a genetic link in the onset of diabetes, is low among Polynesian and Micronesian populations where a traditional life-style has been maintained, but high among modernized groups.

The social antecedents of obesity are discussed elsewhere in this volume. Large physical size has long been considered an attractive physical characteristic by Samoans, and seems to be associated with high status among Polynesian populations in general. Connelly and Hanna (1978), however, found that fatness is more likely to be a consequence of achieving high status rather than its antecedent.

Samoan communities living in the United States contain a high proportion of young people; in California, the median age of Samoans enumerated in the recent national census was 19.1 years (Janes and Pawson in press) compared to over 30 for the state's population as a whole. Because there are an estimated 10 000 Samoans living in this state under the age of 20, problems of excessive weight gain among the young deserve special attention. Adiposity is generally divided into two main clinical manifestations, adult onset, or hypertrophic obesity, and childhood onset, or hypertrophic/hyperplastic obesity. The latter form, in which fat cells grow both in size and number, may be linked to a lower risk for developing cardiovascular and metabolic disease, since metabolic impairment is linked to the size and fat content of the adipocyte, rather than to the absolute number of fat cells (Karam 1979). The early onset of obesity among Samoan children living in the United States may therefore eventually lower their risk of developing certain types of obesity-related disorders as they grow older.

The degree of obesity among certain segments of the young Samoan population, however, at least in California, has no precedent in the literature. Unpublished data show that between the ages of 10 and 18 years, the median weight of Samoan children born and raised in California falls close to the 95th percentile for weight among U.S. children. Such widespread obesity among Samoan youth will likely hinder their integration into a society that places undue value on slimness. In addition, since most young Samoans have never been to Samoa, and may feel little obligation to follow the traditional values of fa'aSamoa, their large physical size may present more problems of social adjustment than it does for their elders.

Chapter 12

The Diet and Nutrition of Contemporary Samoans

JOEL M. HANNA
DAVID L. PELLETIER
VANESSA J. BROWN

A major theme of the Samoan studies has been the changes in health and disease that accompany the processes of modernization and migration. As is clearly demonstrated throughout this volume, the patterns of disease and associated risk factors change as Samoans become exposed to varying degrees of modernization and as they migrate to areas outside of the Samoan archipelago. Western Samoans, living in the continuation of traditional Samoan culture, show diseases and morbidity patterns compatible with other traditional peoples, while American Samoans and Samoan migrants to the United States are now manifesting the morbidity characteristic of modern populations. Hypertension, changed blood lipids, obesity, and diabetes predict a future of modern diseases, even though most adult Samoans were born in Samoa and have lived a significant portion of their lives within the traditional pattern.

The causes and factors contributing to this rapid change in morbidity are interrelated and complex, however, extensive research in modern populations has implicated dietary change as a principal component (Morgen and Caan 1979; Clements 1970) The dietary contribution has also been investigated in studies of other modernizing Pacific populations (Coyne 1981; Zimmet 1979; Henkin et al. 1970). Although the roles of specific nutrients have not been clearly linked to the clinical manifestations of these diseases, dietary cholesterol, saturated fats, refined sugar, dietary fiber, sodium, and excessive caloric intake may play some part.

HISTORICAL STUDIES OF THE SAMOAN DIET

The precontact diet, which formed the basis for the recent traditional diet is unrecorded, but it probably contained most of the major foods carried by the original migrants to the Samoan archipelago and was similar to that of other Polynesian peoples (Bindon 1984b). Taro, yams, coconut, bananas, and breadfruit were the major staples. Domestic pigs and chickens were available, but these were probably reserved for special occasions and did not provide a significant contribution to the everyday diet, especially the diet of women and children who were largely excluded from eating these foods (Malcolm 1954). Protein was provided by exploitation of reef resources. In most respects, the diet is similar to that of horticulturalists throughout the world, with two major exceptions. First, the combination of ocean and reef resources, together with tree and root crops, assured an adequate supply of protein and energy, which reduced the likelihood of malnutrition common in other horticultural populations. Second, due to the heavy use of coconut as a food and condiment, the diet was probably higher in saturated fat than is typical for other horticulturalists. Among contemporary Pukapukans, another Polynesian people, over 36 percent of the average caloric intake is derived from coconut fats with other sources increasing the lipid intake even more (Prior et al. 1981). Indeed, precontact Samoans, and other Polynesians, may have consumed a greater proportion of their diets as saturated fats than paleolithic hunters (Eaton and Konner 1985).

By the time systematic observations began there had been a significant penetration of imported foods into the diet of even the most traditional Samoans. References to imported foods are contained in early dental survey lists of commonly consumed foods (Williams 1939; Ferguson 1934). Early nutrition studies by Holmes (1954) and Malcolm (1954) that described aspects of diet and child feeding practices in Western and American Samoa, respectively, indicated that purchased items such as tinned meat and flour were a normal part of the diet. The major staples were the traditional taro and breadfruit, and dishes containing coconut cream; the local stores reported a large sales volume of several imported foods, such as sugar, bread, rice, canned meat, and lard, that were used as supplements to the locally produced dietary items. Holmes also noted that the diet was adequate with respect to types of foods consumed, but that weanling children were not receiving a sufficient amount of protein and recommended that supplementation be established for young children. Malcolm made similar recommendations and documented the growth problems of several children.

A more comprehensive study was reported by Wilkins (1965), who surveyed several urban and rural villages in Western Samoa employing a food weighing technique. This is apparently the first quantitative study of the Samoan diet. She found that the mixture of traditional and imported foods used in all villages was affected by urbanization. Traditional foods were more characteristic in rural areas and imported foods were more frequently consumed in urban areas. For example, there was a decline in the use of the starchy roots and fruits that provided 64 percent of the calories in the rural areas to 38 percent in the urban areas. A reverse gradation was observed in the use of refined sugar: rural villagers consumed 18 percent of their calories in the simple sugars and urban residents 36 percent. This trade of complex for simple carbohydrates is typical of modernization and has a potential for adverse health effects (Morgen and Caan 1979; Clements 1970). Caloric consumption according to Wilkins was 3166 for rural villages as compared to 2304 Kcal/day for urban areas. Wilkins did not see any adult malnutrition, but also commented upon protein malnutrition in weanling children.

This mixture of traditional and Western foods is typical of the contemporary Samoan diet. Along with the traditional staples, many Western foods—flour, sugar, tinned meat, and tinned fish—introduced generations ago have become incorporated in the diet, and are now considered to be Samoan and are universally consumed. It seems likely that all Samoans are consuming a composite diet that differs only in degree. Those families more exposed to modern society, having a cash income or having access to purchased items are incorporating more imported items in the diet. This traditional diet has probably been consumed by a large number of the Samoans studied in this research program.

The adequacy of the diet seems evident since Samoans appear to be relatively free of nutritional diseases (see Chapter 5; Wilkins 1965); indeed, it may be an optimal diet. In other populations, shifting to a modern diet is frequently accompanied by an increase in stature. This does not occur among Samoans, who show no increase in stature with dietary change (Baker and Hanna 1981), but rather show a substantial gain in body fat (Pawson, see Chapter 11; Bindon and Baker 1985; Bindon 1982), and so the change to a modern diet is not necessarily beneficial. A major exception to the apparent dietary sufficiency of the traditional Samoan diet may have been the protein deficiency reported in infants and young children. It has been commented upon by several observers (Lichton et al. 1983; Neave 1969; Wilkins 1965) and documented in several cases (Wilkins 1965 Malcolm 1954). It may still exist, but since it affected only some infants it may have largely escaped dietary study.

SAMOAN PROJECT DIETARY STUDIES

Three separate studies investigating diet have been conducted as part of the Samoan project. The populations studied have varied, ranging from traditional Western Samoa to migrants resident in Hawaii. Pelletier (1984) studied a rural village in Western Samoa and a more modernized set of subpopulations living and working in Apia. Brown et al. (1984) studied residents of Manu'a, American Samoa, and a sample of residents of Hawaii who are of Samoan ancestry. Bindon (1982, 1984b) has studied residents of Manu'a, Tutuila, and Hawaiian immigrants. Thus, the existing rural-to-modern-to-migrant dietary spectrum has been sampled to some degree.

Studies in Western Samoa

Pelletier (1984) has described a study of diet that he conducted in Western Samoa as part of a larger investigation including a physical activity analysis and a study of blood lipids. Four different groups of men were sampled, including rural villagers, active urban workers, and two groups of sedentary urban workers differing in degree of leisure activity. All subjects were men between 18 and 40 years of age.

The rural village sample consisted of 43 men from Salea'aumua. The modernized samples were chosen from employees of businesses in Apia, initially on the basis of active or sedentary occupations. The group of 73 active workers was selected from occupations including auto mechanics, construction laborers, and copra laborers. The sedentary occupations were further subdivided on the basis of physically active and inactive leisure-time activities. The group of 86 sedentary workers included bank tellers, office clerks, and department store workers. Anthropometric characteristics for these groups are presented in Table 12.1.

The dietary assessment employed a 24-hour recall interview as the primary source of information of the men's dietary patterns. Size models

Table 12.1 Age and Anthropometric Characteristics of Three Groups of Western Samoan Men

Trait	Salea'aumuans (N = 43)		Active workers (N = 73)		Sedentary workers (N = 86)	
	Mean	SD	Mean	SD	Mean	SD
Age (yr)	25.5	6.0	25.4	3.9	24.8	4.0
Height (cm)	170.0	5.9(a)	169.8	5.1(a)	172.5	5.7(b)
Weight (kg)	71.5	8.1(a)	74.4	9.1(a)	80.1	11.6(b)
BMI (kg/m^2)	24.7	2.3(a)	25.8	2.7(b)	26.8	3.4(c)
Skinfolds (mm)						
Triceps	8.6	2.7(a)	10.8	4.5(b)	14.3	6.1(c)
Forearm	4.8	0.6(a)	5.4	1.2(b)	5.7	1.4(b)
Calf	7.6	2.6(a)	9.3	4.2(a)	11.6	5.0(b)
Subscapular	13.6	3.9(a)	16.5	5.8(b)	20.2	8.1(c)
Midaxillary	8.8	2.9(a)	10.7	4.7(a)	14.8	7.8(b)
Suprailiac	15.1	6.4(a)	17.1	7.5(a)	23.0	10.7(b)
Three limb (sum)	21.1	5.3(a)	25.5	9.2(b)	31.6	11.4(c)
Three trunk (sum)	37.5	12.4(a)	44.3	17.1(b)	58.2	25.3(c)
Six sites (sum)	58.6	17.7(a)	69.8	25.4(b)	89.8	35.4(c)

Note: Groups sharing a letter in common are not significantly different at $p < .05$; Tests of statistical significance are based on log-transformed values.

were recalled or standard recipes, reconstructed from interviews with selected informants, were used. Food composition was computed from food tables published by the USDA and a variety of other sources. A total of 296 recalls were collected from 105 men, with 86 percent providing the three recalls and the remaining men providing one or two. Pelletier also recorded the weekday/weekend reports so that the influences of Sunday feasting could also be estimated.

Caloric intake

The estimated levels of energy intake for these Western Samoan men are shown in Table 12.2. The caloric intakes of active workers and villagers are comparable and are higher than reported by sedentary workers, reflecting an expected differential due to daily activity. The level of consumption in excess of 4000 Kcal reported by these active Western Samoans is comparable with that recorded for very active individuals elsewhere (Weiner 1977).

When the caloric intake of Sunday is compared to that reported for weekdays, there is a clear tendency toward a higher consumption on Sundays (Table 12.3). While the tendency is most pronounced in sedentary workers and in villagers, it is also evident in active workers. The reported excess intake on Sunday over weekdays—2300 Kcal for sedentary workers and 1500 Kcal for villagers—may be a significant factor in accumulation of fat in some individuals, specifically the sedentary workers who show the greatest increase on Sunday are also those with the greatest thickness of skinfolds (Table 12.1).

Macro- and micronutrients

The composition of the diet in the three Western Samoan samples is given in Table 12.2. The overall intake of macronutrients as a percent of total calories is approximately 10 percent protein, 60 percent carbohydrate, and 30 percent fat irrespective of location. Most fat intake is saturated (18% of total calories), while unsaturated fat is only 1.5 percent. Overall there are minimal nutrient differences between villagers and active workers, although their vitamin A and C intakes are somewhat lower. Sedentary workers show a higher protein and sodium intake with lower potassium intake (Pelletier 1985). Villagers exceed the others in iron intake. When weekday/Sunday diets were compared, the higher intake of two traditional foods—taro and coconut—was evident in all groups. This resulted in a high intake of saturated fat (coconut oil), higher iron and, among employed men, lowered sodium intake. As a rule the Sunday intakes were more similar among groups than weekday intakes.

These observations can be summarized as follows:

1. Caloric intake is higher among active workers and villagers than among sedentary workers.
2. There are no significant differences in macronutrient composition of

Table 12.2 Summary of Average Nutrients in Diet of Samoan Adults

	RDA[a]	Vil[b]	Act	Sed	Manu'a A	Manu'a B	Tutuila	Haw A	Haw B
N		94	159	226	16	15	91	28	11
Men									
Energy (Kcal)	2700	4192	4300	3794	3110	3192	2671	3153	2320
Protein (gm)	56	82	83	70	117	114	108	133	89
Fat (gm)	—	130	138	115	104	81	89	99	60
CHO (gm)	—	599	650	473	361	534	352	389	345
Calcium (mg)	800	725	763	705	1108	1875	795	763	718
Iron (mg)	10	37	27	24	27	23	20	22	16
Vitamin A (IU)	5000	3327	5371	5765	5218	6537	4762	4017	4739
Vitamin C (IU)	60	64	85	92	187	462	210	187	175
Riboflavin (mg)	1.6	1.0	1.3	1.3	1.9	1.7	1.7	1.9	1.3
Thiamine (mg)	1.4	1.6	1.3	1.3	1.3	2.1	1.4	1.7	1.2
N					21	27	163	35	17
Women									
Energy	2000				2429	3095	2457	2324	2055
Protein	44				103	105	102	103	93
Fat	—				101	80	81	78	69
CHO					280	514	326	297	271
Calcium	800				845	1876	795	699	674
Iron	18				22	21	18	23	15
Vitamin A	4000				2588	4609	4606	3550	4533
Vitamin C	45				148	305	210	162	135
Riboflavin	1.2				1.3	1.8	1.5	1.4	1.4
Thiamine	1.0				1.0	1.8	1.3	1.2	1.1

[a] After Carroll et al. 1983.
[b] After Pelletier 1984: Vil—villagers; Act—active workers; Sed—sedentary workers. Manu'a A, Haw A — Manu'a and Hawaii samples of Brown et al. Manu'a B, Tutuila, Haw B — Manu'a, Tutuila, and Hawaii samples of Bindon (1982).

Table 12.3 Survey Samples: Total Daily Energy Intake (Kcal) on Sundays, Weekdays, and Weekly Average[a]

	Villagers (N = 43)			Active workers (N = 72)			Sedentary workers (N = 86)		
	Mean	SD	Median	Mean	SD	Median	Mean	SD	Median
Sundays[b]	5294	1755	4937	4351	1084	4575	5693	2489	5093
Weekdays[c]	3859	1669	3560	4263	1527	4121	3326	1199	3247
Weekly average[d]	4064	—	3756	4301	—	4186	3664	—	3511

[a] Sunday and weekday samples sizes are, respectively, 10 and 33 for villagers, 10 and 62 for active workers, and 17 and 69 for sedentary workers. Statistical tests are based on Wilcoxon two-group analysis of variance (two-tailed test).

[b] The difference between Sundays and weekdays is significant in the village and sedentary worker groups ($p < .05$).

[c] Sedentary worker median is less than that of active workers ($p < .05$) and villagers ($p = .08$).

[d] Weekly average was calculated as the weighted average of the Sunday and weekday means or medians, with weights of 1/7 Sundays and 6/7 weekdays.

the diet between villagers and active workers, but there are several differences between these and sedentary workers. Such differences are not large and not statistically significant.

3. Changes in diet on Sundays has a large influence on energy intake but only slight influence on the composition of the diet.

Household food economics

To determine the economics of food use among these three groups a list of major imported staple foods was compiled from questionnaire responses. The staples on this list represent foods most frequently purchased by Samoans. The estimated number of calories provided to individual households per week from items on this list of staples was then computed and is summarized in Table 12.4. The table is further subdivided to show the number of calories consumed of each staple item by household and the percentage received as remittance from other households. While this table deals only with items on the list of staples it does provide an insight into rural and urban food economics. As is seen in Table 12.4 the village households derive a significantly smaller portion of their calories from these staple items (on either a per capita or absolute basis) as compared to the households of employed men; that is, the villagers are more reliant upon foods that are locally produced. The employed households purchase all of these staple foods (97%), while villagers received one-third by remittance (31%). This pattern of remittances describes a difference between the respective economies. Even though less cash is available to villagers, substantial quantities of purchased foods are consumed. Pelletier found that access to purchased foods through remittance offsets the urban-to-rural differential in available cash and, as a result of this pattern of remittance, villagers actually have buying power comparable to their employed neighbors. While villagers gain imported items—sugar, flour, bread, rice, and meat—the employed households received their major remittances in the form of traditional foodstuffs from rural relatives. When questioned about the pattern of remittances the villagers did not report receiving any traditional foods, while the employed men reported receiving contributions in that form of 300 to 400 Kcal/day. The system of food remittances then is a reciprocal relationship providing village households with purchased foods and employed households with traditional foods. The system thus reduces the absolute cash requirements for rural villagers and provides employed families with traditional foods.

Pelletier (1984) also calculated correlation coefficients between measures of body size (weight, BMI, skinfold thickness) and measures of energy intake and macronutrient intake in the diet. These did not reveal any consistent relationship. When subgroup averages were compared, a general pattern was evident. The villagers and active workers had higher levels of energy intake, higher levels of expenditure, and lower measures of body fat than did the sedentary workers. This was true regardless of

Table 12.4 Survey Samples: Estimated Caloric Value of Staple Food Items Reportedly Used by the Households

	Villagers			Active workers			Sedentary workers		
	Kcal[a,b]	Rem (%)	Total (%)	Kcal	Rem (%)	Total (%)	Kcal	Rem (%)	Total (%)
Sugar	16 675	27.8	42.5	22 785	2.4	21.1	20 043	2.8	17.5
Flour	3 316	19.6	8.5	10 215	0.0	9.5	9 391	1.3	8.2
Rice	3 314	30.3	8.5	7 173	5.3	6.7	8 434	0.0	7.4
Bread	4 240	41.5	10.8	25 068	0.0	23.3	26 218	0.0	22.9
Fish	878	0.0	2.2	2830	14.3	2.6	3 253	12.8	2.8
Can fish and meat	9 158	43.0	23.3	33 976	5.4	31.5	40 970	3.9	35.9
Lard	1 656	10.4	4.2	5 726	0.0	5.3	4 959	0.0	4.3
Total	39 327	31.0	100.0	107 774	2.9	100.0	114 296	2.4	100.0

Calories per Capita per Day from Carbohydrate Staples, Protein, and Fat Staples and Total Staples[b]

	Mean	Mean	Mean
Carbohydrate	374	957	1091
Protein/fat	165	462	647
Total	539	1419	1738

Source: After Pelletier (1984:134).

[a] "Kcal" indicates mean calories per household per week. "Rem (%)" indicates the percent of each food's calories remitted to the household from relatives living elsewhere. "Total (%)" indicates the percent of staple calories provided by each food.

[b] Except for sugar, villagers, active, or sedentary workers (*p* < .05). Statistical tests were based on Wilcoxon two-group analysis of variance (two-tailed test).

the degree of leisure time activity of the sedentary workers. Thus while intragroup correlations failed to show an association between energy intake, activity, and body fatness, the intergroup comparisons do show a good association between intake and expenditure.

Pelletier concludes:

> Urban employment is associated with energy intake which generally corresponds to levels of energy expenditure during the week, but adherence to the traditional practice of feasting on Sundays results in an energy excess which does not appear to be compensated during the week, as was found among the village agriculturalists. By contrast, the macronutrient composition of the diet does not show large variations with residence and occupation in this age group of men. (1984:319)

Studies in American Samoa and Hawaii

Weighed intake studies

Brown et al. (1984) have described a study of food consumption for two Samoan populations in order to investigate the relationship of diet and migration. One population was resident on the island of Ta'u, one of the outlying Manu'an group. The other sample was of the Samoan migrants living on Oahu, Hawaii, and is representative of several elements of that migrant population.

The Ta'u sample consisted of 120 adults and children of both sexes, residents of the village of Fitiuta and represent 24 percent of the permanent residents of that community. Ta'u is considered by Samoans to be one of the most traditional areas of American Samoa and is geographically removed from the main island of Tutuila. A close inspection of Ta'u shows that many aspects of Americanization and modernization have already arrived. At the time of the study there was a paved road, a few trucks, electricity, air service, and several sundry stores selling imported foods. Despite such evidence of change there is still a strong element of traditional Samoan subsistence agriculture and fishing (see Gage, Chapter 2). At the time of this study, the men of Ta'u were largely involved in traditional agriculture and fishing, while the women seemed more involved in an American life-style (Hornick 1979). The sample from Ta'u is native Samoan in many respects, but also well along on the path toward modernization.

The other sample was drawn from a list of participants in a large survey of Samoan migrants living on Oahu, Hawaii (Hanna and Baker 1979). The 162 adults and children were selected from the survey population to represent a sample of residents from the two major migrant communities on Oahu. About one-half were from a rural community on the north shore of the island, which is a rural Samoan enclave. The other migrant community was resident in public housing areas of Honolulu and were not residentially segregated. See Hecht et al. (Chapter 3) for a complete description. All of the migrant adults and most of the children were born in Samoa.

Identical measurements and procedures were employed in the study of each population. The investigator and assistants arrived early in the morning, before breakfast, and remained throughout the day until after the last meal in the evening. During this time all food consumed by each individual was weighed. Food was weighed at all meals and if the individual traveled away from home, the investigator or an assistant went along. All snacks were included, and in some cases candy wrappers or fast food containers were retained as evidence of food consumption. If an individual was away from home without an observer present, an immediate recall was requested upon their return. School children were visited at school for meals. On Ta'u, where men were working on distant plantations, portions taken to work were weighed prior to departure and any leftover measured upon return. For each subject of the study, food consumption was measured on two different days, so that from 2 to 5 days were required to collect data from the selected members of an entire family.

The quantity and quality of food consumed was analyzed into components by a computer program based upon published food tables. The USDA *Agricultural Handbook No. 8* was the primary source of nutrient information (U. S. Dept. of Agriculture 1982). The items in this table were supplemented and replaced by published analyses of Pacific foods where appropriate. In the case of complex dishes with several ingredients, a recipe was collected at the time of preparation and nutrient values were assigned on the basis of quantity consumed by each individual. Thus in a dish composed of three ingredients, each of five consumers were assumed to eat one-fifth of each of the ingredients.

Anthropometric data by age, sex, and location are provided in Table 12.5. There were no significant differences in height between the Ta'u and Hawaii adult samples, and the Samoans do not appreciably differ from U.S. mean values for age and sex categories (Table 12.5). For children there are minor differences. Ta'u girls are generally shorter than their Hawaiian counterparts, and boys from the two sites differ in a nonconsistent fashion. Samoan residents in Hawaii were generally heavier than the American Samoan sample, a phenomenon previously reported (Baker and Hanna 1981; McGarvey and Baker 1979). As can be seen by comparing adult values presented in Table 12.6 with published RDA values (Adams and Richardson 1981), the Samoans in Ta'u and Hawaii have consistently higher than RDA values calorie intake. This high level of energy intake is most pronounced in older women and least in older men, regardless of location. Daily energy intakes for natives and migrants are similar within sex and age groups, indicating that migration is not followed by increased caloric intake in spite of the increased availability of new and nontraditional foods in the new setting. Adult

Table 12.5 Heights and Weights of Ta'u and Oahu Samoans Compared to U. S. Norms

Age	Height (mm)					Weight (kg)		
	U.S.[a]	Ta'u	N	Oahu	N	U.S.	Ta'u	Oahu
			Male					
7-10	1321	1346	17	1321	26	28.1	30.1	38.1
11-14	1575	1448	11	1524	12	44.9	39.5	52.6
15-18	1753	—		1702	3	65.8	—	87.1
19-22	1753	1778	3	1753	4	72.1	83.9	85.7
23-50	1727	1702	16	1702	28	76.2	84.8	94.8
51+	1727	1676	13	1676	9	72.1	72.1	83.9
Total N			60		83			
			Female					
7-10	1321	1321	18	1347	16	28.1	30.1	35.8
11-14	1575	1448	7	1499	16	45.8	42.6	50.1
15-18	1626	1524	2	1600	1	54.4	55.3	64.4
19-22	1626	1600	2	1600	4	59.9	72.1	70.1
23-50	1600	1600	21	1575	35	70.3	87.1	95.7
51+	1575	1575	10	1575	7	68.9	78.5	81.6
Total N			60		79			

[a] U. S. values from Adams and Richardson 1981, and Fulwood, et al. 1981.

Table 12.6 Daily Energy Intake by Sex and Age (Kcal)

Age	Ta'u	N	Oahu	N	U.S. RDA[a]	U.S.[b]	Age[b]
			Male				
7-10	2344	17	2612	27	2400	2183	(9-11)
11-14	2542	11	2642	12	2700	2430	(12-14)
15-18	—	0	3425	3	2800	2817	(15-17)
19-22	2482	3	3007	4	2900	3040	(18-24)
23-50	3110	16	3153	28	2700	2506	(25-54)
51+	2682	13	2667	9	2400	1950	(55+)
			Female				
7-10	2004	18	2082	16	2400	1857	(9-11)
11-14	2485	7	2286	16	2200	1813	(12-14)
15-18	2224	2	1713	1	2100	1731	(15-17)
19-22	1866	2	1993	4	2100	1687	(18-24)
23-50	2429	21	2324	35	2000	1553	(25-54)
51+	2410	10	2352	7	1800	1348	(55+)

[a] U. S. recommended daily intake values (Adams and Richardson 1981).
[b] Estimates of daily intake for U. S. males and females of similar age, after Carroll et al. 1983.

Samoan mean intake values compare with the 75th percentile for U.S. males and the 90th percentile for U.S. females (Adams and Richardson 1981). Note that estimates used in the comparative tables for the U.S. sample are for similar, but not identical age groups, and that the mean values for U.S. dietary intake are based upon a large dietary survey by questionnaire (Carroll et al. 1983). We recognize that RDA values for U.S. whites may not be strictly applied to Samoans (Beaton 1985), however, since Samoans are of similar stature (Table 12.5), these do provide a useful reference.

The caloric intake of the adult sample was divided on the basis of Body Mass Index (BMI = kg wt ÷ m² ht), which is a convenient, easily derived index that correlates well with body fat (Bray 1979) and has been shown useful for Polynesian populations (Evans and Prior 1969). The index for normal body build ranges from 20 to 26. An index of 30 is equal to about 30 percent above desired weight and is associated with a 25 to 33 percent increased risk to life (Bray 1979). The following categories were thus assigned: 20 to 26 normal, 27 to 30 overweight, over 30 obese. Using these categories as criteria, there were no major differences in either caloric consumption or in blood pressures between categories. This is in agreement with the work of Pelletier previously discussed.

Dietary contents
A summary of intake of protein, carbohydrate, and fat by age and sex is given in Table 12.7. All age groups of both sexes receive an excess of USRDA levels of these nutrients, regardless of place of residence. There seems to be no clear variation in the residential pattern of nutrient consumption.

Micronutrients
Mean intake levels for Oahu and Ta'u in all age and sex groups were at or above RDA for thiamin, riboflavin, niacin, and vitamin C, except for low caloric intake subsamples. There were some indications of potential deficiencies in calcium, vitamin A, and iron (Table 12.2).

A reduced calcium intake was particularly evident in young adults (19-22 years) at all survey locations, except men on Oahu (Table 12.8). The vitamin A deficit intake pattern was ambiguous. Iron intake was somewhat reduced in females, which was most evident in young women and adolescent girls.

With respect to dietary components and micronutrient intake as measured by this study, there were no major deficiencies nor did migration have an important impact upon dietary intake.

Food selection
The 10 most frequently selected food items found in the various investigations are listed in Table 12.9. This total percent of the diet contributed by these items in the Brown et al. (1984) study suggest that

Table 12.7 Protein, Carbohydrate, and Fat Intake by Age and Sex

Mean Dietary Protein (gm)

	Males			Females		
Age	Manu'a	Hawaii	U.S.	Manu'a	Hawaii	U.S.[a]
7-10	78	94	81	68	75	67
11-14	93	118	90	72	79	65
15-18	—	122	111	87	51	63
19-22	118	77	56	65	77	64
23-50	117	133	98	103	103	61
51+	115	112	78	105	83	53

Mean Dietary Fat (gm)

	Males			Females		
Age	Manu'a	Hawaii	U.S.	Manu'a	Hawaii	U.S.[a]
7-10	90	98	88	70	67	75
11-14	101	105	99	94	78	74
15-18	—	162	116	108	49	73
19-22	46	120	124	91	61	64
23-50	104	99	104	101	78	65
51+	109	84	80	101	74	53

Mean Dietary Carbohydrate (gm)

	Males			Females		
Age	Manu'a	Hawaii	U.S.	Manu'a	Hawaii	U.S.[a]
7-10	312	352	276	284	266	283
11-14	311	340	299	341	326	244
15-18	—	460	330	222	268	205
19-22	297	351	321	205	267	297
23-50	361	389	261	280	297	172
51+	310	365	210	279	363	159

[a] Adams and Richardson 1981.

Table 12.8 Mean Calcium Intake (mg)

	Males			Females		
Age	Manu'a	Hawaii	U.S.	Manu'a	Hawaii	U.S.[a]
7-10	895	1203	800	1106	802	800
11-14	1197	1324	1200	897	1014	1200
15-18	—	1672	1200	1467	268	1200
19-22	606	1134	800	550	549	800
23-50	1108	762	800	845	669	800
51+	966	711	800	1033	906	800

[a] Adams and Richardson 1981.

Table 12.9 The 10 Foods Most Frequently Selected by the Various Samoan Populations and Their Contribution to Total Food Intake

Villagers[a] food	%	Active workers[a] food	%	Sedentary[a] food	%	Manu'a A[b] food	%	Manu'a B[c] food	%	Tutuila[c] food	%	Hawaii A[b] food	%	Hawaii B[c] food	%
Taro	39	Taro	24	Taro	23	Banana	7	Banana	14	Banana	13	Rice	7	Banana	12
Palusami	9	Beverage	9	Beverage	9	Sugar	2	Breadfruit	28	Breadfruit	5	Sugar	2	Breadfruit	3
Pork	6	Banana	9	Mutton	8	Oil	4	Fresh fish	6	Fresh fish	4	Milk	3	Fresh fish	3
Beverage	6	Mutton	8	Palusami	7	Tin fish	5	Taro	4	Taro	6	Juice	2	Taro	3
Alcohol	6	Palusami	7	Pancake	6	Onion	1	Corned beef	6	Corned beef	3	Banana	2	Corned beef	2
Tin fish	6	Pancake	6	Bread	5	Tea	—	Cocoa	2	Cocoa	4	Bread	2	Cocoa	4
Drinking nuts	5	Coconut cream	5	Banana	5	Rice	5	Bread	5	Bread	7	Soda	2	Bread	5
Coconut cream	5	Bread	5	Pork	5	Tin beef	5	Butter	3	Butter	4	Oil	1	Butter	2
Fresh fish	4	Chicken back	5	Beef	5	Coconut cream	10	Sugar	3	Sugar	3	Taro	3	Sugar	4
Pastry	3	Tin fish	3	Rice	5	Breadfruit	4	Rice	2	Rice	4	Soy sauce	1	Rice	9
Totals	89		81		78		41		73		53		25		47

[a] Pelletier 1984.
[b] Manu'a and Hawaii samples of Brown et al. 1984.
[c] Manu'a, Tutuila, and Hawaii samples of Bindon 1982.

Ta'u Samoans had a less varied diet than migrants to Hawaii. Some 40 percent of the total Ta'u energy intake was supplied by these 10 items, while the 10 most frequently selected items yielded only 22 percent in Hawaii. The differential in item selection clearly reflects a traditional-to-modern diet continuum, and is in agreement with the effects of change already described by Bindon (1984b). Green banana was most frequently consumed in Manu'a, while rice was most popular in Hawaii. Coconut cream, a rich liquid produced from fresh coconut pulp, although utilized less frequently than green banana on Ta'u, provided the most calories for any single item, contributing a full 10 percent of the energy. Breadfruit was last on the list, contributing only 2 percent. In both communities, white sugar was the second most frequently selected item, providing about 2 percent of the energy. The prominence of milk, canned fruit drinks and soda, and white bread illustrate the Americanization of the diet in Hawaii.

This study has suggested that there are dietary changes with migration and modernization, but these are mostly in the variety of foods consumed. There was no appreciable change in caloric consumption nor was there any great change in the nutrient quality of the diet. It is noteworthy that caloric consumption was even higher than observed in contemporary middle-class Americans (Carroll et al. 1983).

Recall studies

A 24-hour recall study of Samoans in American Samoa and Hawaii was conducted by Bindon (1981, 1982, 1984b). Since this study has been fully described in the literature, only a brief summary is included here. The samples were drawn from three Samoan communities—Manu'a, Tutuila, American Samoan and Western Samoan migrants to Hawaii, yielding 42, 254, and 28 respondents, respectively, representing both adult men and women. The instrument employed was a 24-hour recall questionnaire administered as part of a series of questionnaires concerning various aspects of Samoan life. Respondents were requested to recall the types and amounts of foods consumed during the past 24-hour period preceding the interview and during the interview were asked prompting questions about beverages, snacks, and cooking methods. The responses on the questionnaire were coded and converted into macro- and micronutrients in a procedure similar to that described in the previous studies.

Analysis emphasized two aspects of Samoan dietary habit changes that resulted from modernization and migration (Bindon 1982), and an evaluation of the dietary status of the sample (Bindon 1984b). Bindon suggested that processes of modernization and migration lead to a substantial change in the variety of foods consumed. The number of choices was largest in Tutuila and least in Manu'a, with the Hawaiian migrants in an intermediate position. Manu'a was highest, Tutuila intermediate, and Hawaii the lowest with respect to the amount of food produced by the household as compared to the amount purchased; and with respect to the amount of traditional foods consumed as compared to Western foods. Thus, modernization followed a predicted pattern, with the greatest impact on the migrants to Hawaii and the least impact on the rural Manu'an residents. Bindon (1982) also provided an extensive list of items consumed by the three sample populations.

The Manu'a sample reported the greatest caloric intake, Tutuila intermediate, and Hawaiian migrants the lowest (Table 12.2). Indeed, the migrants were below USRDA values by several hundred calories per day. There was an associated problem with micro- and macronutrients that correlated with degree of modernization. Calcium, iron, and thiamine deficiencies were most pronounced among the migrants and least in Manu'a.

As a result, Bindon recommends that Samoans need more calcium, iron, thiamin, and riboflavin, while they should cut down on some specific types of Western foods—fatty meats, hot dogs, canned cured beef, etc. He also recommends an increase in physical activity. The general process of modernization and migration, he concluded, seems to have led to the following dietary changes: (1) an increase in the variety and quality of foods consumed, including a heavy reliance upon Western foods of questionable nutritional value; (2) a reduced reliance upon traditional Samoan foods, such as breadfruit, banana, taro, and coconut; (3) a reliance upon introduced foods for basic nutrition. These include rice, bread, fresh beef, and canned fish as new staple items (Bindon 1982).

Bindon (1982) and Brown et al. (1984) studied members of the same populations, Manu'a and immigrants to Hawaii. The resulting caloric and nutrient estimates have similarities and differences (Table 12.2), the differences being more evident in Hawaii than in Manu'a. Reasons for this differential are not clear—individual subjects and methods of data collection were different; however, Bindon's small sample from Hawaii (27 adults) may be a contributing factor. It is also possible that some of the differences may be socioeconomic, since Bindon's sample was drawn from a largely unemployed sample (Bindon, personal communication). There is reason to believe that these factors provided some limitation because of the small number of food choices recorded in his questionnaires. Only 81 different items were recorded in Hawaii, as compared to the 116 he recorded in Tutuila and the 191 recorded by Brown et al. in Hawaii (1984). Thus, it is possible that Bindon's small sample was not completely representative of the Samoans in Hawaii, especially in the variance in nutrients associated with food selection. A second factor contributing to the variation between the studies is the period of the breadfruit season during which the sample was collected. Bindon's data were collected during the productive portion of the season, so 28 percent of the calories in Manu'a were from breadfruit. Brown et al., on the other hand, found breadfruit supplying only 2 percent of the calories during the off-season. Thus the presence or absence of this major food item would be expected to modify the dietary composition and probably explains some of the difference between the Manu'an samples.

Discussion

Before reviewing the results of these nutritional studies there are several
limitations and qualifications that must be considered. First, the investigations described have not systematically sampled the various populations, but rather subjects have been selected casually or as volunteers for
other kinds of investigations. Thus we cannot be certain as to their
representiveness in any of the populations sampled. Second, the survey
in Western Samoa did not include women or children and that in
American Samoa did not include children, so we do not know the nutritional status of those segments of the population. Finally, the various
studies have been undertaken by different investigators using different
techniques, hence they are not easily comparable. Brown et al. (1984)
employed a technique for quantitatively weighing food over 2-day periods;
Pelletier (1984) and Bindon (1981), used a 24-hour recall technique.
There are bound to be errors and biases built into each that makes comparison of the results difficult. Such problems are inherent in all such
studies and cannot be avoided.

The major aims of our dietary research have been twofold; one to
document the influences of modernization on the diet of Samoans, and
second, to relate diet to morbidity and the changing health risks observed
in the Samoan populations studied. We can now make some contributions
in both areas.

Modernization, migration and diet

The influence of modernization on the diet of Samoans has been both long-
standing and pervasive. While the late nineteenth century diet seems to
have been largely breadfruit, bananas, yams, taro, coconut, and whatever
reef resources were available (Bindon 1982), during the first two decades
of this century there was an influx of imported foods so that by the 1920s
Richie (1927) noted that large amounts of tinned meats, sugar, and flour
were being consumed as supplements to the regular Samoan diet. This
consumption of imported items continued and nontraditional foods have
penetrated into the more remote rural areas of American and Western
Samoa (Pelletier 1984; Wilkins 1965; Holmes 1954; Malcolm 1954). A
list of the more frequently consumed items is given in Table 12.9. Most
of these foods seem to have been part of the Samoan diet for the better
part of this century and were probably part of the normal daily diet of all
of the individuals discussed in this volume.

Given the nature of the precontact diet and associated food restrictions
the import of several items are of particular importance with respect to
nutritional status of certain parts of the population. In traditional Samoa
pork, chicken, and ocean fish were not readily available and their
consumption was limited. These items were reserved for men and were

usually consumed only on special occasions. With the importation of tinned meats more protein became available and it did not fall under the traditional pattern of distribution. Thus more protein was now consumed by women and children (Malcolm 1954). Similarly, the importation of tinned and dry milk probably had a major impact on the health of postweaning infants (Holmes 1954), perhaps reducing the levels of infant malnutrition that were frequently reported. Finally, the recent increase in the importation of traditional foods such as ocean fish and fresh meats probably modified indigenous consumption patterns to a significant degree.

The amount of energy contributed by these imported items seems to be variable with estimates ranging from about 17 percent in Manu'a to near 20 percent in Western Samoa to more than 66 percent in Hawaii. It seems likely that availability of imported foods and the energy that they contribute should be dependent upon the degree of participation of the particular household in the cash economy as well as upon the degree of modernization of the village. Economic studies (Lockwood 1971) suggest that families with members working for wages or receiving significant remittances in cash are more actively involved in the modern economy, therefore have greater access to imported foods. Pelletier's study of food remittances has shown that this is not necessarily so. In the local exchange system in which imported foods become available by remittance, some families do not require cash; indeed he found that imported food proportions of rural and urban diets were similar regardless of the presence of wage earners in the household. This may assure universal access to the imported foods that have become staples.

Another aspect of modernization has been an increase in the diversity of foods available to Samoans. The precontact diet was very limited in choice due to the small number of foods available and the seasonality of some of the major crops. The introduction of sugar, tinned meat, and fish provided a larger degree of choice. In contemporary rural Samoa the variety of items consumed ranges from 64 to 101 in Manu'a, to 116 in Tutuila, to 191 in Hawaii (Brown et al. 1984; Bindon 1982). While the specific identification of food items may have varied by investigator, these data also support an increased variety of choice according to location.

Table 12.9 provides a different view of the dietary diversity following modernization. In Western Samoa the 10 most frequently selected items contribute about 90 percent of the total calories consumed by villagers and about 80 percent of those consumed in the Apia area. In American Samoa the contribution of the 10 items varies from 40 to 72 percent, while on the island of Tutuila, it is about 54 percent. For the Hawaiian migrants the 10 most frequently selected items account for only 22 percent of total calories. The influence of modernization on food diversity is quite evident, but so too is the result of migration. Comparing Tutuila with Hawaii shows a definite Americanization of the diet, with breadfruit and corned beef replaced by milk, canned beverages, and soy sauce. The Hawaiian sample provides a contrast to Western Samoa where taro and other native dishes are major staples.

The significance of the increase in food choices is not clear; however, laboratory studies (Himms-Hagen 1984; Rothwell and Stock 1979) have shown that an increase in available dietary choices leads to an increase in food consumption. Under these controlled conditions a cafeteria diet has also been linked to an increase in body fat, so that the increase in fat observed among Samoans may in part be tied to increased dietary choice.

On the other side of the energy equation, modernization of occupation seems to lead to an absolute decrease in energy intake. Sedentary workers consume fewer calories per day than villagers or active workers (Table 12.2). This caloric reduction seems related to a reduced level of physical activity in the Samoan samples. While caloric reduction has also been observed in other Pacific populations exposed to modernization, including Micronesians (Labarthe et al. 1973; Henkin et al. 1970), Fijians (Coyne et al. 1984), and French Polynesians (Darlu 1984), Samoans seem unique in that there do not seem to have been substantial changes in the quality of their diet. In other studies there were substantial nutritional changes such as less fiber, less unsaturated fat, and less protein (Darlu et al. 1984); a change to imported items with less nutritive value that also reduced vitamin and nutrient intake (Henkin and Dickinson 1972); and a substantial increase in red meat, dairy products, and poultry at the expense of fish and complex carbohydrates (Henkin et al. 1970). Our observations as well as those of Wilkins (1965) suggest that the reduction in consumption has been nutritionally less significant.

The process of migration seems to have much less influence on the diet than the process of modernization. There was little difference in dietary quality or quantity when Manu'an and Hawaiian samples were compared in the Brown et al. study (1984), although the poorer Samoans in Hawaii studied by Bindon (1982) reported a diet that may be considered somewhat poorer in some aspects. The comparable quality diet reported by Brown et al. occurred in spite of the increased variety of foods utilized. We can only speculate as to how the quality of the diet has been maintained in the face of substantial cultural change, but the early arrival of European foods is probably involved. The list of Western foods given in Table 12.10 is compatible with the vast majority of food items on the shelves of supermarkets in Honolulu or San Francisco so that Samoans have regularly consumed many of these items prior to migration.

Health and diet

The major health risks and problems that we have found to characterize Samoans are also closely tied to the process of modernization. These include the interrelated problems of obesity, hypertension, and diabetes,

Table 12.10 Foods Currently Consumed by Samoans[a]

Traditional foods	Western foods
Taro	Bread
Breadfruit	Rice
Banana	Flour
Coconut	Sugar
Fish	Butter
Shellfish	Milk and Cream
Eggs	Tinned meat and fish
Greens	Fresh beef and pork
Papayas	Mutton
Fresh fruits	Canned fruits
Pork	Cocoa
Poultry	Lard
	Beverages (soda and fruit drinks)
	Alcoholic drinks
	Vegetable oil

[a] This list is taken from foods appearing in the surveys summarized in this paper, as well as from Wilkins (1965), Holmes (1954), and Malcolm (1954).

each of which has been in past studies partially associated with diet (Morgen and Caan 1979).

For the Samoan population obesity seems to play a central role. The prevalence of obesity is high, it seems closely related to hypertension (Chapter 15) and may also be correlated to high levels of death from diabetes (Chapter 5). The dietary studies of Pelletier (1984), Bindon (1984b), and Brown et al. (1984) have attempted to link daily food consumption to obesity, but have not been successful. Blood pressure, BMI, and individual dietary intakes have failed to show any consistent relationships. As noted, the underlying relationships are probably weak, and cross-sectional studies are unable to detect them due to large individual variation in daily food intake. A longitudinal study might demonstrate such linkages, but it is also possible that the linkages that relate to developing obesity in Samoans are via energy expenditure rather than intake. Pelletier (1985) and Bindon (1981) have both suggested that levels of physical activity are more important than energy intake in Samoan obesity. According to this concept, levels of food intake, levels of physical fitness, and levels of physical activity are closely parallel such that obese sedentary workers have the lowest levels of fitness, lowest levels of energy expenditure, and are fattest. Other investigators in the Pacific have made similar suggestions (Coyne et al. 1984; Darlu et al. 1984; Pollock 1974; Labarthe et al. 1973) and the observations of Brown et al. (1984) lend further support to this concept. The men from Manu'a who were actively engaged in plantation agriculture at the time of the study averaged lower BMI than the migrants to Hawaii who were not so actively employed. In both sites the women were less active and resembled each other more closely in terms of BMI and energy intake.

Another important observation concerns the weekday/Sunday differential in food intake. Pelletier (1985) estimates that excessive Sunday food consumption is balanced by weekday energy expenditures in villagers but not in employed men. The villagers may actually be in a daily caloric deficit during the week, while the others are more or less in equilibrium. If this is not a methodological artifact, the obesity that comes with modernization may be partly the result of an interplay between traditional Samoan weekend feasting and a reduced daily activity level during the week. Consumption of food and expenditure of energy are matched during the week, but the excessive consumption on Sundays gradually accumulates in the absence of a slight caloric deficit during the week. This hypothesis could be tested if groups of Samoans who do not engage in Sunday feasting are identified.

Diabetes can also be linked to obesity, and in Samoans its increasing prevalence parallels the increasing obesity in modernizing populations (Baker and Crews, see Chapter 5; Hornick and Hanna 1982). At the physiological level resistance to insulin activity is directly related to a down regulation of insulin receptors in the obese, and a reduction in weight restores glucose tolerance (Arky 1982). Thus, a population with a high level of obesity may also have a high level of diabetes. In this context Pelletier's argument for physical activity as a major factor takes on additional meaning, for physical activity and insulin activity are directly related, regardless of the level of obesity. Laboratory studies have shown that forced bed rest will reduce insulin activity and reduce glucose tolerance in healthy, active men (Arky 1982).

One dietary health factor that we have not investigated is possible malnutrition in infants. Richie (1927) reported high levels of marasmus in Samoan infants, reaching 30 percent in some areas. Other observers have commented on infant malnutrition in some detail (Lichton et al. 1983; Neave 1969; Wilkins 1965; Holmes 1954; Malcolm 1954). This is typically reported to occur at the time of weaning when the child is placed on a high-starch diet with large amounts of sugar and fruit but little protein. The symptoms include low weight for age, dermatosis, reduced serum protein levels, edema, and a host of associated problems (Lichton et al. 1983; Neave 1969; Wilkins 1965). Our nutritional studies did not include this age group, but in light of the results presented in Chapter 10, this reported malnutrition deserves further investigation.

Chapter 13

Work in Contemporary and Traditional Samoa

LAWRENCE P. GREKSA
DAVID L. PELLETIER
TIMOTHY B. GAGE

Work can be defined as "those activities performed by a group in order to meet its social and biological requirements" (Thomas 1975:59). All work thus ultimately involves an interaction between the biological work capacities of workers and the requirements of those essential tasks that they must perform in order to ensure that a sufficient quantity of essential resources are acquired. The utility of this definition lies in the explicit linking of these domains. Traditionally, each would be separately evaluated against arbitrary standards. For example, an individual might be evaluated as physically unfit, while a task might be classified as being of moderate intensity. An important additional question, however, is whether an individual, regardless of absolute level of physical fitness, is physiologically capable of performing specific essential tasks. This interrelationship between physical work capacity and work requirements will be termed *work capability*. Thus, a physically fit worker could potentially have a low work capability if his tasks were highly strenuous, while a relatively unfit individual could have a satisfactory work capability if his tasks were of low intensity. Within the adaptive domain of work, any change that acts to decrease work capability, either by modifying work capacity, work requirements, or both, can be considered maladaptive, while any change that acts to increase work capability can be considered adaptive. Given this theoretical framework, the ultimate aim of this study is to evaluate the effect of the complex biological and social changes associated with exposure to a modern way of life on the work capacity, work requirements, and work capability, and thus biological fitness, of Samoans.

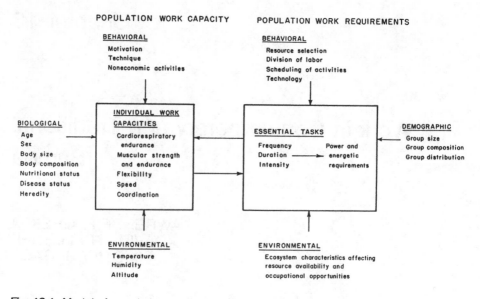

POPULATION WORK CAPACITY   POPULATION WORK REQUIREMENTS

BEHAVIORAL   BEHAVIORAL
 Motivation   Resource selection
 Technique   Division of labor
 Noneconomic activities   Scheduling of activities
                                            Technology

BIOLOGICAL   INDIVIDUAL WORK   ESSENTIAL TASKS   DEMOGRAPHIC
 Age   CAPACITIES   Group size
 Sex   Cardiorespiratory   Frequency   Power and   Group composition
 Body size   endurance   Duration →   energetic   Group distribution
 Body composition   Muscular strength   Intensity   requirements
 Nutritional status and endurance
 Disease status   Flexibility
 Heredity   Speed
                   Coordination

ENVIRONMENTAL   ENVIRONMENTAL
 Temperature   Ecosystem characteristics affecting
 Humidity   resource availability and
 Altitude   occupational opportunities

Fig. 13.1 Model of population work capacity. (Adapted from Thomas 1975)

MODEL OF WORK IN A POPULATION CONTEXT

Many of the factors affecting work capacity and work requirements, and
by extension work capability, are summarized in Fig. 13.1 in a population
model adapted from one originally developed by Thomas (1975). There
are two primary units to the model. The first consists of population work
capacity, or the sum of the work capacities of all individuals in the
population and the variables affecting their work capacities. It should be
emphasized that work capacity is defined as the *physiological* capacity to
perform work. The second component of the model consists of those tasks
that must be performed if a group is to acquire the essential resources
that are needed to meet their biological and social requirements and the
factors that either modify the physiological requirements of these tasks or
place constraints on which individuals will perform them. Each of these
units of the model are discussed in greater detail below.

Physical work capacity is a complex entity, consisting minimally of
cardiorespiratory endurance, muscular strength and endurance, flexibility,
speed, and coordination (Karvonen 1974). This chapter focuses on only
one of these components, cardiorespiratory endurance, for pragmatic
reasons relating to the availability of measurement techniques and the
need to interrelate measures of work capacity and work requirements into
indices of work capability. Although not the sole determinant of work
capacity, cardiorespiratory endurance is generally an important, and often
the most important, determinant of an individual's ability to perform work
(Shephard 1980).

The best overall index of cardiorespiratory endurance is generally considered to be maximal oxygen consumption (VO_2max), or the maximum amount of oxygen that can be utilized by the body's muscles during maximal exercise involving large muscle groups (Buskirk and Taylor 1957). In order to control for body size while comparing work capacity between individuals, maximal oxygen consumption is generally expressed as a function of body weight (ml/kg/min). For some purposes, however, such as the assessment of the work capability of individuals, maximal oxygen consumption is usually expressed in absolute units (l/min). Since both methods of expressing maximal oxygen consumption are used in this report, we will use the following convention to distinguish between them. VO_2max will be used to refer to measures of absolute oxygen consumption [VO_2max (l/min)], while aerobic capacity will be used to refer to weight adjusted maximal oxygen consumptions [VO_2max (ml/kg/min)].

Since energy output is a function of oxygen utilization, VO_2max determines an individual's maximal energy output and thus sets an upper energy limit on the types of work that can be performed. Few individuals are required to work at maximal levels, of course, but a worker's VO_2max is still important since it determines the relative work intensity, or the ratio of submaximal to maximal VO_2, at which work is performed. The importance of this relationship is seen in the fact that the physiological strain associated with submaximal work is positively related to its relative work intensity. Thus, an individual with a low VO_2max will undergo greater physiological strain during the performance of a given task than will an individual with a high VO_2max. One consequence of this relationship is that the fitter individual can perform a given task for a longer period of time without the onset of fatigue (Astrand and Rodahl 1977). These relationships are used in a later section of this chapter as the basis for several indices of work capability.

Aerobic capacity, as are other indices of work capacity, is affected by a variety of factors. These factors have been classified into four broad categories in Fig. 13.1. First, there are numerous biological factors that affect aerobic capacity. For example, aerobic capacity generally decreases with age, is larger in males than females, is positively related to nutritional and disease status, is negatively related to body fatness, and may have a fairly high heritability (Spurr 1983; Bouchard et al. 1981; Shephard 1978; Astrand and Rodahl 1977; Edholm 1967). Second, aerobic capacity is affected by several environmental parameters, particularly temperature, humidity, and altitude (Astrand and Rodahl 1977). Third, there is a strong positive relationship between aerobic capacity and habitually performed activities, both economic and noneconomic (Shephard 1978 Astrand and Rodahl 1977). Finally, behavioral factors also influence work performance, by modulating work intensity for example (Spurr 1983; Edholm 1967).

The second component of the model presented in Fig. 13.1 consists of the variables affecting population work requirements, or the variety of tasks that must be performed if the essential resources that are needed to meet the social and biological requirements of a society are to be acquired. Essential tasks can be defined in terms of their frequency, intensity, and duration, the combined effects of which determine the power and energy requirements that must be met by workers.

The various parameters affecting work requirements are summarized in Fig. 13.1 under three general categories. Demographic parameters determine the minimal level of biological and social resources that are necessary for population maintenance and also set a limit on the number of individuals of specific age/sex groups who are available to perform essential tasks. Behavioral factors that influence work requirements include resource selection, the division of labor, the scheduling of activities, and technology. The latter is particularly important in a population undergoing social changes that include an increased reliance on modern technology, since such changes are generally associated with a decrease in work requirements (Edholm 1967). Finally, work requirements are also affected by environmental parameters. In self-sufficient agricultural societies these are primarily physical and biotic environmental factors that influence nutrient productivity, such as soil fertility and rainfall. The relevant environmental parameters are much more subtle and society-specific when one considers societies that are linked to the world economy. A critical environmental parameter in contemporary Samoa, for example, appears to be its geographic location. The deep water port on Tutuila and the proximity of American Samoa to tuna fishing areas were important stimuli for the development of tuna canneries and thus the specific types of occupations found in such factories. On the other hand, the relative isolation of Samoa from commercial centers has probably hindered other types of economic development.

The two major components of the model, work capacity and work requirements, have been discussed separately, both for convenience and because, for some purposes, they are usefully considered as separate entities. However, we are ultimately interested in examining these components at a higher level of analysis, or one at which the interaction between work capacity and work requirements, or what we have termed work capability, is the focus of interest. In particular, given that Samoans have undergone substantial socioeconomic changes in recent years and these changes have affected numerous determinants of both work capacity and work requirements, what are the combined effect of these changes on work capability? It is appropriate at this time to consider a possible criticism of our ultimately biological definition of work capability. Some would argue that actual work performance, and thus work capability, is determined more by behavioral than by biological factors (e.g., Neff 1985; Rivers and Payne 1982). We are not discounting the importance of behavioral factors; however, we contend that tasks do have requirements that can be defined in physiologically meaningful terms, that individuals do have a limited physiological capacity to perform work, and that the interaction between these two parameters is at least partially responsible for determining whether specific individuals are capable of performing them.

THE WORK OF PRECONTACT AND CONTEMPORARY SAMOANS

Before attempting to evaluate the effect of exposure to a modern way of life on the work capability of Samoans, it is necessary to describe its impact on the two components of work capability—work capacity and work requirements. As a first step in that direction, the work require- ments of Samoans living in three settings are described in the following section. First, we attempt to reconstruct the basic components of the work schedule of Samoa prior to the large-scale socioeconomic changes of recent decades, as our best estimate of work in precontact Samoa. Second, we describe the work pattern of contemporary American Samoa. This pattern is representative of the situation in American Samoa from the early 1960s on, when a U.S. government program of economic and technological aid to American Samoa was instituted. Finally, we describe the work patterns of Samoans working within the United States. The term United States will be used to refer to individuals working in one of the 50 states. When applicable and available, the energy requirements of activities are provided, standardized for a 65-kg man. Where possible, information is provided on the economic activities of both men and women, but usually detailed information is available only on men.

Work in precontact Samoa

Early explorers described Samoa as a land of plenty in which all were easily fed and with the only major hazards being an occasional hurricane or drought (Stair 1897; Turner 1884). Unfortunately, the early explorers did not generally describe traditional Samoan subsistence patterns in detail, making it necessary to utilize more recent historical and ethnographic data to reconstruct precontact work patterns. The extent to which such studies can be used for this purpose is, of course, open to question, as there are obvious limitations. For example, although several researchers have studied villages that were largely self-sufficient, even these villages purchased some nutrient resources. Also, some individuals now produce excess food for sale in the market and there are a few individuals specializing in animal husbandry. In addition, it is quite likely that the relatively few villages in which detailed studies of subsistence patterns have been conducted are not representative of all villages with respect to factors such as environmental and seasonal variation (described in Chapter 2) that can influence the extent to which different foods are utilized. This is particularly a problem for the data on the energy cost of various tasks, much of which is only available from one study (Gage

1982). Even with these limitations, however, the available data probably provide a reasonable model of the broad outlines of precontact work patterns. Based on comparisons of contemporary (Gage 1982; Holmes 1974) and early postcontact (Watters 1958; Kraemer 1902) descriptions of subsistence patterns, the only major change in agricultural practices seems to have been the substitution of steel for stone tools. The actual tasks that must be performed do not appear to have changed greatly. It therefore seems reasonable to use data extracted from studies of traditional villages that were largely self-sufficient to construct a general model of precontact work patterns. Although the pertinent data were derived from studies dating from the late nineteenth century to the recent past and may not be applicable to many contemporary villages, the ethnographic present is utilized in the following account of traditional subsistence practices for the sake of simplicity.

Traditional subsistence in Samoa is based primarily on fruit cultivation and bush-fallow agriculture and secondarily on the exploitation of marine resources (Holmes 1974; Kraemer 1902). The division of labor is based on sex and social status (Holmes 1974; Mead 1969). In general, women are responsible for activities centered in the village, such as household tasks, the production of mats and clothing, and the exploitation of reef and lagoon resources (Holmes 1974). Women in contemporary Samoa also prepare nontraditional foods and may weed plantations (Shore 1982). Men perform all other food production and preparation activities (Lockwood 1971). All males, except perhaps those of very high social status, participate to some extent in subsistence activities, but the most strenuous work appears to be performed by young (less than 30 years) untitled men, either as a group in the 'aumaga, the untitled men's association, or as individuals within their 'aiga (Holmes 1958; Keesing 1934b). Titled men, or matai, supervise the work of the untitled men, perhaps while actively participating in the work themselves, and are also sometimes craft specialists (Gage 1982; Holmes 1974).

A wide variety of plants are cultivated by Samoans, but the bulk of their traditional diet is provided by a few fruit and root crops. The primary staples are breadfruit, taro, and bananas, while coconuts, giant taro, and yams are secondary staples (Holmes 1974). Breadfruit, as in many other Polynesian societies, is the preferred food and the primary component of the diet during the approximate 6 months of the year when it is available. Taro is the primary component of the diet when breadfruit is not available (Holmes 1974; Watters 1958a).

Although fruit crops provide a significant proportion of the nutrient intake of Samoans, little work, none of which is strenuous, is required for their production or collection. As an indication of the minimal demands of fruit cultivation, the average 'aiga of about eight individuals in rural American Samoa invests a total of only about 30 min and 185 Kcal per day in the production and harvesting of breadfruit and bananas (Table

2.6). The same crops provided 66 percent of the caloric content of the diet during breadfruit season.

The root crops, on the other hand, not only provide, on average, a smaller proportion of daily caloric intake (maximum of 20% in rural American Samoa) but were considerably more costly in terms of both time and energy. For example, in the same rural American Samoan village noted above, the daily time and energy investments in taro were, on average, about three and four times as much, respectively, as for fruit cultivation (Table 2.6). Due to the distances of the plantations from villages and the steep slopes on which they were located, the trip to and from the plantation is probably the most strenuous *habitual* task associated with traditional agriculture, with an energy cost of about 7 Kcal/min on the trip to the plantation and about 4 Kcal/min on the return trip (Greksa 1980). The single most strenuous task associated with plantation agriculture is undoubtedly the clearing of a new plantation. Even with steel tools the energy cost of this work is about 10 Kcal/min, or heavy work by most criteria. However, since a plantation is abandoned only after 2 to 4 years (Farrell and Ward 1962; Watters 1958a), this strenuous work is infrequent. An average energy cost for all activities associated with plantation agriculture, weighted for the relative amount of time spent in each activity (Gage 1982), is about 7.1 Kcal/min if travel time is not included and 6.5 Kcal/min if travel time is included (Greksa 1980). It should be emphasized that these (and all following) energy costs are standardized for a 65-kg man.

The available data on time expenditure in Samoan traditional agriculture are summarized in Chapter 2. It was concluded that productive males in precontact Samoa probably needed to expend no more than 6-12 hours of work per week in agricultural labor. Even if one assumes a highly conservative estimate of 15-20 hours of agricultural labor per productive male per week, this is not a great deal of work. No data are available on the energy cost of marine exploitation in Samoa but few of the available strategies appear to have been strenuous or time consuming (Chapter 2; Lockwood 1971; Watters 1958a).

In summary, to the extent that the available data provide a valid model of precontact subsistence patterns, the work associated with traditional Samoan subsistence was occasionally strenuous but, on the whole, was neither time nor energy costly. A majority of the diet for at least half of the year was obtained in many villages from fruit crops, a food source with minimal time and energy requirements. Plantation agriculture, although more demanding than fruit crop cultivation, nevertheless required only a moderate energy and time investment by productive males. It should be remembered that the category *productive male* is a culturally, and not biologically, defined category. With respect to subsistence activities, particularly the more strenuous activities, this category consisted primarily of young adult males in traditional Samoa.

Work in contemporary American Samoa

The most detailed information on the diversity and distribution of
occupations in modern American Samoa at the time of the study was provided by a census conducted by the East-West Center in 1974 (Levin and
Pirie 1974). The following discussion is limited to adults 20 years and
older, excluding 85 adults whose ages could not be reliably determined
during the census.

There were 6379 adult males and 6482 adult females residing in
American Samoa in 1974. Of these, 74 percent of adult males and 36
percent of adult females were economically active. The distribution of
workers by age and sex is presented in Table 13.1. This pattern did not
differ greatly between districts and the most economically active individuals in both sexes are those between 30 and 49 years of age.

Only 31 females reported any involvement in agriculture in American
Samoa in 1974 and therefore our description of the distribution and
frequency of agriculturalists is restricted to males. Only 405 males, or 6
percent of all adult males, reported that they devoted any time to
agriculture (Table 13.2). This figure includes men who had a full-time
position in the wage labor economy and only worked part-time as agriculturalists. The proportion of active agriculturalists varied by district,
with the proportion being considerably higher in Manu'a, the most
traditional area of American Samoa, than in the three Tutuila districts.
In addition, there is a tendency of men over 40 years of age to be
somewhat more likely than younger men to work as agriculturalists. This
trend is, of course, in the opposite direction to that described for
precontact Samoa, perhaps reflecting a greater desire by older men to
maintain traditional values. Indirect evidence for this hypothesis is
provided by calculating the proportion of men in each age group who
worked solely as agriculturalists (4% of the adult male population) from
the 1974 census. In this case, there was less variation between age
groups in each district. For example, in Manu'a the proportions for the

Table 13.1 Percentage of Economically Active Individuals in American Samoa, by
Age and Sex

Age	Males	Females
20 - 29	68.3	42.9
30 - 39	86.6	46.2
40 - 49	82.8	37.9
50 - 59	73.9	17.2
60 +	37.4	6.0
Total	74.0	36.3

Source: Adapted from Levin and Pirie 1974.

Table 13.2 Percentage of American Samoan Males Participating to Any Extent in Agriculture, by District of Residence and 10 Year Age Groups

Age	Western District	Central District	Eastern District	Manu'a	Total
20 - 29	6.2	0.4	0.7	35.2	3.5
30 - 39	8.6	0.1	1.0	42.7	5.2
40 - 49	12.8	1.2	4.6	41.7	8.4
50 - 59	10.3	0.9	4.8	67.5	10.8
60+	9.2	1.4	2.6	40.0	8.4
Total	8.9	0.6	2.4	46.4	6.3

Source: Adapted from Levin and Pirie 1974.

five age groups shown in Table 13.2 are .05, .09, .03, .04, and .05, respectively. In other words, it appears that older men were more likely than younger men to perform agricultural labor in addition to their full-time employment as wage laborers.

The census provided data on two categories of time expenditure by agriculturalists: less than 35 hours and greater than or equal to 35 hours per week. Given the discussion in the previous section on the time requirements of subsistence agriculture, a cutoff point of 10 to 20 hours per week would perhaps have been more appropriate. Of all agriculturalists, 29 percent worked for 35 or more hours per week. This percentage did not vary greatly by age within each district, but it did vary between districts. In particular, approximately 50 percent of all agriculturalists in the three Tutuila districts worked for 35 or more hours per week, while the corresponding percentage in Manu'a was only 1 percent. These data probably reflect the production of crops for both the market and the 'aiga in Tutuila and primarily for the 'aiga in Manu'a.

Most American Samoans, or 96 percent of all economically active males and almost all economically active females, were employed in the money economy in 1974. Expressed differently, 71 percent of all adult males and 36 percent of all adult females worked in the money economy. A large proportion of these individuals, or 56 percent of all employed males and 49 percent of all employed females, worked for the government of American Samoa. The remainder worked for private businesses, with the largest private employers being the fisheries, which employed 14 percent and 32 percent, respectively, of the male and female work force.

The distribution of the 1974 work force by age, sex, and seven occupational categories is presented in Table 13.3. The average energy costs of occupations that are reported below were obtained from Durnin and Passmore (1967) and Katch and McArdle (1983). The professional and clerical/sales categories, which include sedentary and semisedentary occupations with energy requirements of 1.7 to 2.5 Kcal/min, are self-

Table 13.3 Percentage of Economically Active American Samoans in Seven Occupational Categories, by Age and Sex

Age	Professional	Clerical/ Sales	Service	Machine trades	Structural work	Fishery	Miscellaneous
			Males				
20 - 34	27.1	15.4	12.5	10.7	13.7	9.0	11.7
35 - 54	30.4	7.9	12.2	10.1	22.7	6.2	10.4
55 +	45.7	6.8	19.2	6.4	12.6	2.7	6.6
Total	30.4	11.3	13.0	10.0	17.5	7.2	10.6
			Females				
20 - 34	28.1	33.5	12.1	1.6	0.0	21.8	2.9
35 - 54	33.5	10.9	17.1	1.3	0.0	21.8	2.9
55 +	53.0	10.6	22.7	6.1	0.0	7.6	0.0
Total	30.8	24.6	14.2	1.6	0.0	26.0	2.7

Source: Adapted from Levin and Pirie 1974.

explanatory. The service category includes individuals such as maids, cooks, tailors, and barbers. Most of the occupations within this category, have energy requirements of 2.5 to 3.5 Kcal/min. Individuals such as mechanics and metal workers are included in the machine trades category, while the structural work category includes carpenters, welders, electricians, painters, and general construction workers. Most of the occupations within these two categories have energy requirements of 3.5 to 4.0 Kcal/min. The fishery category includes all individuals involved in the processing and canning of fish, excluding office personnel. No data are available on the energy costs of these positions. All other manual laborers are included in the miscellaneous category. Thus, with few exceptions, the occupations available in modern American Samoa in 1974 were not particularly strenuous, only varying in energy requirements from light to moderate intensity (Karvonen 1974).

As can be seen in Table 13.3, a large proportion of the work force, or 43 percent of males and 55 percent of females, had sedentary or semisedentary occupations (professional and clerical/sales categories). Not unexpectedly, the proportion of the work force in these categories tends to increase with age, primarily in the professional category. Nevertheless, many older men were also employed as manual laborers. Although few of their positions are particularly strenuous, they are among the most strenuous occupations reported.

Work in the United States

Occupational data on Samoans residing in the United States are available from the U.S. 1980 census (U.S. Bureau of the Census, 1983a, 1983b, 1983c). Since 81 percent of all Samoans live in two states, Hawaii and California, we present the available data for the United States as a whole and also individually for these states. Data by age group are not available. The census defines the labor force as those individuals (16 years and older) who are either employed or are actively seeking employment. As indicated in Table 13.4, approximately 74 percent of all male and 47 percent of all female Samoans were classified as belonging to the labor force in 1980. These values are 1 to 3 percent below the equivalent values for the entire U.S. population. Labor force participation rates for Samoans in the United States are comparable to those for American Samoa (74% in males and 36% in females). However, since 16- to 19-year-olds, many of whom are probably not actively seeking work, are included in the U.S. figures but not in those for American Samoa, it is likely that labor force participation is actually somewhat lower in American Samoa than in the United States. The employment rate among the Samoan civilian labor force in the United States (both sexes) was 90 percent (Table 13.4), or about 3 percent less than for the entire U.S. population for the same time period. The level of employment of Samoans working in the United States can also be evaluated by comparing labor

Table 13.4 Samoan Work Force in the United States

	United States		Hawaii		California	
	Males	Females	Males	Females	Males	Females
Persons 16 years and older	11 550	11 189	3735	3938	5358	5236
Labor force	8484	5216	2545	1485	3911	2698
Percentage of persons 16 + years	73.5	46.6	68.1	37.7	73.0	51.5
Civilian labor force	7141	5151	2180	1476	3638	2687
Employed	6476	4622	1985	1298	3194	2401
Military	1343	65	365	9	373	11

Source: U. S. Bureau of the Census 1983a, 1983b, 1983c.

force participation rates between Samoans and the eight other Asian and Pacific Islander groups listed in the census (Japanese, Chinese, Filipinos, Koreans, Asian Indians, Vietnamese, Hawaiians, and Guamanians). Labor force participation was lower in Samoan females than in any of these other groups, while only one group (Vietnamese) had a lower percentage of adult men in the labor force.

Almost 16 percent of the Samoan male labor force in the United States were members of the military in 1980 (calculated from Table 13.4). With the exception of Guamanians, this is a considerably higher rate of military participation than in the other Asian and Pacific Islander groups listed in the census. The equivalent value for the entire U.S. male labor force is 0.2 percent. This high rate of military service by Samoans is not unexpected, given their long history of military enlistment (Franco 1984).

The distribution of the U.S. Samoan civilian work force, by sex and six occupational categories, is presented in Table 13.5. The breakdown of occupations in the 1980 census was not as precise as in the 1974 American Samoa census, but the categories provided in Table 13.5 are as similar as possible to those used in Table 13.3 for American Samoa. One major difference, however, is that men involved in the fishing industry in the United States were included in the miscellaneous category. Also, due to the limited occupational opportunities in American Samoa, there is probably a greater diversity of occupations within each category in the United States than in American Samoa.

The primary difference between the distributions of occupations in American Samoa and the United States lies in the percentage of men with sedentary and semisedentary occupations (professional and clerical/sales). Such occupations account for 45 percent of all male workers in American Samoa, but for only 29 percent of the men working in the United States The source of this difference lies primarily in the professional category, which accounts for 31 percent of the American Samoan civilian male work force, but only 14 percent of the U.S. Samoan civilian male work force

Table 13.5 Percentage of Economically Active Samoans in Various Occupational Categories in the U. S., by Sex

	United States		Hawaii		California	
	Males	Females	Males	Females	Males	Females
Professional	13.9	15.9	10.3	17.0	12.4	15.8
Clerical/sales	14.3	37.8	9.6	33.2	16.9	40.5
Service	17.3	23.2	24.7	34.7	13.8	17.1
Machine trades	18.7	13.0	11.3	8.4	24.6	14.3
Structural work	14.0	6.8	13.4	2.7	15.5	9.4
Miscellaneous	21.8	3.3	30.8	4.0	16.8	2.9

Source: U. S. Bureau of the Census 1983a, 1983b, 1983c.

(Tables 13.3 and 13.5). Whether this pattern represents selective migration or an inability of men with professional experience in American Samoa to compete in the United States is unknown. Whatever the explanation, a larger proportion of Samoan males employed in the United States in 1980 worked in manual labor occupations than was true in American Samoa in 1974. The percentage of women with professional occupations was also considerably lower in the United States than in American Samoa (16% versus 31%). In this case, however, there was a compensating increase in the clerical/sales category among women working in the United States, so that the overall proportion of women with sedentary and semisedentary occupations was similar between American Samoa and the United States (55% versus 54%). The proportion of the Samoan civilian labor force in the United States with professional and clerical/sales occupations is considerably less than for the entire U.S. population, where about 43 percent of all employed males and 67 percent of all employed females have sedentary or semisedentary occupations.

Summary of work in different settings

Given the nature of the data, it is only possible to make general comparisons between the work of precontact and contemporary Samoans. Nevertheless, several important patterns can be detected. First, there has been a change in the demographic structure of the work force, with older men becoming increasingly economically active relative to younger men. Second, there has been an increase in the amount of time devoted to economic activities. The most economically active group in precontact Samoa, young adult males, spent a maximum of 15-20 hours per week in economic activities, while the modern work week is generally about 40 hours. It is likely that the first figure refers to actual working time, while the latter does not. Even if contemporary Samoans spend less than 40 hours per week actually working, they are nevertheless largely constrained from using this time for other purposes. This trend even applies to many contemporary agriculturalists, due to the increased time demands of providing food for both the 'aiga and the market. Third, although the available data on energy expenditure in each setting are necessarily rough estimates, there appears to have been a decrease in average occupational energy expenditure. Due to the relatively moderate requirements of traditional Samoan subsistence, however, this decrease appears to have been only moderate, and perhaps less than often envisaged for such a transition. No data are available on changes in leisure activity levels between precontact and contemporary Samoa. Assuming leisure activity levels also decreased, which seems likely, the decrease in total activity level may have been somewhat greater than indicated by only the economic data, which thus provide a conservative estimate of the decrease in total activity level. Finally, there is some pertinent work-related variation among contemporary Samoans. In particular, for whatever reason, employed Samoans living outside of American Samoa are more likely to have manual labor occupations than Samoans working in American Samoa.

WORK CAPACITY OF SAMOANS

Work capacity, as evaluated by aerobic capacity [VO_2max (ml/kg/min)], was measured in three samples of Samoan men; tests were conducted in Hawaii in 1975, in American Samoa in 1978, and in Western Samoa in 1982. The samples, methods, and results of each study are summarized below. Greater detail can be found in the original sources (Western Samoa: Pelletier 1984; American Samoa: Greksa and Baker 1982; Greksa 1980; and Hawaii: Lukaski 1977).

The Hawaiian sample included 30 adult males (30-59 years) residing on Oahu and categorized as active (manual laborer) or inactive (office employee or unemployed). There were no detectable leisure activity differences between the subgroups in Hawaii and their mean length of residence in Hawaii was 12 years. The American Samoa sample consisted of 34 adult males (18-30 years) residing on Tutuila, categorized as office employees, manual laborers, or agriculturalists. The Hawaiian and American Samoan samples were given progressive maximal exercise tests on a treadmill ergometer. Heart rate was monitored with an electrocardiograph and expired air was collected in Douglas bags and analyzed with electronic analyzers or a Scholander apparatus.

The Western Samoan sample consisted of 97 men (20-35 years) partitioned into four categories on the basis of occupation and leisure activity levels. These groups are (1) village agriculturalists (VILL); (2) men with active (manual labor) occupations and with sedentary leisure activities (ACT/SED); (3) men with sedentary occupations and active leisure activities (SED/ACT); and (4) men with sedentary occupations and sedentary leisure activities (SED/SED). These subjects were given a progressive submaximal step test, with heart rate determined by auscultation. Oxygen consumption was not measured. Instead, submaximal data were used to predict a work load associated with a predicted maximal heart rate of 220 minus age. These data were then used to predict aerobic capacity, using the formula of Nagle et al. (1965). Predicted aerobic capacity was adjusted by analysis of covariance to control for significant differences between testing sites in ambient temperature.

Before discussing the results of these studies, it is useful to consider the validity of predicted aerobic capacities. It is not possible to accurately predict the aerobic capacity of specific individuals. According to Shephard (1980), such predictions may be confounded by two possible sources of error. First, there are inherent errors in the process of predicting aerobic

Table 13.6 Anthropometric and Exercise Characteristics of the Work Capacity Samples

	N	Age (yr) X̄	SD	Stature (cm) X̄	SD	Weight (kg) X̄	SD	Sum of skin-folds (mm) X̄	SD	Maximal heart rate (beat/min) X̄	SD	VO$_2$max[a] (ml/kg/min) X̄	SD
Western Samoa													
Villager													
ACT/SED	23	25.4	4.0	169.4	6.8	69.9	4.5	57.5	17.6	220-age		50.4	7.7
SED/ACT	29	25.8	4.3	169.5	4.5	72.9	8.0	66.2	22.8	220-age		56.2	9.3
SED/ACT	23	23.0	2.2	173.1	6.3	78.7	12.1	79.0	34.0	220-age		47.8	6.9
SED/SED	22	23.6	3.1	173.2	5.2	78.4	10.8	88.5	27.4	220-age		41.1	8.3
American Samoa													
Total sample	34	23.6	3.5	172.6	4.8	77.4	12.1	95.7	29.7	185.0	9.5	38.9	5.9
Agriculture	10	24.9	3.7	171.8	5.4	77.4	15.9	86.0	33.7	183.5	6.6	41.2	5.9
Manual labor	14	23.0	3.1	173.1	4.6	78.2	8.7	100.7	27.5	184.0	9.3	35.8	4.6
Office employee	10	23.0	3.6	172.7	5.0	76.3	13.0	98.5	29.2	188.0	12.1	40.8	6.1
Hawaii													
Active	7	30 - 39		168.6	5.6	86.3	9.8	—	—	188.0	9.3	39.4	2.9
Active	8	40 - 49		172.3	4.8	83.8	9.3	—	—	182.0	5.4	38.1	1.7
Inactive	7	30 - 39		172.5	4.0	98.9	7.7	—	—	188.0	8.2	32.5	5.0
Inactive	5	40 - 49		166.6	5.9	87.5	15.0	—	—	185.0	8.2	32.7	3.7

Source: Pelletier 1984; Greksa 1980; Lukaski 1977.
[a] VO$_2$max in Western Samoa is predicted from submaximal data and statistically adjusted for ambient temperature.

capacity based on submaximal data, even if submaximal VO_2 is measured. Second, additional errors may accrue if submaximal VO_2 must also be estimated. Factors such as differential levels of anxiety, interindividual variation in the mechanical efficiency of stepping, and maximal heart rates that differ from their predicted value are possible sources of error. If, as appears to be generally true, however, the error of prediction is random and there are no systematic biases, it is valid to use such data for internal comparisons between subsamples (Shephard 1980). It is because there were systematic differences between activity categories in the ambient temperature of the testing sites that aerobic capacity in the Western Samoan sample was adjusted for this variable. There do not appear to have been any other systematic differences between subsamples (other than those built into the research design). Thus, there is no reason to doubt the validity of internal comparisons within the Western Samoan sample. In addition, for reasons to be discussed shortly, it is likely that comparisons between American Samoa and Western Samoa are also reasonably valid.

Anthropometric and exercise characteristics of each sample are described by activity category in Table 13.6 and by age in Fig. 13.2. Western Samoan males tend to be considerably fitter than American Samoan males of the same age. Also, aerobic capacity in the American Samoan and Hawaiian samples, both of which were measured using similar maximal exercise protocols, is very low. Mean aerobic capacity in these Samoans is even less than in Easter Islanders, a population cited for its low level of physical fitness (Ekblom and Gjessing 1968), although

Fig. 13.2 VO_2max in Samoans and selected populations.

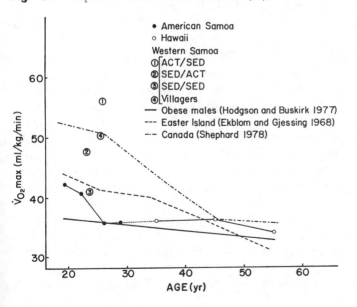

the difference between these groups decreases with age. A simila:
pattern can be observed when comparing Samoans with Canadian:
(Shephard 1978). From the late twenties on, which is the approximate
age range during which many American Samoan and Hawaiian male:
display rapid increases in body fatness, the age pattern of aerobic
capacity in these samples of Samoans is very similar to that predicted for
obese males by Hodgson and Buskirk (1977).

Exposure to a modern way of life in Samoans is associated with ar
increase in body fatness and, as argued earlier, a moderate decrease in
average occupational energy requirements. One would expect these
trends to be associated with a decrease in aerobic capacity. The relative
importance and potential magnitude of these trends on the aerobic
capacity of Samoans can be estimated by examination of within and
between sample variation in fatness and activity level. The effects of
body fatness on the aerobic capacity of Samoans is investigated first
followed by an examination of the effects of changes in occupationa
activity patterns.

Body fatness was evaluated in the American Samoan and Western
Samoan samples with the sum of six skinfolds (triceps, subscapular.
midaxillary, suprailiac, forearm, and medial calf). Equivalent data are
not available for Hawaii. These data were normalized with a square-root
transform prior to analysis. American Samoans tend to be fatter thar
Western Samoans (Chapter 10) and the same is true for the exercise test
samples (Table 13.6). In addition, there tends to be greater variation in
body fatness in the American Samoan sample than in the Western
Samoan sample. Therefore, in order to estimate the effects of an increase
in body fatness with exposure to a modern way of life on the aerobic
capacity of Samoans, we will first investigate the relationship between
fatness and aerobic capacity in American Samoans.

The relationship between the transformed sum of skinfolds and aerobic
capacity in the American Samoan sample is portrayed in Fig. 13.3.
There is a highly significant ($p < .001$) negative relationship between
these variables, with 50 percent of the variation in aerobic capacity
explained by differences in body fatness. Based on the regression
equation provided in Fig. 13.3, the difference in predicted VO_2max
between individuals with sums of skinfolds of 49 and 169 mm (square
roots of 7 and 13), or between the leaner and fatter men in the sample.
was about 17 ml/kg/min. Also included in Fig. 13.3 are the mean values
for the four Western Samoan activity categories. In this case, however,
aerobic capacity in the Western Samoan sample was predicted using a
maximal heart rate of 185 beat/min (rather than 220 minus age) in order
to increase comparability with the American Samoan data. With the
exception of the ACT/SED group, the mean aerobic capacities of these
groups are within the range of those for American Samoans with a
similar level of body fatness.

The general similarity between American and Western Samoan males

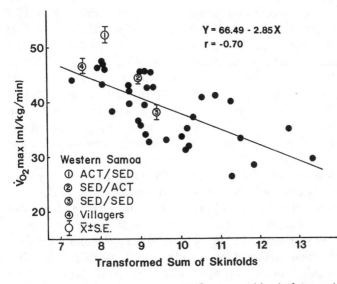

Fig. 13.3 Relationship between VO₂max and body fatness in Samoans.

in aerobic capacity, after controlling for body fatness, provides some support for the validity of the Western Samoan predicted aerobic capacities and also suggests that one cause of the difference in aerobic capacity between these samples is a difference in average body fatness. However, since body fatness is itself influenced by activity level, as indicated by the general correspondence between occupational category and sum of skinfolds in American and Western Samoan males, and since American Samoan males have considerably larger sums of skinfolds (by up to 50%) than Western Samoan males with equivalent occupations (Table 13.6), it is likely that there are also overall activity level differences between Western and American Samoan males. In other words, differences in both fatness and activity level are probably responsible for the differences in aerobic capacity between American and Western Samoan males.

We next investigate the relationship between activity and aerobic capacity within each sample. The potential effect of activity differences on aerobic capacity are clearly evident in the Hawaiian sample, where there is a 5-7 ml/kg/min difference between active and inactive individuals of the same age. Since the Hawaiian sample includes no men performing traditional activities and a major goal is to estimate the effect of a transition from traditional to wage labor occupations on aerobic capacity, the analyses focus on the samples of American and Western Samoan males.

The relationship between aerobic capacity and activity is examined first in the American Samoan sample. Within this sample, maximal heart rate did not differ significantly between occupational groups, indicating that the men in each group were working at similar levels of maximal exertion. Regression analyses suggested that the rate of decrease with age in aerobic capacity differed considerably between the agriculturalists and wage laborers, decreasing significantly ($p < .05$) in the office employees (- 1.22 ml/kg/min/yr) and manual laborers (- .91 ml/kg/min/yr), but not in the agriculturalists (- .66 ml/kg/min/yr). This is the pattern one would predict; however, comparison of these regression equations indicated that aerobic capacity did not differ significantly between agriculturalists and office employees and that it was significantly higher in both groups than in manual laborers ($p < .05$).

One possible explanation for the unexpectedly high aerobic capacity (relative to the other groups) of the office employees is a difference between groups in leisure activity. In addition, the relationship of aerobic capacity with age in office employees and agriculturalists may have been unduly affected by a few outliers. In particular, the regression relationship for office employees was affected by an older male with a relatively high aerobic capacity, while that of the agriculturalists was depressed by two older males with relatively low aerobic capacities. Removal of these outliers had a significant effect on the analyses. For example, mean aerobic capacity increased to about 44 ml/kg/min in the agriculturalists and decreased to about 40 ml/kg/min in the office employees. In this case, aerobic capacity was significantly greater in agriculturalists than in office employees ($p < .01$). These findings still do not conform totally to expectations, since aerobic capacity in office employees was greater than in manual laborers, even after excluding outliers. As expected, however, aerobic capacity was greater in agriculturalists than in both of the wage labor categories. Thus, these data suggest that the transition from traditional agriculture to modern wage labor occupations was associated with a decrease in aerobic capacity of, at most, 4-5 ml/kg/min. A relatively small decrease in aerobic capacity due to changing occupational activity levels is not entirely unexpected, given that we have estimated that there were only moderate activity level differences between precontact and contemporary Samoa. However, the Western Samoan data, which are examined next, suggest that the American Samoan data may underestimate the effect of this transition on aerobic capacity. This may be partially due to the limitations of utilizing average occupational energy expenditure to estimate habitual activity level. Perhaps more important, there may simply be less within sample variation in activity level than is true for fatness, such that the impact of activity level differences on aerobic capacity are overwhelmed by differences in body fatness.

The general pattern of aerobic capacity in the Western Samoan wage labor groups corresponds to expectations, with aerobic capacity highest in the ACT/SED group, intermediate in the SED/ACT group, and lowest in the SED/SED group (Table 13.6, Fig. 13.3). Mean aerobic capacity was significantly greater in the ACT/SED group than in the other two wage labor groups ($p < .05$). In addition, aerobic capacity was significantly greater (by about 7 ml/kg/min) in the SED/ACT than in the SED/SED group, indicating the potential importance of leisure activity upon aerobic capacity. Mean aerobic capacity in the VILL group was significantly greater than in the SED/SED group (by about 9 ml/kg/min), but signficantly less (by about 6 ml/kg/min) than in ACT/SED males ($p < .05$). The latter difference does not seem to fit expectations; however, the ACT/SED group may not be representative of most manual laborers in Western Samoa. This group was specifically selected to include some of the more strenuous manual labor occupations, in order to accentuate activity level differences within the Western Samoan sample. There also may have been leisure activity differences between groups and perhaps also differences in the nature of the work, other than average energy expenditure, which influenced aerobic capacity.

In conclusion, aerobic capacity is very low in those groups of Samoans who have been most affected by a modern way of life (American Samoans, Hawaiians, and the SED/SED group from Western Samoa). The low levels of physical work capacity in American Samoan males appears to be due to both an increase in body fatness and a decrease in activity level, but primarily due to the former. The maximal effect on aerobic capacity of increases in body fatness due to exposure to a modern way of life can be estimated from the regression equation in Fig. 13.2 by determining the average difference in aerobic capacity between the leanest and fattest American Samoans. The average difference between these groups is about 17 ml/kg/min, which is physiologically highly significant. For example, if the mean aerobic capacity for all American Samoans was increased by this amount, it would be in the mid-50s, or similar to that found in most reasonably unacculturated groups (Shephard 1980).

The maximal effect on aerobic capacity of changes in occupational activity patterns due to exposure to a modern way of life can be estimated by comparing aerobic capacity between the VILL and SED/SED groups in the Western Samoan sample. Mean aerobic capacity was about 9 ml/kg/min higher in the VILL than in the SED/SED group, suggesting that decreases in occupational activity level with exposure to a modern way of life have not had as great an effect on the aerobic capacity of Samoans as increases in body fatness. The estimation of the maximal possible effect of increases in body fatness on aerobic capacity took into account, however, the full range of variation in fatness in the American Samoan sample. A comparison of mean aerobic capacity between the VILL and SED/SED groups is only a comparable technique if one assumes that there is little or no variation in activity level within these occupational categories. That is a reasonable assumption for the SED/SED group, but may not be for the VILL group. In other words, some agriculturalists may be more active than others. Therefore, in order to make the estimate of the maximal possible effect of changing occupational activity patterns on aerobic capacity comparable to that for body fatness, it is necessary to attempt to take such variation into account. If accurate daily energy expenditure data were available for the agriculturalists, one could use an approach similar to that used for fatness, or regress aerobic capacity against daily energy expenditure and then predict average aerobic capacity in the most active agriculturalists. Since this is not possible, this value might be conservatively approximated by an aerobic capacity of one standard deviation above the mean of the VILL group. Comparing this value with the mean of the SED/SED group indicates that the maximal effect of changing occupational activity patterns on aerobic capacity is about 17 ml/kg/min, or similar to the potential maximal effect of body fatness on aerobic capacity.

It is difficult to estimate the combined maximal possible effect of increases in fatness and decreases in activity level on the aerobic capacity of Samoans exposed to a modern way of life, since both trends are interrelated. Our best estimate is obtained by determining the difference between the mean-plus-one standard deviation in the VILL group and the average for the fattest American Samoans, which is about 28 ml/kg/min. This is probably a high estimate, due to difficulty in adequately controlling for the interaction between fatness and activity. Nevertheless, it provides some indication of the magnitude of the combined effect of changes in body fatness and activity level on the work capacity of Samoans.

WORK CAPABILITY

Measurement of work capability

Work capability refers to the ability of individuals to perform specific essential tasks. Thus, individuals with low work capabilities cannot easily perform essential tasks while individuals with high work capabilities can. One reasonable strategy for evaluating work capability is to relate the energy requirements of essential tasks to the levels of energy expenditure that an individual can maintain without the onset of fatigue (Greksa in press). Thus, the greater the ability of individuals to perform essential tasks without undue physiological strain, the higher their work capability, and vice versa.

The relationships between VO_2max, duration of a task, and the maximal level of work that an individual can perform without undue physiological strain were first systematically examined in laboratory studies by Astrand (1960) and later by Wyndham and colleagues (1966). Additional work by Bink (1962) and Bonjer (1971, 1968, 1962) resulted in the derivation of a formula describing the relationship between these variables. This formula is:

$$V_1 = \frac{3.756 - \log_{10}(t)}{3.1} \times VO_2max,$$

where V_1 is the highest submaximal VO_2 (l/min) that an individual of a given VO_2max (l/min) can sustain for a period of t minutes without the onset of fatigue. V_1 can easily be converted to Kcal/min by assuming a respiratory exchange ratio of 0.83, in which case each liter of consumed oxygen has an energy value of 4.83 Kcal. For example, an individual with a VO_2max of 2.50 l/min could work without undue strain for 2 hours at a submaximal VO_2 of 1.35 l/min (about 6.5 Kcal/min), or at a relative work intensity of 54 percent. Work for longer periods of time or at a higher submaximal VO_2, or essentially at a higher relative work intensity, would result in debilitating fatigue and thus require the cessation of work or a decrease in work intensity. V_1 will be referred to as *sustainable* VO_2.

Sustainable VO_2 and the average energy requirements of tasks can be used to calculate several useful indices of work capability. Before describing these indices in detail, we first discuss their limitations. The calculations for sustainable VO_2 assume that work is continuous while actual work conditions are generally intermittent. Also, the average energy requirements of a task may be less important in determining work capability, at least under some conditions, than other aspects of the energetics of work, such as the energy cost of the most strenuous acts that must be performed (Astrand and Rodahl 1977). Even given these limitations, there is evidence to suggest that the sustainable VO_2 predicted by this formula is reasonably accurate and meaningful. For example, Astrand (1967) determined that construction workers worked at an average relative work intensity of 39 percent over an 8-hour day. The formula predicts a relative work intensity of 35 percent, which is a slight underestimate but is nevertheless reasonably close to the actual value.

Even given a valid index of work capability, a major drawback in evaluating work capability by this method is that VO_2max must be measured in a large number of workers. This is no simple matter since maximal exercise tests require sophisticated equipment and are time-consuming. However, since VO_2max is highly correlated to body size, body composition, and activity level, it is possible to derive regression equations from relatively small exercise test samples that can be used to predict VO_2max in larger survey samples (Shephard et al. 1971). Such equations do not permit the accurate prediction of VO_2max for individuals, but are useful for the comparison of mean predicted VO_2max between groups (Shephard et al. 1971). This is the approach taken in this study. Prediction equations were derived from the exercise data on the American Samoan (18-30 years) and Hawaiian (30-60 years) samples described earlier. Since exercise data are available only for males, all calculations are restricted to this group. In each case, activity level was controlled by deriving separate regressions for each occupational category. A best-fit regression procedure was utilized, with the input data being age and anthropometric indices of body size and composition. The correlation coefficients for the regressions based on the American Samoan sample varied from 0.76 to 0.98, while those for the Hawaiian sample varied from 0.61 to 0.89. These equations were used to predict VO_2max in 18- to 60-year-old Samoan males in American Samoa and Hawaii who had been examined during several surveys and who were employed in occupations for which average energy requirement estimates are available. Estimates of work capability in Western Samoa were restricted to the village agriculturalists (VILL) and the manual laborers (ACT/SED) in the Western Samoan exercise sample, using their own predicted VO_2max's.

Two indices of work capability were calculated. First, we determined the ratio of occupational energy requirements to sustainable VO_2. Since sustainable VO_2 represents the maximal level of work that can be maintained for a given time period without the onset of fatigue, the larger the ratio, the lower the work capability of the individual. Individuals with ratios of 1.0 or greater are particularly stressed by their work. Second, we calculated the proportion of individuals with occupational energy requirements greater than their sustainable VO_2. This measure identifies men who are clearly stressed by their work. It should be noted that average occupational energy requirements have been presented in this report standardized for a 65-kg man. For these calculations, of course, individualized estimates of occupational energy requirements, adjusted for the weight of the subject, were utilized.

Finally, in order to calculate sustainable VO_2, information on the normal duration of more-or-less continuous work is necessary. No such data are available for Samoans and therefore separate calculations of sustainable VO_2 were made for two potential work periods, or for 2 and 8 hours. It is not unreasonable to assume a 2-hour work period to represent a minimal level of work that must be performed fairly frequently by most workers while 8 hours represents the opposite extreme, or a work period that is probably rare but nevertheless must be performed on occasion.

In the following sections, we first describe the work capability of contemporary Samoans employed in the wage labor economy in American Samoa, Western Samoa, and Hawaii. Finally, we investigate the work capability of Samoan agriculturalists.

Work capability in contemporary Samoans

The analysis of the work capability of contemporary Samoans is based on Samoans residing in Western Samoa, American Samoa, and Hawaii, and focuses on those individuals employed as manual laborers in the wage labor economy. Manual laborers comprise 60 percent of the American

Samoan work force and 72 percent of the work force of Samoans in the United States. Men with sedentary occupations are not considered since such men, not unexpectedly, are not physically stressed by their work, due to its low energy requirements. One would not expect biological factors to influence work performance to a great extent in men with such occupations (Spurr 1983).

Mean indices of work capability for the American and Western Samoan samples of manual laborers are included in Table 13.7. Focusing first on American Samoa, these data suggest that some manual laborers are stressed by their work. For example, the ratio of occupational energy requirements to sustainable VO_2 increases for the total sample of manual laborers (18-60 years old) from 0.58 for work of 2 hours duration to 0.91 for work of 8 hours duration. In other words, the *average* American Samoan manual laborer must work at 58 percent of sustainable VO_2 for work periods of 2 hours and at 91 percent for 8 hours of work. Also, within each age group of American Samoan manual laborers, the proportion of men unable to perform their work without undue physiological strain increases as the duration of the task increases. For the total sample, none are incapable of working without fatigue for 2 hours, but this value increases to 21 percent for 8 hours of work. It may seem

Table 13.7 Mean Indices of Work Capability in Contemporary Samoan Males

Duration of work (hr)	N	Ratio of occupational energy requirements to sustainable VO_2		Proportion with occupational energy requirements $>$ sustainable VO_2	
		2	8	2	8
American Samoa					
Manual laborers					
18 - 30	83	0.63	0.99	0.00	0.48
30 - 45	128	0.54	0.84	0.00	0.04
45 - 60	74	0.59	0.92	0.00	0.20
18 - 60	285	0.58	0.91	0.00	0.21
Unemployed					
18 - 30	24	0.53	0.83	0.00	0.17
30 - 45	17	0.54	0.84	0.00	0.12
45 - 60	36	0.59	0.92	0.00	0.25
18 - 60	77	0.56	0.87	0.00	0.19
Western Samoa					
ACT/SED	29	0.51	0.62	0.00	0.00
Hawaii (18 - 60 years)					
Manual laborers	68	0.58	0.90	0.00	0.18
Unemployed	49	0.60	0.93	0.02	0.31

logical that the proportion should increase as the work duration increases, but this would not occur in a physically fit sample.

The impact of larger VO_2max's on work capability is clearly demonstrated by comparing the work capability of American and Western Samoan manual laborers (Table 13.7). An average occupational energy requirement of 3.5 Kcal/min (for a 65-kg man), the approximate average for American Samoan manual laborers, was assumed for the ACT/SED group. All indices of work capability are considerably smaller in this group than in American Samoans of a comparable age (< 30 years). For example, none of the Western Samoans would be unable to perform their work for 8 hours while the corresponding figure for American Samoan manual laborers is 48 percent. In order for the indices of work capability to be as high in Western Samoan workers as in American Samoan workers, one would have to assume occupational energy requirements of 5.5 Kcal/min for the Western Samoans, or nearly twice as strenuous as the work of the average American Samoan manual laborer.

Surprisingly, both indices of work capability are somewhat larger in 18- to 30-year-old American Samoan workers than in older workers, especially those 30 to 45 years of age, indicating a slightly lower work capability in younger men (Table 13.7). This is unexpected since VO_2max tends to decrease with age (Astrand and Rodahl 1977). The simplest explanation for this finding, but one that cannot be evaluated, is that it relates in some way to the use of two exercise test samples to predict VO_2max. An alternative hypothesis, however, that might at least partially explain this finding is that, since VO_2max decreases with age, some of the less fit manual laborers are selected out of the manual labor work force, either into a less strenuous occupation or into unemployment, as they age and their occupations become increasingly strenuous in relation to their capabilities.

Although the causes for individual unemployment in American Samoa have not been analyzed, it is likely that some men are unemployed by choice. There is no reason to assume that such men would be less fit, on average, than employed men; however, the inclusion among the unemployed of older men who are unemployed because of their low work capabilities should be reflected in the indices of work capability. Therefore, the work capability of unemployed American Samoan males was evaluated assuming average occupational energy requirements of 3.5 Kcal/min (for a 65-kg man). These data are also included in Table 13.7. In comparison with the data on manual laborers, it can be seen that the indices of work capability are consistently lower in unemployed than in employed 18- to 30-year-olds, indicating a somewhat greater work capability in the former. However, the indices of work capability in unemployed 30- to 60-year-olds are either similar to or greater than in employed 30- to 60-year-olds. For example, 20 percent of the 45- to 60-year-old manual laborers are unable to work for 8 hours duration without fatigue, but the corresponding value for the unemployed men is

25 percent. The corresponding values for 30- to 45-year-olds are 4 percent and 12 percent, respectively. These patterns are consistent with the hypothesis that, due to their low work capabilities, some older manual laborers may be selected out of the work force.

It was demonstrated in the previous section that a number of American Samoan manual laborers have a low work capability. All manual laborers are capable of working for 2 hours duration without undue strain, but work of longer duration would be limiting for a significant number of men. In order to examine how this might affect their relative performance when competing with manual laborers in other societies, a comparison to published data from Sweden was made. We chose to use data collected by Astrand (1967) on the work capability and work requirements of 14 bricklayers and carpenters between the ages of 30 and 60 years. Astrand monitored heart rate throughout an 8-hour work day and used average heart rate for this period to predict average oxygen consumption. The average energy expenditure of these men was about 5.0 Kcal/min, or greater than that of most American Samoan manual labor occupations.

As noted earlier, Astrand found that these men worked at an average relative work intensity of 39 percent over an 8-hour work day, rather than the 35 percent predicted by the Bonjer formula. Therefore, calculations of the ability of American Samoan manual laborers to perform the work of these construction workers were made using both estimates of sustainable VO_2. These analyses provide striking evidence for the potential difficulty of many American Samoans to work as manual laborers outside of American Samoa. For example, assuming a sustainable VO_2 based on 39 percent of VO_2max and occupational energy requirements equivalent to those presented above for the construction workers, the ratio of occupational energy requirements to sustainable VO_2 in 30- to 45-year-old American Samoan manual laborers was 1.04 and 75 percent of the men would have sustainable VO_2's less than their occupational energy requirements. The equivalent values for 45- to 60-year-olds were 1.17 and 97 percent. The corresponding values, assuming a sustainable VO_2 that is based on the Bonjer formula, were 1.36 and 100 percent for 30- to 45-year-olds and 1.17 and 100 percent for 45- to 60-year-olds.

Based on such data, it is tempting to suggest that a large number of American Samoan migrants with manual labor occupations are stressed by their work and that some may be excluded from the workplace due to their low work capabilities. However, these relationships could be modified by a variety of factors. For example, since activity level and physical fitness are positively related (Fig. 13.1), the physical fitness, and thus work capability, of migrant men who were able to perform their work, even if minimally, should improve over time. Also, there are many manual labor occupations that are less strenuous and more intermittent in nature than construction work. In addition, Parsons (1982) utilized blood group and anthropometric data to demonstrate the existence of selective migration among Samoans. It is not clear whether migrants are physically fitter than nonmigrants, but there is some support for this possibility (Greksa 1980).

For all of the reasons stated above, we therefore also examined work capability in Samoan migrants in Hawaii. As elsewhere in the United States there is a high rate of unemployment among Samoans in Hawaii. For example, considering only Hawaiians, Guamanians, and Filipinos, the groups probably most similar to Samoans in occupational opportunities, 2.4 to 5.6 percent fewer of Samoan males of 16 years and older are in the labor force and 1.4 to 4.4 percent fewer who desire to work are employed. The majority of employed Samoans are, as noted earlier, employed as manual laborers (Table 13.5). This may appear, in itself, as evidence against a low work capability limiting the occupational opportunities of Samoans; however, this pattern may primarily reflect the educational and training background of migrant Samoans. The important question is whether some men are only able to perform their work with difficulty and whether others are actually excluded from the workplace due to their low work capabilities. In a sense, this involves estimating whether the high unemployment rate of Samoan migrants can be partially explained by biological factors or whether it can be totally explained by social factors.

Work capability was assessed in 117 adult migrant males (18-60 years), 68 manual laborers and 49 unemployed men. Precise occupational descriptions were unavailable for many of the manual laborers. We therefore estimated indices of work capability assuming average occupational energy expenditures of 3.5 Kcal/min, or the approximate average occupational energy requirements for American Samoan manual laborers. The results of these analyses are summarized in Table 13.7. Assuming equivalent work durations and occupational energy requirements, the indices of work capability are similar between manual laborers and unemployed men residing in American Samoa and Hawaii. These data suggest, however, that migrants would have a lower work capability than sedentes if work duration and/or work requirements were greater in Hawaii than in American Samoa, which seems possible. For example, if one assumes a work duration of 2 hours and energy requirements of 5.0 Kcal/min, the ratio of occupational energy requirements to sustainable VO_2 in Hawaiian manual laborers would be 0.83, and 10 percent would have occupational energy requirements greater than sustainable VO_2. For 8 hours of work these indices would increase to 1.29 and 96 percent, respectively. In other words, as noted earlier, the potential exists for a low work capability in migrants. Finally, the indices of work capability are similar between manual laborers and unemployed men in Hawaii, suggesting that the latter group does not include men of such low work capability that they have been excluded from the work force. In other words, these data do not support the suggestion that the high rate of unemployment of Samoans in Hawaii has a physical fitness basis.

Work capability in Samoan agriculturalists

Finally, the available data suggest that both physical work capacity and work requirements decreased with the transition from traditional agriculture to a wage labor economy in Samoa. Since work capability is determined by the relationship between these parameters, an interesting question is whether these trends also resulted in a decrease in work capability, or, in other words, whether there were differential changes in work capacity and work requirements. Therefore, the work capability of traditional Samoans was estimated by evaluating the capability of modern Samoan agriculturalists to perform plantation agriculture, the most strenuous component of the traditional Samoan food production system.

These analyses were restricted to 18- to 30-year-old agriculturalists (30 men in American Samoa and the VILL group from Western Samoa), since this is the age group that performed most of the agricultural labor in traditional Samoa. Work periods were assumed to be of 2 hours duration. This is probably the lower end of the distribution of actual work periods, if one includes travel time, but this work duration will facilitate comparison with the data on contemporary manual laborers. The average energy cost of plantation agriculture, including travel time, was calculated in a previous section to be about 6.5 Kcal/min. However, in order to adjust for the fact that this estimate was based on periods of continuous work while actual agricultural labor is intermittent (Gage 1982), an estimate of 6.0 Kcal/min was used for these calculations.

Based on these assumptions, the ratio of occupational energy requirements to sustainable VO_2 in American Samoan agriculturalists was 0.80 and energy requirements were greater than sustainable VO_2 in 13 percent of the men. Among the physically fitter Western Samoan agriculturalists, the ratio of occupational energy requirements to sustainable VO_2 was 0.70, while energy requirements were greater than sustainable VO_2 in none of the men. Even assuming four continuous hours of strenuous agricultural work, only 13 percent of the VILL group would have energy requirements greater than sustainable VO_2.

Thus, if the Western Samoan agriculturalists are considered to be representative of precontact Samoans, which is probably a conservative assumption, subsistence work would not have been stressful for the large majority of males. Given the low work capability of contemporary manual laborers, especially those residing in American Samoa, this suggests that exposure to a modern way of life may have resulted in a decrease in work capability. This is not at all clear, however. If work durations of only 2 hours are assumed for both manual laborers and agriculturalists, the indices of work capability presented above are similar to or higher than those for 18- to 30-year-old American and Western Samoan manual laborers (Table 13.7). This comparison, however, does not include the 40 percent of the contemporary American Samoan work force with sedentary occupations who, because of the low energy requirements of their work, have high work capabilities, and also does not include older men, who form a significant proportion of the contemporary work force but not of the precontact work force. Finally, the estimates of work capability for traditional Samoa are based on the most strenuous component of the food production system. Thus, in general, these data suggest that, assuming a work duration of only 2 hours for precontact and contemporary Samoans, there has been little change with modernization in work capability, at least in young adult males. In other words, the magnitude of the decrease in work requirements between precontact and contemporary Samoa was similar to the decrease in physical work capacity. On the other hand, although an increase in assumed work duration is not likely to result in a further decrease in the work capability of sedentary employees, work durations of greater than 2 hours would lead to a significant decrease in the work capability of manual laborers, resulting in their having lower work capabilities than agriculturalists. Thus, although exposure to a modern way of life does not currently appear to be associated with a decrease in the work capability, and thus adaptive status, of contemporary Samoans, the potential for such a pattern exists. Of course, these conclusions apply only to the domain of work. The effect of changes in activity level, body fatness, and physical fitness on other domains, such as health, and on overall biological fitness, is another matter.

Chapter 14

Blood Lipid Studies

DAVID L. PELLETIER
CONRAD A. HORNICK

The concentration of cholesterol and triglycerides in whole plasma and in certain lipoprotein fractions are potentially important intermediate variables in the relationship between modernization and increased incidence of cardiovascular diseases (CVD). Along with blood pressure and cigarette smoking, these characteristics have been identified as major risk factors for CVD and are known to vary in response to changes in many of the environmental and personal characteristics often associated with modernization. In the particular case of Samoans, other chapters in this volume examine differences in dietary intake, activity patterns, and body composition in relation to modernization, all of which may be related to variation in these blood lipids. This chapter describes the levels of plasma total- and HDL-cholesterol and triglycerides for adult Samoans of both sexes in two areas of American Samoa and Hawaii; it also examines variation in these blood lipids in relation to diet, activity patterns, and body composition using data from the same locales and from specialized studies of men in these areas and in Western Samoa. The overall aim is thus to examine the extent to which modernization may be associated with changes in blood lipids among Samoans, and to examine the extent to which these may be related to changes in body composition, activity patterns, and dietary intake.

The effect of modernization on plasma total-cholesterol and triglycerides has previously been examined in a variety of world populations. In these studies modernization (defined by rural/urban residence or access to nontraditional foods, wage labor, education, or health services) is generally found to be associated with higher levels of plasma cholesterol and triglycerides. Examples are available from South Africa (Walker and Arvidsson 1954), Guatemala (Scrimshaw et al. 1957),

327

Greenland Eskimos (Bang et al. 1971), Puerto Rico (Garcia-Palmieri et al. 1972), Solomon Islands (Page et al. 1974), and many others (Knuiman et al. 1980). Most studies have cited high fat and cholesterol diets as a possible explanation for these trends, although quantitative dietary information has not always been collected. It is also difficult to distinguish the possible effects of dietary change from those of decreases in physical activity, increases in body fatness, and other possible changes associated with modernization.

Other factors capable of influencing plasma lipid levels in modern environments are investigated by another set of comparative studies in populations whose traditional life-style includes a diet high in total and saturated fats. Examples include several groups of East African pastoralists (Murray et al. 1978; Day et al. 1976; Lowenstein 1964), Micronesians from Palau (Labarthe et al. 1973), and Polynesians from the Cook Islands (Prior et al. 1966). These populations maintain relatively low levels of plasma lipids in their traditional settings compared to their urbanized counterparts, despite consuming diets that are hyperlipidemic in experimental studies. In these cases it has been suggested that high activity levels and/or low levels of body fatness associated with traditional life-style may prevent the rise in plasma lipids typically seen in experimental studies, and that changes in these characteristics with modernization may account for the higher lipid levels in the urbanized groups. Quantitative data on dietary intake, activity patterns, and body composition have not been systematically collected in support of this suggestion, however. Nor has the interaction of body composition, plasma lipid level, and physical activity always been analyzed. In addition, these studies have generally focused on fasting plasma total-cholesterol, which is now known to consist primarily of two fractions (LDL and HDL), which have opposing influences on CVD risk in prospective studies and which bear different relationships to diet, activity, and body composition (Krauss 1982; Schwartz et al. 1982; Castelli et al. 1977).

Differences in body composition, dietary intake, and indicators of habitual activity patterns have been studied in Samoan populations differing in exposure, to examine the possible association each may have with plasma lipid levels in the context of modernization.

INDIVIDUAL PROJECTS

As shown in Table 14.1 data on plasma lipids were collected during several field seasons in Western and American Samoa and Hawaii. The lipids measured and techniques used in each season varied, with total-cholesterol measured in all studies, HDL-cholesterol measured in all except the first two Hawaiian studies, and triglycerides measured in all except the Western Samoan and specialized Tutuila samples.

The study by Hornick (1979) focused on adults of both sexes and

Table 14.1 Summary of Samoan Blood Lipid Studies

Year of study	Location	Age/Sex	Primary contrasts	Lipids measured[a]	Author
1975	Hawaii	30-59 M	Occupational Activity	TC, TG	Lukaski (1977)
1975/77	Hawaii	12-74 M/F	Descriptive	TC, TG	Klapstein (1978)
1978	Hawaii	> 20 M/F			
1978	Tutuila	> 20 M/F	Residence	TC, TG,	Hornick
1978	Ta'u	> 20 M/F		HDL-C	(1979)
1978	Tutuila	> 20 M	Leisure Activity	TC, HDL-C	Hornick (1979)
1982	W. Samoa	20-35 M	Residence, Occupation, Leisure Activity	TC, HDL-C	Pelletier (1984)

[a]TC = total-cholesterol; TG = triglycerides; HDL-C = HDL-cholesterol.

included 108 residents of the village of Fitiuta on the island of Ta'u, American Samoa, 153 residents and employees of villages and businesses on Tutuila, and 107 residents of rural and urban areas of Oahu, Hawaii. Klapstein (1978) studied 94 adults of both sexes living on Oahu who were volunteers from a larger anthropometric survey. These groups are used here to examine the overall effects of modernization on plasma lipids by contrasting the mean values on Ta'u with those on Tutuila, while any additional effect due to migration to Hawaii is examined by comparing migrants with the Tutuila sample.

The other studies shown in Table 14.1 focused on select groups of men chosen primarily for contrasts in activity patterns. These include an opportunity sample of 30 men in high- and low-activity occupations selected from residents of rural and urban communities on Oahu (Lukaski 1977) and 30 members of a rowing team on Tutuila who are contrasted with the general Tutuila sample surveyed by Hornick (1979). The Western Samoan samples consist of 27 male full-time residents of a rural Upolu village (Salea'aumua) and three groups of men employed in urban Apia: 32 men in physically active occupations who have sedentary leisure-time activities; 24 men in sedentary occupations who also have sedentary leisure activities; and 24 sedentary workers in Apia who play sports in their leisure time. Together, these specialized studies are used to examine the effects of physical activity on plasma lipids in the three different environmental settings on Western and American Samoa and Hawaii. In addition, data are available on the anthropometric characteristics of all these men, while quantitative data on dietary intake, activity patterns, or

work capacity are also available on some of these men or on other
residents of the same locations.

Comparison of techniques

The study of Samoan migrants by Lukaski (1977) measured serum total-
cholesterol and triglycerides in frozen serum and employed the methods of
Block and Jarett (1969) for triglycerides and Abell et al. (1952) for
cholesterol. This study and the others that measured triglycerides
requested that subjects fast overnight before blood was drawn. Klapstein
(1978) measured total-cholesterol from frozen plasma using Hycell Stable
Cholesterol Reagent, and triglycerides were measured with the Technicon
Auto-Analyzer based on Kessler and Lederer (1965). Hornick (1979),
also using frozen plasma, measured triglycerides by the Auto-Analyzer I
method, while total-cholesterol and the cholesterol content of the HDL
fraction were measured by the Dow technique described by Wybenga et
al. (1970). The HDL fraction was separated using the magnesium-
chloride sodium phosphotungstate technique as described by Lopes-Virella
et al. (1977). Pelletier (1984) measured total- and HDL-cholesterol in
fresh plasma using the same techniques as Hornick described above.
Comparison of the results with those obtained by a reference technique
(Zak et al. 1954; Abell et al. 1952), however, showed that the Dow technique gave values 28 mg/dl higher on average than the reference technique,
although there was a high correlation between the two ($r = .97$).
Accordingly, the values for total- and HDL-cholesterol in the Western
Samoan samples have been adjusted to the reference method of Abell/Zak.
Due to the many methodological differences among these studies, comparisons across studies should be evaluated cautiously, while comparisons
within studies can be considered more reliable.

GROUP COMPARISONS

Figure 14.1 shows the mean levels of total- and HDL-cholesterol and triglycerides for men and women in all locales for which data are available.
In general, total-cholesterol is lower among Western Samoan men, but
there is very little additional variation among residents of Ta'u, Tutuila,
and Hawaii. The one exception is the study of select samples of men in
Hawaii by Lukaski (1977), which found total-cholesterol to be higher than
the other Hawaiian and American Samoan groups. Among women, total-
cholesterol does not vary by residence in the three groups studied by
Hornick (Ta'u, Tutuila, and Hawaii), while the Hawaiian women
measured by Klapstein had relatively much lower total-cholesterol levels.
Bearing in mind the differences in methodology and the nature of the
samples, these results indicate a possible tendency toward higher total-
cholesterol levels in American Samoa and Hawaii than in Western

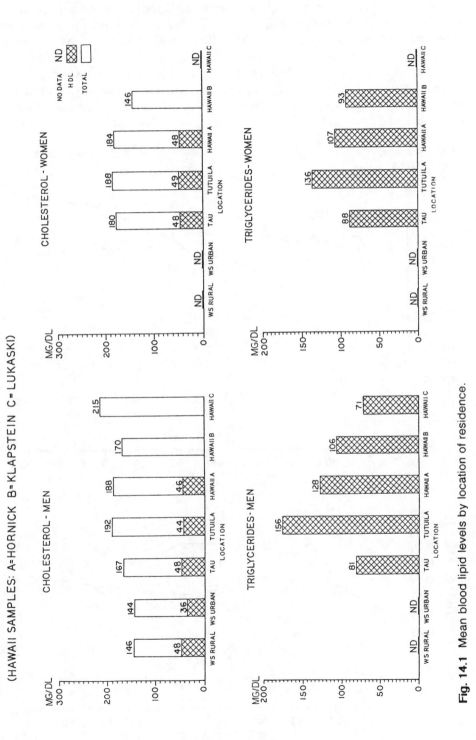

Fig. 14.1 Mean blood lipid levels by location of residence.

Samoa, but they show a relatively limited degree of variation in total-cholesterol levels across subpopulations compared to those found in other world populations (Kagan et al. 1974; Reed et al. 1970; Scrimshaw et al. 1957).

A similar conclusion holds with respect to HDL-cholesterol, which shows no variation by location among women and, with the exception of men employed in Apia, fairly uniform levels among men. It should be

Fig. 14.2 Mean total-cholesterol and triglycerides in U. S. and Samoan men and women by age.

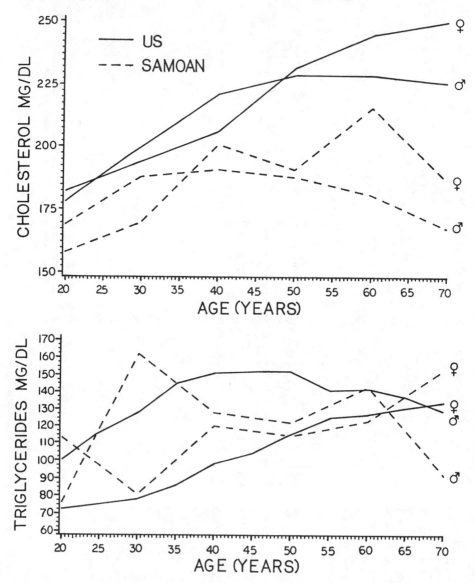

noted, however, that small differences in HDL-cholesterol have a more pronounced effect on the risk of subsequent CVD than do similar differences in total-cholesterol (Castelli et al. 1977). The figure also shows little or no differences between men and women in HDL-cholesterol, unlike the situation in U.S. populations where women typically have levels 10 to 15 mg/dl higher than men (Heiss et al. 1980; Castelli et al. 1977). A similar lack of sex difference in HDL-cholesterol has been noted in New Zealand Maoris and Marquesans (Darlu et al. 1984; Beaglehole et al. 1979).

Plasma triglycerides are higher in Tutuila men and women than in Ta'u or Hawaii. There is substantial variation in triglycerides among the three samples of men and two samples of women in Hawaii, however, making it difficult to draw further conclusions regarding the differences between Ta'u and Hawaii.

Age associations

Figure 14.2 compares the levels of total-cholesterol and triglycerides of the combined groups surveyed by Hornick to U.S. values at various ages. The data from Ta'u, Tutuila, and Hawaii have been combined in this figure because of the small sample sizes at some ages in certain locales. Total-cholesterol levels of Samoans are well below those of the United States at all ages and in both sexes. Samoan women show an increase with age, as do U.S. women, but Samoan men have fairly constant or decreasing values through middle age. These differences in age trends are such that Samoan men have mean total-cholesterol levels that are 30-58 mg/dl lower than U.S. men after age 40, and Samoan women have mean values 30-63 mg/dl lower than U.S. women after age 50. Considering the levels of obesity and blood pressure among residents of these same locales noted in Chapters 11 and 15 and the moderately high fat intake noted in Chapter 12, these total-cholesterol results are somewhat unexpected and are examined further in the next section.

In contrast to the patterns of plasma total-cholesterol, the triglyceride levels (Fig. 14.2), although erratic at various ages, are generally closer to those found in the United States. In this case, however, the average values for Samoan residents in various locales differ markedly from each other and from the United States at comparable ages; specifically (as can be calculated from data presented in Figs. 14.1 and 14.2), average triglyceride levels of Ta'u men are 71 mg/dl lower and those of Tutuila women are 50 mg/dl higher than their comparably aged counterparts in the United States.

Body composition associations

Correlations among individuals
One of the factors consistently found to be related to blood lipid levels is body composition, with higher levels of body fatness being associated with

Table 14.2 Pearson Correlation Coefficients Between Blood Lipids and Age, Body Mass Index, and the Sum of Three Skinfolds (Triceps, Subscapular, and Waist Skinfolds)

	Total-Cholesterol		HDL-Cholesterol		Triglycerides		Age	
	N	r	N	r	N	r	N	r
				Men				
Age	158	-.03	156	.12	123	-.01	—	—
Body Mass Index	156	.22[a]	154	-.38[c]	118	.14	156	-.11
Sum of Skinfolds	152	.18[b]	152	-.30[c]	120	.18[b]	152	-.17[b]
				Women				
Age	193	.24[c]	194	-.08	142	.21[a]	—	—
Body Mass Index	183	.18[a]	185	.04	137	.20[a]	200	.10
Sum of Skinfolds	189	.14[c]	190	.04	139	.17[b]	206	.02

Source: From Hornick 1979.
[a] $p < .01$.
[b] $p < .05$.
[c] $p < .001$.

higher total-cholesterol and triglycerides but lower HDL-cholesterol (Gordon et al. 1982; Glueck et al. 1980). Table 14.2 shows the Pearson correlation coefficients between these three lipids and age, body mass index (weight/height^2) and the sum of skinfolds. In men and women total-cholesterol and triglycerides are positively correlated with body mass index and skinfold thickness. In women these lipids are also positively correlated with age, but the latter is not associated with the measures of body mass or fatness, suggesting that age and body composition may have independent effects on these lipids in women. In men there is no correlation between age and blood lipids (as seen also in Fig. 14.2) and thus the effects of body fatness on total-cholesterol and triglycerides in men also appears to be independent of age. These results are consistent with those of other epidemiological studies in showing a positive association between body fatness and plasma cholesterol and triglycerides.

   The table also shows that HDL-cholesterol is negatively correlated with body fatness among men, as expected, but not among women. The reason for these results among women are not clear. Beaglehole et al. (1979) reported a weak but statistically significant correlation ($r = .138$) between HDL-cholesterol and body mass index among Maori women, similar to that found in other studies. Since the correlations in this case are based on women from three quite different environments, it is possible that variation in other factors with known influences on HDL-cholesterol (e.g. alcohol intake, cigarette smoking, physical activity, diet, medications, etc.) may be obscuring the relationship with body fatness. For instance, Heiss et al. (1980) found no correlation between HDL-cholesterol and body mass index among women taking gonadal hormones, although women in the same study not taking hormones showed the expected negative correlation. One of the factors with possible relevance in the present case is in plasma insulin. Insulin shows a strong negative correlation with plasma HDL in all groups studied. Moreover, this relationship remains significant when age, place of residence, and relative obesity based on skinfolds are controlled for. On the other hand, when insulin levels are controlled for, the correlation between adiposity and HDL in the men becomes insignificant (-.181) suggesting that this relationship may be mediated by insulin (Hornick and Fellmeth 1981).

The influence that variation in body mass index or body composition may have on plasma lipid levels is further illustrated in Fig. 14.3, which shows the mean total-cholesterol levels of various populations and sub-populations plotted against the mean body mass index. The figure includes data from 12 Samoan individual samples, as well as 11 samples from other Pacific studies and 22 samples from developed countries, most of which took part in the Seven Countries study (Keys 1970). In most cases the points refer to men with an average age of 35-45 years, with the notable exceptions being the three Western Samoan samples studied by Pelletier (1984) where the average age is 25 years.

Correlations among populations

Figure 14.3 shows a clear increase in total-cholesterol with increasing body mass index among the Samoan groups ($r = .81$). Similar trends are evident for the 11 other Pacific groups ($r = .72$) and the 22 samples from

Fig. 14.3 Mean total-cholesterol in relation to body mass index in various world populations.

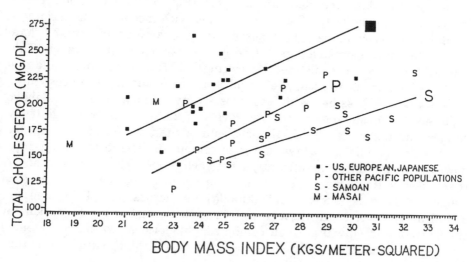

developed countries ($r = .65$). It may thus be that Samoans living in the more modernized environments of Hawaii and Tutuila are characterized not only by somewhat higher total-cholesterol levels but also higher body mass index relative to the residents of Ta'u and Western Samoa. This is consistent with the hypothesis that for Samoans, as well as other populations, some of the variation in plasma cholesterol levels may be due to differences in body fatness and/or factors associated with it among residents of contrasting environments (Baker and Hanna 1981; Garcia-Palmieri et al. 1972).

The figure also reveals that Samoans, and to a lesser extent other Pacific populations, have much lower total-cholesterol levels at any level of body mass index than the levels found in developed countries. Unlike the Tarahumara and the Japanese in Japan, who also have low total-cholesterol levels, however, the Samoans and other Polynesian groups in this figure have relatively low total-cholesterol levels with moderate fat intakes, a high proportion of which is saturated (see Chapter 12). Similarly, the figure shows that the rural Masai have total-cholesterol levels within the developed country range, and that the urban Masai are on the high side of the range, when their relatively low body mass index is taken into consideration. This further highlights the unusual character of the Samoan results, since it has been suggested that the rural Masai maintain low plasma cholesterol levels despite their hyperlipidemic diet through an efficient negative feedback system for suppression of endogenous cholesterol synthesis (Biss et al. 1971). From this figure it would appear that Samoans and other Pacific people may also possess a physiology characteristic that results in low plasma cholesterol levels relative to body fatness and dietary fat intake.

Interactions with residence

Based on the data for Samoan samples (Fig. 14.3) variation in body fatness may be one of the factors associated with modernization related to variation in blood lipids. This possibility is examined in greater detail using the anthropometric and blood lipid data from surveys on Ta'u, Tutuila, and Hawaii conducted by Hornick (1979). Since the same lipid analysis techniques were applied in all locales, stronger inferences can be drawn than is possible with the cross-study comparisons used above. Ta'u men have significantly lower levels of total-cholesterol and triglycerides and higher levels of HDL-cholesterol than the Tutuila men alone or than the Tutuila and Hawaiian men combined. The latter two groups, however, do not differ significantly from each other. The table also shows that these differences in blood lipids are paralleled by differences in weight, body mass index, and skinfold thicknesses; the Ta'u men are significantly lower in all three variables than either Tutuila men alone or the Tutuila and Hawaiian men combined. Modernization, as represented by the sociocultural differences between Ta'u and Tutuila, is thus

Table 14.3 Age, Anthropometric, and Plasma Lipid Measurements of Men and Women from Ta'u, Tutuila, and Hawaii

Variable	Ta'u			Tutuila			Hawaii		
	N	Mean	SD	N	Mean	SD	N	Mean	SD
Men									
Age (yr)	46	53.0	14.8[a]	58	41.0	16.9[b]	58	43.0	14.5
Height (cm)	45	170.6	7.0	55	172.2	5.6	58	169.9	14.5
Weight (kg)	45	76.8	13.3[a]	51	87.7	21.6[b]	58	90.4	19.0
BMI[c] (kg/m^2)	45	25.7	5.0[a]	51	29.6	10.0[b]	58	29.3	9.0
Skinfolds[d]	45	18.4	8.3[a]	51	27.6	12.7[b]	58	24.6	11.3
Triglycerides (mg/dl)	32	81.3	29.7[a]	37	156.2	40.0[b]	53	127.7	68.
Cholesterol (mg/dl)	46	167.4	36.9[a]	50	191.9	40.4[b]	61	188.0	43.4
HDL-Cholesterol (mg/dl)	46	48.4	8.0[a]	57	44.3	8.0[b]	62	45.9	
Women									
Age (yr)	54	45.0	14.1[a]	104	37.0	15.6[b]	49	42.0	14.9
Height (cm)	54	162.9	5.8[a]	102	158.2	11.6[b]	45	159.7	5.7
Weight (kg)	54	83.3	17.5	99	83.7	18.7	45	83.4	24.3
BMI[c] (kg/m^2) (1)	54	31.0	5.0	99	32.0	14.0	45	30.0	12.0
Skinfolds[d]	54	35.7	10.2	102	37.6	13.1	45	32.0	12.3
Triglycerides (mg/dl)	44	88.5	34.5[a]	56	136.5	93.5[b]	43	107.0	52.7
Cholesterol (mg/dl)	49	180.2	40.3	96	188.3	76.8	48	184.3	41.6
HDL-Cholesterol (mg/dl)	45	47.8	8.8	101	49.1	10.5	48	48.5	7.1

Source: From Hornick 1979 and Hornick and Hanna 1982.

[a] Ta'u mean differs significantly from Tutuila and Hawaii combined.

[b] Ta'u mean differs significantly from Tutuila mean.

[c] BMI = Body Mass Index.

[d] Mean of triceps, subscapular, and waist skinfolds.

associated with altered lipid levels and increases in body fatness; however, in the absence of further increases in body fatness, migration to Hawaii is not associated with further change in blood lipid levels. It should also be noted that dietary surveys in these same locales found total fat intake to be higher among Tutuila men (30% of calories) than on Ta'u or Hawaii (both with 23% of calories), thus making it difficult to distinguish the possible effects of body composition from dietary intake on blood lipids of these three groups of men (see Bindon 1984b).

The pattern of blood lipids in the three groups of women is different from that seen in the men and is not entirely consistent with the macro-environmental characteristics or the personal characteristics for which data are available. The fairly uniform levels of total- and HDL-cholesterol in women from all three locales does conform to the similarity in body fatness; however, the dietary data collected on a different set of women in each locale, indicate greater fat consumption by women in Tutuila and Hawaii (30% of calories) than by Ta'u women (23%). On the basis of that evidence, higher levels of plasma cholesterol and triglycerides would have been expected among Tutuila and Hawaiian women compared to Ta'u, but the only observed difference was the higher levels of triglycerides in Tutuila compared to Ta'u. The reasons for these discrepancies among women are not readily apparent from the data available, but may involve inaccuracies in measuring total body fat from skinfold data in relatively obese groups such as these. Additional possibilities include unmeasured differences in activity patterns, seasonal differences in dietary intake, or the fact that blood samples and dietary data were obtained from different individuals during separate surveys. It should be noted, for instance, that the percentage of calories provided by total fat on Ta'u during the breadfruit season was 23 percent for adults of both sexes (Bindon 1984b) compared to 33 percent of men and 38 percent for women when breadfruit was not in season.

Physical activity effects

Physical activity is another factor that may be related to changes in blood lipids during modernization. Transition from an agricultural to a wage economy may lead to changes in physical activity through changes in both occupational and leisure activity patterns. Several studies of Samoans in different locales have found associations between work capacity and occupational and leisure activities (Chapter 13). Since blood lipids were also measured on several of those groups, it is possible to examine the extent to which changes in physical activity with modernization may also have an effect on blood lipid levels.

Occupational activity in Hawaii
The study by Lukaski (1977) examined differences in serum total-cholesterol and triglycerides among 30- to 49-year-old Samoan migrants

Table 14.4 Anthropometric, Work Capacity and Serum Lipid Data for Men in Active and Non-active Occupations in Hawaii

| | 30- to 39-year olds | | | | 40- to 49-year-olds | | | |
| | Active (N = 7) | | Nonactive (N = 7) | | Active (N = 8) | | Nonactive (N = 5) | |
Variable	Mean	SD	Mean	SD	Mean	SD	Mean	SD
Height (cm)	168.6	5.6	172.5	4.0	172.3	4.8	166.6	4.7
Weight (kg)	86.3	9.8	98.9	7.7[a]	83.8	9.3	87.5	11.9
BMI[b] (kg/m^2)	30.4	—	33.2	—	28.2	—	31.5	—
Body fat (%)	25.2	2.9	27.5	1.9	23.0	3.7	26.0	1.3
VO$_2$ (ml/kg/min)	39.4	2.9	32.5	5.0[a]	38.1	1.7	32.7	2.9[a]
Triglycerides (mg/dl)	66.6	24.9	96.1	61.1	60.9	33.1	61.0	23.7
Cholesterol (mg/dl)	192.5	32.3	230.5	26.2[a]	209.3	14.1	231.2	43.2

Source: From Lukaski 1977.
[a] Active significantly different from nonactive ($p < .05$).
[b] BMI = Body Mass Index.

to Hawaii who differed in occupational activity levels as assessed by job titles (manual laborers and landscapers versus clerical workers and the unemployed). Skinfold measurements were taken at ten sites from which an estimate of percentage body fat was calculated according to the formula by Allen et al. (1956) and a maximal exercise test was performed to estimate work capacity. Active workers are lighter and have a lower body mass index and percentage fat than nonactive workers, though the only statistically significant difference is in the weight of 30- to 39-year-olds (Table 14.4). Active workers in both age groups have significantly higher work capacity than nonactive workers. Serum total-cholesterol and triglycerides are both lower among active 30- to 39-year-olds than their nonactive counterparts, but only the cholesterol difference reaches statistical significance. Among 40- to 49-year-olds there are no differences in triglyceride levels, and the cholesterol levels are again lower among active workers, though not at a significant level.

These results on migrants to Hawaii therefore suggest possible differences in blood lipid levels among men differing in physical activity as a result of their occupations. However, the results are based on small sample sizes and thus have low power to detect statistically significant differences. Since the groups also differ in weight and body mass index, it is not possible to separate the effects of those factors from possible effects of physical activity. In addition, since there is considerable evidence that body fatness and physical activity may have opposing influences on HDL-versus LDL-cholesterol (Dufaux et al. 1982; Schwartz et al. 1982; Heiss et al. 1980), it is possible that differences in LDL-cholesterol between these groups may be partially offset by differences in the opposite direc-

Table 14.5 Comparison of General Tutuila Sample of Men with Members of the Rowing Team

Variable	Rowing Team			Tutuila Men			
	N	Mean	SD	N	Mean	SD	p
Age (yr)	30	32.6	11.2	58	41.0	16.9	.015
Height (cm)	30	174.3	5.1	58	172.2	5.5	.133
Weight (kg)	30	86.4	14.0	51	87.7	21.6	.785
BMI (kg/m^2)	30	28.4	4.0	51	29.6	10.0	.402
Cholesterol (mg/dl)	30	176.5	37.2	50	191.9	40.4	.068
HDL-Cholesterol (mg/dl)	30	51.3	7.1	57	44.3	8.0	.001

Source: From Hornick 1979 and 1982.

tion in HDL-cholesterol, with the result that total-cholesterol levels may show little variation.

Leisure activity in American Samoa

As part of the larger survey conducted in American Samoa, Hornick (1979) examined the anthropometric and blood lipid characteristics of 30 Tutuila residents who were members of a rowing team and had trained together for 6 weeks prior to measurement. These men were presumed on the basis of general observations to be similar to the general Tutuila sample in most respects other than their training regimen (Hornick and Hanna 1982). They were compared with a sample of men from the general Tutuila population studied during the same period (Table 14.5). The rowers were significantly younger than the survey sample, but did not differ significantly in height, weight, or body mass index. Total-cholesterol was lower among the active men, similar to the results found by Lukaski among migrants to Hawaii. HDL-cholesterol was higher among the active men, consistent with other cross-sectional epidemio-logical studies (Dufaux et al. 1982). It is also noteworthy that the ele-vated levels of HDL-cholesterol among the rowers did tend to reduce the differences in total-cholesterol between the two groups: if the two groups had similar HDL levels they would have a 22 mg/dl difference in total-cholesterol instead of the 15 mg/dl difference that was observed.

BODY COMPOSITION, PHYSICAL ACTIVITY, AND DIET IN WESTERN SAMOA

The study in Western Samoa also examined the possible relevance of variation in physical activity to blood lipid levels and was designed to distinguish the effects of occupational from leisure activity. In order to further distinguish those effects from other characteristics, data were also collected concurrently on the same men on body composition, dietary intake, alcohol and cigarette use, and other personal characteristics. In addition, the presumed differences in physical activity among various groups were confirmed by quantifying daily energy expenditure and estimating work capacity through submaximal exercise tests. The study focused on four select groups of men in order to limit variation due to age, sex, and extremes of obesity.

One of the groups consists of full-time residents of the Salea'aumua village who were engaged in traditional agriculture and fishing as their primary occupation. These are compared to men employed in physically active and sedentary occupations in the urban Apia area. All samples were restricted to 20- to 35-year-old men, and the Apia samples were restricted to those men reporting the most sedentary leisure activities in order to examine the effects of occupational activity with minimal confounding by leisure activity. The effects of leisure activity itself were examined by selecting a fourth group representing those sedentary workers who had been playing organized sports (primarily rugby, soccer, and field hockey) for about 4 weeks prior to study. The dietary intake, activity patterns, and estimated work capacity of these men is described in other chapters, but selected results will be included below as necessary to interpret the variation in blood lipids.

Table 14.6 presents the mean age, anthropometric characteristics, estimated work capacity, and plasma lipids of men in three of the four samples from Western Samoa. (The exercising sedentary workers are presented separately for clarity.) The mean height, weight, body mass index, and sum of six skinfolds is lowest among villagers, intermediate among active workers, and highest among sedentary workers. The villagers and active workers differ significantly from the sedentary workers in all these characteristics, but do not differ significantly from each other in these comparisons; they do differ significantly in body mass index and fatness in the larger samples from which these men were chosen (Pelletier 1984). These results are thus similar to those found in migrants to Hawaii by Lukaski (1977), in that body mass and fatness varies with occupation type and as such represents a potentially confounding factor on plasma lipid comparisons. The Western Samoan results further show that villagers practicing traditional subsistence weigh less and have smaller skinfold thicknesses than employed men.

Also similar to the Hawaiian results, occupation type is associated with significant differences in work capacity, with the sedentary workers having significantly lower levels than villagers or active workers. These differences are found both before and after adjustment for variation in body fatness among the groups, thereby strengthening the suggestion that habitual activity patterns, in addition to body fatness, are responsible for the observed differences in work capacity. The results of the 24-hour recalls of diet and activity also indicate significantly higher energy intake and expenditure among villagers and active workers than sedentary

Table 14.6 Age, Anthropometric Characteristics, Estimated Work Capacity, and Plasms Lipids in Western Samoan Men

Variable	Villagers			Active workers Sedentary leisure			Sedentary workers Sedentary leisure		
	N	Mean	SD	N	Mean	SD	N	Mean	SD
Age (yr)	27	25.4	4.0	32	25.8	4.3	24	23.6	3.1[a,b]
Height (cm)	27	169.4	4.5	32	169.5	4.5	24	173.2	5.2[a,b]
Weight[d] (kg)	27	69.9	7.3	32	72.1	7.6	24	79.4	10.4[a,b]
BMI[d] (kg/m^2)	27	24.3	2.4	32	25.1	2.4	24	26.4	2.4[a,b]
Skinfold Sum[d,e]	27	56.2	16.9	32	65.0	23.0	24	93.5	24.2[a,b]
Estimated VO_2 Max[f]	23	48.7	7.5	29	54.9	9.0	22	43.3	7.1[a,b]
Total cholesterol (mg/dl)	25	145.7	18.4	29	142.1	22.9	23	151.6	27.2
HDL-cholesterol (mg/dl)	25	48.1	10.4	29	38.0	5.5	23	35.2	7.1[a]
Non-HDL-cholesterol (mg/dl)	25	97.6	21.3	29	104.2	21.7	23	116.4	27.2[a,b]

Source: From Pelletier 1984.

[a] Sedentary workers differ significantly from villagers ($p < .05$).
[b] Sedentary workers differ significantly from active workers (all $p < .05$ except non-HDL-cholesterol, where $p = .05$).
[c] Active workers differ significantly from villagers ($p < .05$).
[d] Adjusted for age by analysis of covariance.
[e] Skinfold sites include triceps, forearm, medial calf, subscapular, midaxillary, and waist; values shown are in millimeters, but significance of differences is based on log-transformed data.
[f] Estimated from submaximal step-tests; values are in ml/kg/mn and are adjusted for temperature and the sum of six skinfolds.

342

workers, with the differences in energy expenditure arising from occupational activities only (Pelletier 1984).

In light of these significant differences in body fatness, work capacity, and energy expenditure, it is interesting to note that plasma total-cholesterol levels do not vary significantly among these three groups. This finding contrasts with those from American Samoa and Hawaii, in which the more active men tended to have lower total-cholesterol levels. HDL-cholesterol, however, does vary among the groups, with the villagers having significantly higher levels than either employed group. When HDL-cholesterol is subtracted from total-cholesterol to estimate the concentration of cholesterol in all other plasma fractions combined, the results show that villagers and active workers have lower levels than sedentary workers, with differences at least approaching significance ($p < .06$). It thus appears that when variation in HDL-cholesterol is controlled, the remainder of total-cholesterol does vary in a manner consistent with the differences in physical activity and body composition among these groups and is consistent with the findings from American Samoa and Hawaii.

In an attempt to distinguish the possible influences of physical activity, body fatness, dietary intake, and alcohol and cigarette use, the blood lipid data were examined in relation to these variables through simple comparison of group means and through multifactor analysis of variance and covariance. First, the possible influence of dietary intake is examined in Table 14.7, with data obtained from the 24-hour recalls from each man. The three groups have remarkably similar nutrient intakes, with roughly 10 percent of calories derived from protein, 60 percent from carbohydrate, and 30 percent from total fat, with a P/S ratio (polyunsaturated to saturated fats) of 0.15 to 0.18 and cholesterol intakes ranging from 54 to

Table 14.7 Nutrient Composition of the Diet for Three Groups of Western Samoan Men

	Villagers	Active workers Sedentary leisure	Sedentary workers Sedentary leisure
Number of recalls:	73	86	67
Nutrient	Median	Median	Median
Protein (% Kcal)	8.5	8.7	10.7[a]
Carbohydrate (% Kcal)	62.0	57.5	58.2
Total fat (% Kcal)	29.6	33.3	31.6
P/S ratio[c]	0.18	0.16	0.15
Cholesterol (mg/1000 Kcal)	54.2	64.2	86.7[b]

[a] Sedentary workers differ significantly from villagers or active workers ($p < .05$).
[b] Sedentary workers > active workers > villagers (all $p < .05$). [Statistical tests based on Wilcoxon Two-Group Analysis of Variance (two-tailed tests).]
[c] P/S = polyunsaturated/saturated fatty acids.

Table 14.8 Mean Levels of Plasma Total-, HDL-, and Non-HDL-Cholesterol among Western Samoan Men after Adjustment for the Sum of Three Trunk Skinfolds by Analysis of Covariance

Variable	Villagers Mean	Active workers Sedentary leisure Mean	Sedentary workers Sedentary leisure Mean
Total Cholesterol (mg/dl)	150.3	144.1	146.9
HDL-Cholesterol (mg/dl)[a]	42.9	38.5	36.0[b]
Non-HDL-Cholesterol (mg/dl)	102.3	106.3	111.5

[a] Regression model includes a group × skinfolds interaction term.
[b] Village > sedentary workers ($p = .02$).

87 mg/1000 Kcal. While some of the differences among groups do reach statistically significant levels, it is clear that the differences are very slight in terms of their potential influence on plasma cholesterol levels. Since these data were collected in the one to three weeks prior to blood collection on the same men from whom blood samples were obtained, it appear that the differences in plasma cholesterol levels are probably not a result of variation in the intake of these nutrients.

The reported use of alcohol and cigarettes was examined for potential influence on plasma cholesterol levels by multifactor analysis of variance, with habitual alcohol and cigarette use coded as simple dichotomous (yes/no) variables. Reported alcohol use is not associated with any of the lipid variables, while cigarette smoking is associated with significantly lower HDL-cholesterol levels ($p = .04$) and shows a trend for higher non-HDL-cholesterol levels ($p = .14$). Controlling for cigarette and/or alcohol use has no effect on the group mean cholesterol levels shown above, however, due to the fact that a similar proportion (about 50%) of the men in each sample reported these behaviors. On the basis of this relatively crude control for alcohol and cigarette use, they do not appear to have any large confounding effect on the plasma cholesterol levels.

The relative effects of variation in physical activity and body composition were examined by comparing the mean plasma cholesterol levels among the three groups after controlling for anthropometric variables by analysis of covariance. In these analyses the sum of three trunk skinfolds was the strongest covariate and its effects on the group means are therefore shown in Table 14.8. Adjusting for fatness differences has little effect on total-cholesterol but, as expected, reduces the intergroup variation in HDL- and non-HDL-cholesterol. As a result the HDL-cholesterol of villagers is no longer significantly higher than active workers, but is still significantly higher than sedentary workers; there are no significant differences in non-HDL-cholesterol after fatness-adjustment, although the same trend exists for lowest levels among villagers and highest levels among sedentary workers.

These analyses therefore suggest that some, but not all, of the variation in HDL- and non-HDL-cholesterol among these groups is due to their differences in body fatness. It appears that the differences in HDL-cholesterol originally observed between villagers and active workers is accounted for by the higher body fatness of the latter, while a residual difference persists between villagers and sedentary workers that is potentially attributable to the differences in activity levels. By contrast, all of the significant differences in the non-HDL portion of total-cholesterol appear to be related to differences in body fatness. It should be noted, however, that since body fatness and activity levels are positively correlated with each other, controlling for body fatness involves, to some extent, controlling for activity levels as well. It is therefore possible that analyses such as these may overestimate the effects of fatness by including its covariation with activity levels.

As noted previously, the Western Samoan study also included a group of sedentary workers who were active in their leisure time by virtue of participation in organized sports. Table 14.9 presents the anthropometric, work capacity, and blood lipid data for this group of men and compares it to the three groups discussed above. The sportsmen are anthropometrically similar to their coworkers who do not play sports: they are therefore significantly taller, heavier, and fatter than the villagers and active workers. Their estimated work capacity is significantly greater than their more sedentary coworkers and less than that of the active workers, consistent with their intermediate position in total daily energy expenditure as estimated from 24-hour recalls (Pelletier 1984). The only significant difference in plasma cholesterol levels is in HDL-cholesterol, which is lower among the sportsmen than the villagers. There is the suggestion of lower plasma total-cholesterol relative to their more sedentary coworkers, but this does not reach significant levels ($p = .09$). Thus, despite having a higher estimated work capacity and daily energy expenditure than their sedentary coworkers, the sportsmen do not differ significantly in body weight, body fatness, or plasma cholesterol levels. This may relate to the seasonal nature of their participation in sports, since they had only exercised for about 4 weeks prior to blood collection, or may indicate that changes in plasma cholesterol levels may require changes in body composition together with chronic elevation of activity levels.

OVERVIEW

The variation in blood lipids among Samoans shows some similarities but also some differences compared to the patterns observed in other Pacific populations. Some of these comparisons for men are presented in Table 14.10. Total-cholesterol is higher among all the urban groups in all populations except Western Samoa. The study in Western Samoa

Table 14.9 Comparison of Western Samoan Sedentary Workers Who Play Sports with the Other Three Western Samoan Groups

Variable	Sedentary workers Active leisure			Villagers	Active workers Sedentary leisure	Sedentary workers Sedentary leisure
	N	Mean	SD	Mean	Mean	Mean
Age (yr)	24	23.0	2.2	25.4[a]	25.8[a]	23.6
Height (cm)	24	173.1	6.3	169.4[a]	169.5[a]	173.2
Weight[b] (kg)	24	82.4	11.6	69.9[a]	72.1[a]	79.4
BMI[b] (kg/m^2)	24	27.3	2.8	24.3[a]	25.1[a]	26.4
Skinfold Sum[b,c] (mm)	24	88.9	32.2	56.2[a]	65.0[a]	93.5
Estimated Work Capacity[d]	23	49.0	6.2	48.7	54.9[a]	43.3[a]
Total-cholesterol (mg/dl)	23	140.6	20.1	145.7	142.1	151.6
HDL-Cholesterol (mg/dl)	23	35.2	6.8	48.1[a]	38.0	35.2
Non-HDL (mg/dl)	23	105.4	21.9	97.6	104.2	116.4

[a] Significantly different from sedentary workers with active leisure.

[b] Adjusted for age by analysis of covariance.

[c] Skinfold sites include triceps, forearm, medial calf, subscapular, midaxillary, and waist. Values shown are in millimeters but significance tests are based on log-transformed data.

[d] Estimated from submaximal step-tests. Values are in ml/kg/mn and are adjusted for temperature and the sum of six skinfolds.

Table 14.10 Mean Plasma Cholesterol and Triglycerides in Some Pacific Island
Populations (Men Only) [a]

Population	Ages	Cholesterol (mg/dl) Rural	Urban	triglycerides (mg/dl) Rural	Urban	References
Cooks	20-40	165	182	82	118	Prior et al. 1966b
	40-70	185	204			
Tokelau	20-35	204	—	46	—	Prior et al. 1981
	35-65	218	—	57	—	
Marquesas	16-62	201	191	136	157	Darlu et al. 1984
Maori	20-35	—	218	—	146	Beaglehole et al. 1979
Palau	20-60+	147	171	91	139	Labarthe et al. 1973
Marianas	20-40	182	190	—	—	Reed et al. 1970
	40-60	198	203	—	—	
Solomons	15-70+	119	—	—	—	Page et al. 1974
New Guinea Highlands	20-40	145	—	124	—	Sinnet and Whyte 1973
	40-60	148	—	126	—	
Western Samoa	20-35	166	157	64	84	Zimmet 1980
	35-64	189	183	86	118	
Western Samoa	20-35	146	144	—	—	Pelletier 1984
American Samoa	20-75+	167	192	81	155	Hornick 1979
Hawaii	20-75+	—	188	—	128	Hornick 1979
Hawaii	20-74	—	170	—	106	Klapstein 1978

[a] Where necessary averages have been calculated as weighted averages of several age groups. top
1.37i

reported here and that done earlier by Zimmet (1980), both found total-cholesterol to be similar among rural and urban men. The most striking finding, however, is that with the possible exception of older Maoris, the mean total-cholesterol levels of Pacific Islanders are well below those found in European-derived populations (see also Figs. 14.2 and 14.3). Unlike the Masai, Tarahumara (Connor et al. 1978; Day et al. 1976), or Japanese (Kagan et al. 1974), who also have low levels of total-cholesterol, the low levels in Samoans and other Pacific Islanders are found in conjunction with both high levels of body fatness and moderate to high levels of saturated fat intake in certain populations. Whether this results from the influences of other dietary, life-style, or environmental factors common to these Pacific populations, or from distinctive biological characteristics of these groups is not known with certainty; however, the diverse range of environments and life-styles represented in this table would argue against the existence of common environmental factors that could account for this finding.

In contrast to the situation with cholesterol, plasma triglycerides of urbanized Samoans and other Pacific Islanders are roughly comparable to values found in developed countries (see Fig. 14.2). Triglyceride levels in the rural areas of these populations are generally much lower than the urban areas, in agreement with findings in other world populations (Garcia- Palmieri et al. 1972; Bang 1971).

While the data for comparison of HDL-cholesterol levels are scarcer, the results for Samoans are in agreement with those of Beaglehole et al. (1979) for the Maoris and Darlu et al. (1984) for the Marquesans. HDL-cholesterol was similar in men and women for both of these populations rather than the higher female values typically found in populations of European ancestry (Heiss et al. 1980). In addition, the low levels of HDL-cholesterol reported here for Western Samoan men working in Apia agree with the reports of low HDL-cholesterol levels among Maori adolescents and adults (Beaglehole et al. 1979; Stanhope and Sampson 1977), Marquesans (Darlu et al. 1984), and with the reports of low apoprotein A-1 levels in Western Samoans and New Caledonians (Nestel and Zimmet 1981). It is of further interest to note that the study of Maori adolescents by Stanhope and Sampson (1977) found that the absolute level of HDL-cholesterol and the size of the male/female difference both decrease with increasing Maori ancestry. Similar studies among Samoans and other Pacific populations would be most valuable for examining the possible reasons for the distinctive patterns of total- and HDL-cholesterol found in these populations.

The studies of select groups of men from Western and American Samoa and Hawaii represent attempts to examine the possible relevance of activity patterns to blood lipid variation in each of these locales. In each case comparison of the more active with the less active men provides some evidence that plasma cholesterol may be lower among more active men. Since this result was found in all three environments, it indicates that variation in physical activity as a result of occupation and/or leisure activities may be one of the specific mechanisms by which modernization could affect blood lipids as was suggested by earlier investigators (Day et al. 1976; Labarthe et al. 1973; Prior and Davidson 1966). Moreover, the data from Western Samoa suggest that activity may have an effect on HDL-cholesterol independent of body fatness, diet, alcohol, and cigarette use, while activity and body fatness may have an indistinguishable combined effect on non-HDL-cholesterol as a result of participation in the wage economy.

In addition to the possible effects of physical activity on blood lipids, it is clear from experimental evidence that other factors such as nutrient composition of the diet may play a role in the changes in blood lipid levels with modernization in some cases. It is difficult to discern the influence that dietary composition may have in the case of Samoans, however, because of apparent seasonal changes in dietary composition found on the islands of Manu'a and because of the similarity in dietary composition among the four groups studied in Western Samoa.

Chapter 15

Blood Pressure of Samoans

STEPHEN T. MCGARVEY
DIANA E. SCHENDEL

This chapter presents the blood pressure studies in the context of the overall project objective: to explain biocultural adaptations to the physical and sociocultural environment, past and present, among Samoans residing in areas from traditional Western Samoa to modern California. The chapter is divided into two sections after the general material is presented: (1) *children*, including juveniles and adolescents aged 4-22 years, and (2) *adults*. We emphasize the overall blood pressure patterns among different Samoan residence groups and the relative effects of biological and sociocultural factors on blood pressure variation.

Blood pressure is variously studied depending on perspective and purpose, and during the Samoan research project of almost 10 years, several investigators produced different perspectives and purposes. There are, however, at least three approaches common to human population biology in our blood pressure studies. First, we emphasized the specific Samoan population characteristics of blood pressure and its covariates relative to other populations in a comparative and human variability focus. Second, we emphasized blood pressure as a trait responsive to large-scale environmental changes such as modernization and migration in a human adaptability focus. Third, we emphasized the multiple causes of blood pressure variation and the individual and interactive roles of various biological and cultural factors in a human ecology focus. These three foci are not exhaustive nor mutually exclusive, but constitute the major theme in our research.

HYPOTHESES

The primary hypothesis of the blood pressure research is that modernization of Samoan life is associated with increased levels of blood pressure. Modernization change and its operational definition have been discussed in detail in earlier chapters. It is only important to amplify that we do not regard *traditional* Samoan life as stable, unchanged, or even archetypal. Biocultural adaptation is a process and changes responsive to precontact demography and politics, European contact, Christian missionization, and early U.S. influence all precede the age of most of our subjects and of course our research itself. Modernization is not a single event or influence imposed on a pristine society and culture. This chapter shows the difficulties of measuring modernization easily and assessing its effects on blood pressure. Nevertheless, we tried to detect (1) whether residence-based ecological differences in exposure to modern life *in situ* are related to blood pressure, and (2) whether migration from the more traditional life of Samoa to the modern world outside of Samoa is associated with blood pressure.

The secondary hypotheses pertain to specific individual changes in biology and behavior resulting from exposure to modern life among Samoans. Thus, we are concerned with specific variables and their individual and interactive association with blood pressure. These secondary hypotheses include: (1) age increases in blood pressure are directly associated with exposure to modern life; (2) adiposity is directly related to blood pressure; (3) salt intake, due to modernization of the diet, is directly related to blood pressure; and (4) psychosocial factors due to modernization are related to blood pressure.

METHODS

Research design

The research design differed according to each study but two basic types were used. First, the ecological designs focused on the broad areal comparisons and contrasts of different locales, for example, the difference in mean adult pressure between American Samoa and Hawaii. This design was generally used to address the primary hypothesis. Second, the cross-sectional design focused on the correlates of blood pressure in different locales and on modernization differences after considering other biocultural factors. Most often, this design was used to determine the concomitants of blood pressure variation, for example, does adiposity have a similar association with blood pressure in Samoans living traditionally and urban, migrant Samoans? The overwhelming majority of our research was cross-sectional and addressed the secondary hypotheses by partitioning the blood pressure variance attributable to specific factors.

Populations

Blood pressure was studied among the Samoan communities in the San Francisco Bay area of California, on the island of Oahu, Hawaii, most villages from three regions of American Samoa, and selected villages in traditional Western Samoa.

These populations were chosen to represent the spectrum of Samoan experience from most traditional to most modern: Western Samoa, Manu'a in American Samoa, Tutuila, the large island of American Samoa, the Pago Pago harbor region of American Samoa, Hawaii, and California. Samples were chosen from these locales during the project for both general survey and specific research. Study populations and samples were slightly different for children and adults and they are described at the start of each section on children and adults.

We recruited subjects from a range of communities to represent the diversity of employment, length of residence, education, and area of origin in Samoa. At the time of our studies there were no censuses of Samoans in California or Hawaii. In American Samoa comparisons of our adult study samples with census data, available afterward, show a strong similarity in sociodemographic characteristics and a good representation of the overall population, except in one group of villages in the eastern district of Tutuila island. There our samples are biased toward adults outside of the wage economy (Bindon 1981). In Western Samoa the villages were almost completely recruited, but no comparisons were made with the Western Samoa census materials. We cannot assess representativeness of the samples because random sampling techniques were not used in any location. There may be selection bias, but we do not know in what direction.

Blood pressure measurement

Blood pressure was measured as part of a broad human biological investigation of rapid sociocultural change. It must be made explicit that none of the blood pressure research reported here took place as the sole and primary objective during any years of the project. Not as much time and money were spent on precision in blood pressure measurement and estimating and reducing the error (at least compared to studies explicitly focused on blood pressure). Nevertheless, although the overall objective of the research took precedence, the blood pressure studies were one of the specific goals and were generally marked by a consistent attention to proper measurement. These data cannot be referred to as "casual" blood pressures.

Measurement took place in church halls, community centers, and homes of prominent Samoans in California and Hawaii, as well as in individual homes in Hawaii. In American and Western Samoa measurement took place always in unoccupied guest houses in the villages. Under these conditions it was rarely possible to isolate subjects in a room for the blood pressure measurement and instead a relatively undisturbed corner of a large room, hall, or guest house was used. Blood pressure was measured in the seated position after quietly resting seated for a minimum of 5 minutes and after having refrained from cigarette smoking.

Mercury sphygmomanometers (Baumanometers) were used in all study locales and these were regularly checked and cleaned. Several cuff sizes from child through adult thigh size were used with the mercury sphygmomanometers and these were placed on the upper arm according to standard procedure. In Hawaii in 1975 an aneroid instrument with extra large cuff was used for those with massive upper arm circumference, and no differences were found between it and the mercury manometers.

In Western Samoa in 1979 an aneroid sphygmomanometer was also used by one of the two observers during the village surveys. Unfortunately, a discrepancy between this aneroid instrument and a mercury manometer was detected by several observers when reliability checks were performed. The decision was made to add the consistent 4 mmHg difference to the aneroid determined blood pressures in Western Samoa in 1979.

In Western Samoa in 1982 another aneroid sphygmomanometer was used in one village by one observer working alone. There was a consistent discrepancy between this aneroid device and a mercury manometer against which it was later checked. The regression equation

$$BP = BP + (BP \times .021) + .412$$

was derived from the reliability check. Note this would increase a measurement of 100 mmHg to one of 102.5 mmHg and one of 140 mmHg to 143.5 mmHg.

The first and fifth Korotkoff sounds were used to determine systolic and diastolic pressure, respectively. The second of the two readings one minute apart was recorded to reduce the effect of anxiety on measurement. In California one observer took the measurements. In Hawaii during the two field seasons, ten observers were used, but the majority of the readings were performed by four individuals. In American Samoa in 1976 three observers took blood pressures, with most done by two. In Western Samoa in 1979 two observers took readings, while in 1982 one person took the measurements. All individuals received blood pressure training from the same two individuals, but there were no rigorous inter- or intraobserver reliability checks. A post-hoc check on 2-year reliabilities on 160 subjects in Hawaii showed correlations of better than .5 (McGarvey and Baker 1979), comparable to other large community studies. It is

Table 15.1 Blood Pressure by Age Group of Samoan Boys — Means ($\bar{X}$), Standard Deviations (SD), and Sample Sizes (N)

Age	WeSam79			WeSam82			AmSam78			Hawaii		
	$\bar{X}$	SD	N	$\bar{X}$	SD	N	$\bar{X}$	SD	N	$\bar{X}$	SD	N
						Systolic						
4-5	99.0	7.5	17	111.0	8.9	7	92.8	8.3	5	97.8	7.8	34
6-7	97.3	8.9	30	110.2	8.9	27	91.6	11.1	10	102.5	10.3	39
8-9	104.4	6.2	16	113.7	9.7	15	102.1	6.0	18	104.9	13.3	32
10-11	98.5	6.0	19	118.8	11.8	21	100.1	9.9	13	104.9	13.3	40
12-13	105.0	8.0	19	120.0	7.8	15	113.4	16.3	8	109.3	10.6	36
14-15	109.4	11.0	15	123.9	10.4	17	116.9	12.9	7	113.4	9.6	30
16-17	116.8	10.1	8	131.2	14.5	7	126.7	8.9	6	118.5	8.8	27
						Diastolic						
4-5	63.1	5.9	17	64.6	10.5	7	54.8	7.8	5	64.3	8.6	34
6-7	62.0	5.1	30	69.9	11.3	27	60.1	11.7	10	70.4	8.1	39
8-9	68.4	6.1	16	69.0	8.6	15	67.6	5.4	18	71.0	9.2	32
10-11	66.7	7.6	19	72.3	8.6	21	68.0	10.4	13	71.3	8.6	40
12-13	71.5	6.3	19	75.7	9.9	15	73.5	13.7	8	71.3	8.7	36
14-15	66.9	6.1	15	75.1	9.8	17	71.7	9.9	7	72.9	7.6	30
16-17	74.4	7.5	8	74.5	6.9	7	78.2	12.5	6	77.8	10.1	27

Table 15.2 Blood Pressure by Age Group of Samoan Girls — Means ($\bar{X}$), Standard Deviations (SD), and Samples Sizes (N)

Age	WeSam79			WeSam82			AmSam78			Hawaii		
	$\bar{X}$	SD	N	$\bar{X}$	SD	N	$\bar{X}$	SD	N	$\bar{X}$	SD	N
Systolic												
4-5	98.0	9.1	12	133.6	6.4	6	90.2	9.5	9	96.5	11.9	28
6-7	103.5	7.3	17	114.4	10.3	11	97.6	6.3	9	98.7	10.5	41
8-9	106.1	10.4	23	116.1	10.2	16	103.4	12.0	10	106.7	10.9	42
10-11	102.3	10.9	20	119.5	9.1	19	107.9	5.4	17	107.2	14.0	32
12-13	100.9	7.9	22	117.7	9.9	12	104.8	8.9	10	110.1	10.1	41
14-15	111.5	9.5	11	125.7	14.2	13	107.6	8.9	7	112.7	9.8	23
16-17	115.0	10.2	17	126.2	13.6	10	110.8	13.3	10	110.9	9.6	31
Diastolic												
4-5	66.3	7.7	11	76.3	6.2	6	66.4	5.3	9	65.6	9.4	28
6-7	69.9	8.3	17	70.6	7.5	11	67.1	8.5	9	66.5	6.8	41
8-9	70.7	8.9	23	72.6	12.2	16	67.4	10.3	10	73.5	8.2	42
10-11	67.6	9.3	20	76.6	10.0	19	71.5	5.8	17	70.6	10.6	32
12-13	69.2	7.9	22	72.2	6.0	12	70.1	8.7	10	72.1	8.1	41
14-15	72.9	7.6	11	74.1	8.1	13	67.3	5.8	7	75.3	9.0	23
16-17	73.7	7.6	17	74.7	11.4	10	72.4	11.9	10	76.2	7.1	31

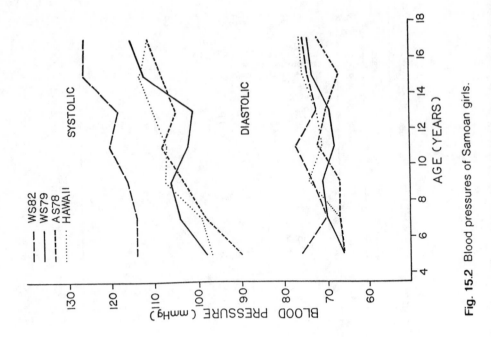

Fig. 15.2 Blood pressures of Samoan girls.

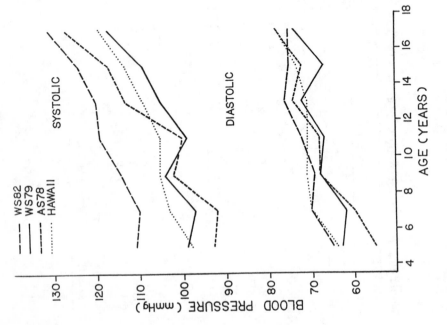

Fig. 15.1 Blood pressures of Samoan boys.

clear that we did not do enough to try to reduce the error of measurement variance in our blood pressure studies. This may affect the ability to detect associations and the precision of hypertension rates.

RESULTS: BLOOD PRESSURE AMONG CHILDREN

Blood pressure patterns in the samples

Young childhood and adolescent blood pressure data were collected in Hawaii in 1975 and 1977, in American Samoa in 1978, and in Western Samoa in 1979 and 1982. Eight communities from four areas of Oahu, Hawaii, were sampled. The American Samoan sample was collected as part of a special study of extended families from Tutuila, the main island of American Samoa, and relatives in Hawaii. The Hawaiian sample excludes the relatives of children of American Samoa so as to eliminate sample bias due to closer genetic relationship than among the other samples of children. The present American Samoan sample may be slightly different from other American Samoan children, due to their selection based on kin in Hawaii, but blood group genetic studies suggest this is not the case for adults (Parsons 1982), so we assume the same for children. Blood pressure data for children measured in Western Samoa in 1979 (WeSam79) and in 1982 (WeSam82) are reported separately because they derive from separate communities on different islands at two different times with different investigators, measurement differences as described, and because as shown next there are wide differences between the two Western Samoan samples in the age group means of blood pressure.

Tables 15.1 and 15.2 present the sample sizes, means, and standard deviations of systolic and diastolic blood pressure by age among the study samples for boys and girls aged 4 to 18. Figures 15.1 and 15.2 illustrate mean systolic and diastolic blood pressure by age, sex, and sample.

Samoan children's blood pressure tends to increase with cross-sectional age in all four study samples. Among boys the cross-sectional age increase tends to be sharper in those 12 years and older, relative to the younger boys, especially in systolic blood pressure. For girls the sharper age increase occurs among those 4 to 11 years. Boys tend to have similar or lower blood pressure than girls in the younger ages but higher values than girls in the adolescent years, with the crossover occurring around ages 12 to 14.

The age/sex blood pressure values among the samples are not very different from one another, except for the WeSam82 sample. Blood pressure of the WeSam82 sample is noticeably higher than the other samples, especially in systolic pressure. Tables 15.1 and 15.2 show that apart from a few 2-year age groups the variance is not dissimilar from the other samples, so we cannot attribute this difference to a few outliers.

Comparing Samoan children's blood pressure with other groups assesses human biological variability and provides clues about genetic and environmental influences, although caution is required due to differences in techniques. Compared to other Polynesian populations such as Tokelauan children aged 4 to 15 years residing in the Tokelauan atolls and in New Zealand (Beaglehole et al. 1977a, 1977b), Samoan children have higher blood pressure values from ages 4 to 12 years and similar values afterward. Compared to another Polynesian group, Maori children in New Zealand (Beaglehole et al. 1975), Samoan blood pressures are lower, except for the anomalous WeSam82 study sample. These comparisons parallel general differences in modernization among Polynesian groups and suggest that children's blood pressure for age increases with greater exposure to modern environments.

Despite the range of mean blood pressures in the Samoan samples, the values fall within the range of juvenile blood pressure in the United States reported by both the task force on blood pressure in children (Task Force 1977), and recent summaries of several pediatric studies (Lee et al. 1984; Fixler et al. 1980). If we accept the WeSam82 sample's measured values and correction, there is no a priori reason to be overly skeptical about the higher blood pressure values, because they fall within the range of studies of U.S. children. This Samoan population variation in blood pressure could be due to unmeasured specific local factors, analogous to U.S. variation attributed to geographic factors.

The cross-sectional age increases in blood pressure among Samoan children resemble patterns observed among many other populations of young children and adolescents. Variation in blood pressure of Samoan children by community of residence resembles community variation in the United States, Europe, and other world populations (Lee et al. 1984). Thus, comparisons with similar and dissimilar ethnic groups suggest that childhood blood pressure differences among the Samoan samples, and between them and various world populations, are probably attributable to environmental variation, perhaps specifically, exposure to factors of modern life. If there are genetic influences they are probably at the familial level rather than the population level.

Blood pressure correlates

Age

The data are cross-sectional, so it is difficult to evaluate age effects; however, analyses show age to be unrelated to young childhood and adolescent blood pressure among Samoans when simple measures of body morphology are included. Table 15.3 shows the partial correlations of blood pressure with age, weight, height, and triceps skinfold thickness across all ages by sex.

Age is only weakly and negatively related to systolic blood pressure in the Hawaiian sample. Weight is the only significant blood pressure cor-

Table 15.3 Partial Correlation Coefficients of Blood Pressure with Age, Weight, Height, and Triceps Skinfold in Samoan Children[a]

Sample	Boys				Girls			
	Age	Weight	Height	Triceps	Age	Weight	Height	Triceps
Systolic								
Hawaii	.07	.33[d]	-.07	.07	-.13[b]	.25[d]	.05	.01
AmSam	.03	.31[c]	.03	.05	-.14	.28[c]	.16	-.08
WeSam79	-.12	.34[d]	.04	.02	-.10	.27[d]	-.12	.05
Diastolic								
Hawaii	-.05	.22[d]	.00	-.01	-.02	.16[c]	-.02	.04
AmSam	.11	.05	.03	.17	-.06	.14	-.04	.14
WeSam79	.09	.04	.05	.05	-.06	.10	.00	.02

Source: Adapted from Schendel 1980, Table 4, p.13.
[a] The coefficients refer to the single variable with the remaining three variables held constant.
[b] $p < .05$.
[c] $p < .01$.
[d] $p < .001$.

relate. This suggests that body morphology changes due to growth and development during these ages, rather than chronological age itself, should be investigated for explanations of blood pressure variations.

Body morphology
The blood pressure and body morphology analyses are presented by sex and sample and in two age groups: ages 4 to 11 years, the young children, and ages 12 years and above, the adolescents. Some of the analyses included the few 18- to 22-year-olds in the adolescent subgroup to represent the social and biological end points of adolescence. All variables are age-adjusted in the two age groups. The specific variables in the analyses are ones used in other studies of children's blood pressure and are representative measures of body mass, subcutaneous fat, muscularity, and skeletal size.

Tables 15.4 and 15.5 show the correlation matrix for age-adjusted blood pressure with the body morphology variables by age group and study sample for boys and girls. Overall, no one variable is more associated with blood pressure consistently across age/sex groups; however, pertaining to sample differences, the WeSam82 sample had far fewer associations, whereas the American Samoan and Hawaiian samples had many more significant associations.

There are some age differences in body morphology correlates discernable in these preliminary analyses. In the WeSam79 sample of boys, fatness measures are not related to systolic blood pressure in the

Table 15.4 Partial Correlation Coefficients of Blood Pressure with Select Anthropometric Characteristics among Samoan Boys[a]

Age group: 4 to 11 years

	WeSam79		WeSam82		AmSam78		Hawaii	
	Systolic	Diastolic	Systolic	Diastolic	Systolic	Diastolic	Systolic	Diastolic
Weight	.33[b]	.21	.38[b]	.35[b]	.51[b]	.30[b]	.25[b]	.10
BMI	.20	.07	.23	.24[b]	.55[b]	.32[b]	.23[b]	.09
Upper arm circumference	.31[b]	.00	.37[b]	.43[b]	.39[b]	.29[b]	.28[b]	.06
Triceps skinfold	-.02	-.01	.08	-.13	.40[b]	.30[b]	.17[b]	.06
Subscapular skinfold	.03	.17	.00	-.14	.45[b]	.38[b]	.24[b]	.09
Midaxillary skinfold	.00	.16	-.10	-.25	.38[b]	.38[b]	.23[b]	.09
Upper arm muscle area	.35[b]	-.01	.33[b]	.44[b]	.17	.13	.22[b]	.03
Height	.30[b]	.21	.34[b]	.33[b]	.08	.03	.17[b]	.09
Biacromial diameter	.26[b]	.29[b]	.39[b]	.37[b]	.31[b]	.30[b]	.28[b]	.22[b]
Biiliocristal diameter	.06	-.07	.15	.02	.22	.16	.36[b]	.27[b]
AP chest	.14	-.01	.30[b]	.17	.03	-.03	.39[b]	.27[b]
Transverse chest	-.06	-.02	.28[b]	.33[b]	.22	.12	.21[b]	.18[b]

Age group: 12 to 22 years

	WeSam79		WeSam82		AmSam78		Hawaii	
	Systolic	Diastolic	Systolic	Diastolic	Systolic	Diastolic	Systolic	Diastolic
Weight	.40[b]	.08	.14	.24	.56[b]	.38[b]	.28[b]	.22[b]
BMI	.25[b]	.12	.19	.28[b]	.60[b]	.46[b]	.26[b]	.22[b]
Upper arm circumference	.32	.00	.18	.21	.56[b]	.53[b]	.12	.27[b]
Triceps skinfold	.20	.13	.05	.07	.43[b]	.37[b]	.10	.12
Subscapular skinfold	.36[b]	.24[b]	.03	.06	.57[b]	.57[b]	.17	.12
Midaxillary skinfold	.25[b]	.25[b]	.16	.09	.52[b]	.49[b]	.14	.16[b]
Upper arm muscle area	.26	-.02	.17	.19	.27	.32	.03	.26[b]
Height	.33[b]	-.05	.04	.10	.24	.06	.10	.07
Biacromial diameter	.33[b]	-.06	.11	.23	.47[b]	.35[b]	.29[b]	.24[b]
Biiliocristal diameter	.23	-.02	-.05	.12	.46[b]	.47[b]	.26[b]	.26[b]
AP chest	.31[b]	.14	.19	.26[b]	.40[b]	.39[b]	.24[b]	.18[b]
Transverse chest	.20	.09	.20	.26[b]	.57[b]	.51[b]	.27[b]	.22[b]

[a] All variables were adjusted for the effects of age.
[b] $p < .05$.

Table 15.5 Partial Correlation Coefficients of Blood Pressure with Select Anthropometric Characteristics among Samoan Girls[a]

	WeSam79		WeSam82		AmSam78		Hawaii	
	Systolic	Diastolic	Systolic	Diastolic	Systolic	Diastolic	Systolic	Diastolic
Age group: 4 to 11 years								
Weight	.31[b]	.28[b]	.14	.13	.33[b]	.24	.29[b]	.15
BMI	.37[b]	.31[b]	.00	.08	.31[b]	.13	.28[b]	.16[b]
Upper arm circumference	.31[b]	.24[b]	.16	.20	.34[b]	.24	.26[b]	.18[b]
Triceps skinfold	.18	.20	.18	.20	.12	.04	.22[b]	.08
Subscapular skinfold	.37[b]	.29[b]	.17	.19	.29[b]	.10	.23[b]	.11
Midaxillary skinfold	.24[b]	.32[b]	.07	.10	.18	.16	.16[b]	.06
Upper arm muscle area	.29	.18	.10	.15	.35[b]	.31[b]	.20[b]	.16[b]
Height	.14	.13	.23	.16	.38[b]	.30[b]	.20[b]	.08
Biacromial diameter	.17	.15	.17	.19	.22	.02	.32[b]	.28[b]
Biiliocristal diameter	.23	.23	.05	.04	.22	.26	.30[b]	.29[b]
AP chest	.03	.07	-.04	.15	.09	.05	.40[b]	.22[b]
Transverse chest	.24[b]	.22	-.13	.07	.33[b]	.23	.25[b]	.19[b]
Age group: 12 to 22 years								
Weight	.49[b]	.18	.24	.15	.41[b]	.28[b]	.32[b]	.23[b]
BMI	.49[b]	.14	.26	.23	.37[b]	.30[b]	.30[b]	.21[b]
Upper arm circumference	.52[b]	.21[b]	.27	.22	.42[b]	.43[b]	.27[b]	.24[b]
Triceps skinfold	.39[b]	.11	.25	.15	.23	.41[b]	.24[b]	.24[b]
Subscapular skinfold	.48[b]	.14	.24	.01	.31[b]	.20	.36[b]	.30[b]
Midaxillary skinfold	.53[b]	.27	.16	.08	.32[b]	.37[b]	.34[b]	.15
Upper arm muscle area	.40[b]	.19	.16	.16	.42[b]	.22	.16	.12
Height	.29[b]	.17	.15	-.08	.19	.02	.12	.12
Biacromial diameter	.33[b]	.24[b]	.32[b]	.13	.41[b]	.38[b]	.26[b]	.21[b]
Biiliocristal diameter	.32[b]	.26[b]	.04	-.13	.36[b]	.31[b]	.22	.20
AP chest	.37[b]	.11	.15	.10	.30[b]	.16	.24[b]	.09
Transverse chest	.30[b]	.06	.29[b]	.06	.34[b]	.21	.28[b]	.20[b]

[a] All variables were adjusted for the effects of age.
[b] $p < .05$.

younger boys but are in the older boys. However, the reverse age pattern of fat and blood pressure correlates is present in the Hawaiian sample. Height is related to blood pressure in young boys from the WeSam82 sample but not in the older boys. The sample differences in body morphology and blood pressure correlates in boys are more striking than any consistency in body morphology associations across samples.

In girls, there is a slight trend in all the samples for more fat measure correlates in older girls than in younger girls and fewer height correlates in older girls than younger girls. As in boys, however, the sample differences in blood pressure and body morphology correlates are more noteworthy than any obvious body morphology trends. For example, the WeSam82 sample has very few blood pressure correlates, whereas the AmSam78 and Hawaiian samples have similar blood pressure correlates in both younger and older girls.

In a preliminary study of subcutaneous fat distribution and blood pressure, greater relative trunk fat is positively associated with blood pressure only in Hawaii, the fattest sample. The index of fat distribution is the difference between age and sex Z-scores for calf and midaxillary skinfold (Stallones et al. 1982). (A positive index indicates more limb fat and a negative index more trunk fat.) In Hawaii boys aged 4 to 11 years years the correlation equals $-.39$ for the index and diastolic pressure, in boys 12 to 22 years $r = -.26$ for the index and diastolic, and in girls 4 to 11 years, $r = -.29$ for systolic; all significant at $p < .05$. This is similar to findings among lean Samoan adults (see below) and normal U.S. adults (Blair et al. 1984) and children (Stallones et al. 1982). Multivariable analyses are, however, needed to determine how the index relates to blood pressure along with body mass and overall fatness. For example, in U.S. children aged 12 to 17 years, the fat distribution index does not relate to blood pressure when weight and fatness are included in multivariable models (Stallones et al. 1982).

Blood pressure variation within samples of Samoan children is primarily associated with body morphology not age. Measures of body mass, fatness, and maturity are also more related to blood pressure in young children and adolescents than age in U.S. studies (Katz et al. 1980; Cornoni-Huntley et al. 1979; Harlan et al. 1979; Voors et al. 1977). Body mass, size, and muscularity are related to blood pressure among young Samoan children aged 4 to 11 years, while overall subcutaneous fat is more related among adolescents. This pattern is more pronounced among girls than boys and may reflect earlier maturation and cessation of growth among girls. The sex difference in age increases of blood pressure is also probably related to different growth patterns. A similar sex difference in age patterns exists among U.S. children and adolescents (Cornoni-Huntley et al. 1979; Harlan et al. 1979). What remains as most striking still is the sample variation in blood pressure and body morphology correlates among these Samoan children. More detailed analyses are needed as well as village, family, and household level data to explain growth and blood pressure relationships in these data. In future studies more detailed biological data are also required.

Modernization

If there are detectable modernization effects on children's blood pressure, we expect that blood pressure would increase from Western Samoa to American Samoa to Hawaii. Table 15.6 presents the results of an analysis of covariance with age, and Table 15.7 one with age and specific body morphology variables, to test whether there are significant differences in blood pressure among the four study samples.

Table 15.6 shows WeSam82 children aged 4 to 11 years have, with one exception, significantly higher blood pressure than the other samples. Among children 12 to 22 years, WeSam82 also has higher systolic pressure than the other samples. The WeSam79 boys under 12 years have lower diastolic blood pressure than the Hawaiian boys. Among the older boys WeSam79 has lower systolic and diastolic pressure than Hawaiian boys, and lower systolic pressure than the American Samoan boys. The American Samoan boys under 12 years have lower blood pressure than the Hawaiian boys. Among girls, both the American Samoan and WeSam79 12- to 22-year-old samples have lower diastolic pressure than the Hawaiian girls.

Table 15.7 shows that differences in age-adjusted blood pressure between WeSam79 children and the American Samoan and Hawaiian samples are eliminated with inclusion of body morphology variables. Differences between the American Samoan and the Hawaiian children are also eliminated, with only one exception—diastolic blood pressure in 4- to 11-year-old boys. The most noteworthy result, however, is that the WeSam82 sample still has significantly higher age-, and body-morphology-adjusted blood pressures than the other samples.

The majority of the blood pressure differences among the WeSam79, American Samoan and Hawaiian sample thus are due to differences in body morphology such as shown in Table 15.8. In general, these results suggest an increase in body size, weight, BMI, and fatness with increasing modernization. However, the blood pressure of the WeSam82 boys and girls is relatively high despite their smaller body size, muscularity, and fatness compared to the American Samoan and Hawaiian sample. Adjusting for these differences only increased the blood pressure differences between WeSam82 and the other samples.

Higher blood pressure in children from the supposedly most traditional area of Samoan residence suggests that operational definitions of modernization need to be more complex and the community studied in 1982 deserves intensive study to explain differences, especially in systolic pressure.

Setting aside the WeSam82 sample blood pressures, it does appear the Hawaiian sample has relatively increased blood pressure. This difference

Table 15.6 Analysis of Covariance of Blood Pressure in Samoan Children: Effects of Age Adjusted

Boys

| | Age group: 4-11 years | | | | | | Age group: 12-22 years | | | | | |
| | Systolic | | | Diastolic | | | Systolic | | | Diastolic | | |
Independent variables:	F	p	R^2	F	p	R^2	F	p	R^2	F	p	R^2
sample	33.24	.0001		8.96	.0001		19.87	.0001		3.01	.03	
age	22.43	.0001	.27	26.19	.0001	.14	91.55	.0001	.34	9.51	.002	.06
Mean adjusted blood pressure:												
1. WeSam79	99.7			64.9			111.7			71.7		
2. WeSam82	113.3			69.6			126.0			72.7		
3. AmSam	97.6			64.1			119.1			74.8		
4. Hawaii	102.5			69.4			117.5			75.5		
	$1^a, 3, 4 < 2, 3 < 4$			$1^a, 3 < 2, 4$			$1 < 3^a, 4 < 2$			$1 < 4^a$		

Girls

| | Age group: 4-11 years | | | | | | Age group: 12-22 years | | | | | |
| | Systolic | | | Diastolic | | | Systolic | | | Diastolic | | |
Independent variables:	F	p	R^2	F	p	R^2	F	p	R^2	F	p	R^2
sample	22.61	.0001		3.81	.01		20.39	.0001		4.51	.004	
age	39.12	.0001	.28	14.18	.0002	.08	19.47	.0001	.19	20.56	.0001	.09
Mean adjusted blood pressure:												
1. WeSam79	103.1			68.9			110.6			72.4		
2. WeSam82	115.8			73.6			123.3			74.6		
3. AmSam	100.9			68.5			109.0			71.3		
4. Hawaii	103.0			69.5			112.3			75.4		
	$1^a, 3, 4 < 2$			$1^a, 3, 4 < 2$			$1^a, 3, 4 < 2$			$1^a, 3 < 4$		

[a] All differences significant at $p < .01$.

Table 15.7 Analysis of Covariance of Blood Pressure in Samoan Children: Effects of Age and Body Morphology Adjusted

	Pairwise comparisons in adjusted sample means			
	Age group: 4-11 years		Age group: 12-22 years	
Adjusted sample means	Systolic	Diastolic	Systolic	Diastolic
Boys				
WeSam79 vs.WeSam82	100.4 vs. 112.4[a]	64.9 vs. 69.6[a]	113.2 vs. 125.5[a]	n.s.
AmSam78 vs.WeSam82	96.3 vs. 114.9[a]	63.0 vs. 71.1[a]	117.3 vs. 127.1[a]	n.s.
Hawaii vs. WeSam82	101.7 vs. 115.2[a]	n.s.	115.6 vs. 128.2[a]	n.s.
AmSam78 vs. WeSam79	n.s.[b]	n.s.	117.2 vs. 113.9	n.s.
Hawaii vs. WeSam79	n.s.	68.2 vs. 66.5	116.0 vs. 114.5	74.8 vs. 72.9
AmSam78 vs. Hawaii	99.9 vs. 101.9	64.0 vs. 69.5[a]	n.s.	n.s.
Girls				
WeSam79 vs.WeSam82	105.4 vs. 113.5[a]	70.0 vs. 72.6	110.9 vs. 122.4[a]	n.s.
AmSam78 vs.WeSam82	101.0 vs. 117.0[a]	68.4 vs. 74.3[a]	107.7 vs. 124.5[a]	n.s.
Hawaii vs. WeSam82	101.1 vs. 119.6[a]	69.0 vs. 74.6[a]	111.4 vs. 125.8[a]	n.s.
AmSam78 vs. WeSam79	n.s.	n.s.	n.s.	n.s.
Hawaii vs. WeSam79	n.s.	n.s.	n.s.	74.6 vs. 73.8
AmSam78 vs. Hawaii	n.s.	n.s.	n.s.	72.7 vs. 75.1

[a] Difference significant at $p < .05$.
[b] n.s. No significant difference between samples in age-adjusted blood pressure means.

Table 15.8 Analysis of Covariance of Select Anthropometric Characteristics of Samoan Children: Effects of Age Adjusted

	Comparisons in adjusted sample means						
	WeSam79		WeSam82		AmSam		Hawaii
Boys							
Weight	37.2		39.2	[a]	44.3	[a]	48.3
BMI	18.3		19.0	[a]	20.1	[a]	21.2
Upper arm circumference	20.0		21.1	[a]	22.4		24.2
Triceps skinfold	7.1	[a]	9.5	[a]	13.2		12.8
Subscapular skinfold	6.5	[a]	9.8	[a]	12.1		11.6
Midaxillary skinfold	4.7	[a]	6.5	[a]	10.1		10.5
Height	137.6		139.6	[a]	144.4		146.3
Biacromial diameter	30.0		29.8	[a]	32.6	[a]	33.6
Biiliocristal diameter	22.0		24.4	[a]	24.0		24.1
Upper arm muscle area	29.1		27.6		28.2	[a]	34.6
Girls							
Weight	39.6		41.3	[a]	46.1	[a]	50.8
BMI	19.2		19.6	[a]	21.1	[a]	22.8
Upper arm circumference	20.7	[a]	21.6		22.3 -	[a]	24.8
Triceps skinfold	11.2	[a]	14.5	[a]	18.1		18.3
Subscapular skinfold	10.4	[a]	13.7	[a]	18.5		18.4
Midaxillary skinfold	8.0		10.0	[a]	17.1		17.2
Height[b,c]	139.1		141.5		143.6		145.0
Biacromial diameter	29.9		29.8	[a]	32.0		32.4
Biiliocristal diameter[b]	23.2		23.0	[a]	25.4	[a]	24.6
Upper arm muscle area	24.4		24.0		22.7	[a]	30.0

[a] Difference between the samples in adjacent columns significant at $p < .01$.
[b] WeSam79, WeSam82 less than Hawaii.
[c] WeSam79 less than AmSam.

and the majority of differences among WeSam79, American Samoa, and Hawaii disappear when body morphology is considered. It is clear that blood pressure differences among these three samples are due to body morphology variation. Increasing body mass size and fatness with modernization are associated with sample blood pressure differences.

Migration

To detect migration effects in children in the Hawaiian sample, blood pressure is associated with certain sociodemographic factors that assess exposure to the new environment: birthplace, length of residence in Hawaii, and residence locale in Hawaii.

There are no blood pressure differences in Samoan children resident in Hawaii by birthplace. Time since migration as a proportion of children's

Table 15.9 Blood Pressure in Samoan Children Residing in Hawaii by Percent of Life Spent in Hawaii, and Place of Residence (blood pressure is adjusted for age and weight)

Percent of life resident in Hawaii	Boys			Girls		
	N	Systolic	Diastolic	N	Systolic	Diastolic
1-24	58	113.0	74.8	74	111.2	74.7
25-49	62	109.6	73.7	66	109.6	72.8
50-74	36	112.0	73.9	33	107.9	70.1
75-99	31	106.9	71.5	42	108.6	75.3
Hawaii-born	78	108.9	72.2	82	106.1	71.1
Within-column comparisons	n.s.[b]		n.s.		n.s.	[a]

Source: Adapted from Schendel 1980.
[a] $p < .05$.
[b] n.s. not significant.

Residence in Hawaii, Island of Oahu	Boys			Girls		
	N	Systolic	Diastolic	N	Systolic	Diastolic
Southwest shore	79	107.0	72.1	86	107.2	73.2
North shore	109	107.0	70.0	121	105.0	73.1
Honolulu	81	107.8	68.8	91	106.0	66.9
Within-column comparisons	n.s.[b]	[a]			n.s.	[a]

Source: Adapted from Hanna and Baker 1979.
[a] $p < .01$.
[b] n.s. not significant.

life and locale in Hawaii are significantly related to blood pressure among Samoan children in Hawaii. Girls who have spent the smallest (1-24%) and the largest (76-99%) proportion of their life in Hawaii have higher age- and weight-adjusted diastolic blood pressure than those resident in Hawaii for 25-75 percent of their life. Boys and girls who live in urban Honolulu have lower age- and weight-adjusted diastolic blood pressure than those resident in rural and suburban areas of Oahu.

The cross-sectional U-shaped relationship between proportion of life resident in Hawaii and girls' diastolic blood pressure suggests that the Hawaiian environment has both acute effects on blood pressure, perhaps related to dislocation, establishing routines in new neighborhoods, and novel aspects of the actual migration to Hawaii, and chronic effects of long-term exposure, perhaps related to diet, activity pattern, growth and development changes, and psychosocial adaptation. Residence differences in Hawaii are more difficult to explain, but higher blood pressure among the semirural predominantly Mormon Church-affiliated community indicates a need for detailed studies of community variation. Religious variation in acculturative behaviors and attitudes toward modern and traditional life may exert detectable effects on blood pressure correlates. Differences between community and individual achievement orientations, attitudes toward the larger Hawaiian and American culture, and social control are important topics for further investigation among families and their children.

Conclusions

The children's blood pressure data support the overall hypothesis of increases due to modernization, if we exclude the seemingly anomalous WeSam82 sample. However, the lesson of the WeSam82 sample is that simple conceptualizations of exposure to modern life are not enough to account for blood pressure variation in one community. The data strongly support a direct role for body morphology in blood pressure variation, and in determining most of the intersample blood pressure variation. That is, the modernization differences in blood pressure among the samples may be attributed to body morphology differences among the samples. The Samoan data are generally similar in the age, sex, and morphology relationships with blood pressure to those in other populations, although sample differences in correlates exist.

A causal model for Samoan children's blood pressure is suggested by the demonstrated modernization differences in body mass, size, and fatness and the blood pressure and body morphology associations. General modernization is prime mover of a variety of processes, such as altered diet and physical activity, whose net effect is increased height, weight, and fatness for age, thereby leading to slightly higher blood pressure. Other factors are weaker, for example, psychosocial aspects of exposure to the modern environment such as those hypothesized among the Hawaiian children.

We need to pursue longitudinal research in a human ecological context to understand how the biology of growth and development and psychosocial coping with new environments affect children's blood pressure in rapidly changing societies, whether in Hawaii or in supposedly traditional Samoan villages.

RESULTS: BLOOD PRESSURE AMONG ADULTS

Blood pressure patterns in the samples

Adult study samples were also chosen to represent Samoan life from traditional to modern: Western Samoa, three areas of American Samoa, Hawaii, and California. The American Samoan villages are divided into three groups based on degree of exposure to modernization: Manu'a—consisting of villages on the islands of Manu'a and those on the north

Table 15.10 Sizes of Adult Blood Pressure Samples by Sex, Number of Sampling Sites, and Years of Study

Area	Men	Women	Number of sites	Years of study
Western Samoa	140	184	2 villages	1979
Western Samoa	96	95	1 village	1982
Manu'a	105	191	9 villages	1976
Tutuila	331	413	23 villages	1976
Pago Pago	262	381	9 villages	1976
Hawaii	249	332	8 communities	1975-1977
California	166	158	3 communities	1979

shore of Tutuila accessible in 1976 by very rough tracks or no roads; Tutuila—the villages on the main island of Tutuila, excepting the isolated north shore villages and those in the Pago Pago group; Pago Pago— villages in the immediate Pago Pago Bay area and nearby, plus two large villages on the south shore with historical, political, and commercial ties to the Pago Pago area and its modernizing influences. Significant differences among these three American Samoan groups in occupation, education, and proportion of English speakers validate the regional grouping (McGarvey 1980; McGarvey and Baker 1979). Throughout the adult section, these group names are used; note the distinction between them and the actual geographic place-names.

Table 15.10 presents the sample sizes by sex, number of village, and community sampling sites, and the years of study for the adult samples. The two Western Samoan samples are treated separately in descriptive analyses for the same reasons described for children.

Tables 15.11 and 15.12 and Figs. 15.3 and 15.4 present and illustrate, respectively, the mean blood pressure by age among Samoan adults.

Systolic and diastolic pressure of both men and women increase with cross-sectional adult age; however, there is a sex difference in the pattern of blood pressure increase by age. Pressure curves of men show sharp rises among 18- to 40-year-olds, followed by no or gradual increase in the more traditional locales and by continuing sharp increases in the more modern locales. By contrast, women show little increase before age 30, and then a sharp rise in later ages to values exceeding men. Notably high values are achieved in women, for example, the mean systolic pressure in 60- to 69-year-old Pago Pago group is 158 mmHg. The sex crossover in blood pressure values with cross-sectional age occurs in the 40- to 49-year-old group, the approximate time of menopause. This crossover has been demonstrated in many modern Western populations.

Comparison of mean blood pressure in the samples shows a trend for lowest values in WeSam79, followed in order by the Manu'a group, Tutuila, Pago Pago, and California. The WeSam82 values are markedly

Table 15.11 Blood Pressure by Age Group of Samoan Men — Means ($\bar{X}$) and Standard Deviations (SD)

Age		WeSam79	WeSam82	Manu'a	Tutuila	Pago Pago	Hawaii	California
				Systolic				
18-19	$\bar{X}$	117.5	130.4	—	121.3	120.5	118.7	—
	SD	7.8	12.1	—	9.7	12.7	13.5	—
20-29	$\bar{X}$	119.4	134.1	123.8	124.4	125.6	129.9	129.5 [a]
	SD	10.3	12.3	8.3	11.5	11.2	15.9	12.9
30-39	$\bar{X}$	118.0	126.0	124.0	127.8	133.7	132.4	131.6
	SD	10.8	9.5	12.8	18.4	16.1	17.0	14.7
40-49	$\bar{X}$	122.3	128.4	124.6	132.7	132.3	129.6	140.5
	SD	12.1	13.7	12.1	22.4	17.8	17.9	18.6
50-59	$\bar{X}$	125.2	135.4	120.2	137.7	145.3	129.6	142.5 [b]
	SD	13.5	21.0	17.4	25.3	21.8	14.7	29.8
60-69	$\bar{X}$	124.6	124.2	131.1	139.6	135.8	135.3	—
	SD	18.4	19.7	18.6	23.9	16.1	24.1	—
70+	$\bar{X}$	—	—	140.1	132.8	154.9	129.1	—
	SD	—	—	9.1	20.2	32.5	15.5	—
				Diastolic				
18-19	$\bar{X}$	72.5	68.1	—	78.0	77.8	76.8	—
	SD	7.8	7.7	—	6.8	11.6	10.3	—
20-29	$\bar{X}$	75.7	74.7	81.9	80.4	80.9	84.4	83.7 [a]
	SD	8.1	11.5	4.6	8.6	8.3	11.8	9.2
30-39	$\bar{X}$	77.8	78.8	82.8	85.8	89.6	89.1	86.9
	SD	9.1	9.4	9.7	11.9	11.5	12.5	8.8
40-49	$\bar{X}$	80.6	84.9	82.5	86.1	88.9	87.6	94.4
	SD	7.9	9.8	8.1	12.5	11.8	11.7	12.8
50-59	$\bar{X}$	82.1	83.0	79.2	89.6	93.5	86.2	90.6 [b]
	SD	9.9	12.8	12.2	16.7	11.6	11.3	11.9
60-69	$\bar{X}$	82.0	69.8	84.3	85.7	85.0	90.7	—
	SD	10.7	3.2	11.1	12.6	9.6	17.3	—
70+	$\bar{X}$	—	—	75.5	86.6	94.1	83.1	—
	SD	—	—	4.4	6.7	18.1	4.2	—

[a] Age group 18-29.
[b] Age group 50 and above.

higher than the WeSam79 sample, but are less than the more modernized samples.

Among men aged 18 to 20 years, the mean pressures are very similar across samples, but the older age means are higher with modernization. The Hawaiian male sample is an exception to this trend due to the considerable variation in area of origin, length of residence, and probable exposure to modernization in the Hawaiian migrants (McGarvey 1980).

For women this trend is not as clear, but there is the same general finding of increasing blood pressure with modernization. This is quite evident in women aged 20 to 40, and in those over 60 years of age.

Table 15.12 Blood Pressure by Age Group of Samoan Women — Means ($\bar{X}$) and Standard Deviations (SD)

Age		WeSam79	WeSam82	Manu'a	Tutuila	Pago Pago	Hawaii	California
				Systolic				
18-19	$\bar{X}$	121.1	123.5	127.3	114.1	115.9	112.8	—
	SD	10.1	4.5	12.7	12.0	10.4	12.5	—
20-29	$\bar{X}$	114.2	119.9	114.7	114.3	116.5	116.9	117.4 [a]
	SD	10.2	10.7	12.4	11.4	11.0	11.2	10.7
30-39	$\bar{X}$	112.7	117.6	120.7	125.0	124.4	122.6	132.1
	SD	10.1	12.7	15.1	18.5	16.5	15.4	15.5
40-49	$\bar{X}$	123.2	130.6	127.8	137.1	140.0	121.1	137.1
	SD	10.9	15.9	20.8	20.1	26.0	18.6	14.2
50-59	$\bar{X}$	131.5	134.8	135.0	140.3	140.1	125.6	145.2 [b]
	SD	23.4	29.4	25.3	26.8	23.6	17.9	21.5
60-69	$\bar{X}$	132.1	139.3	148.6	142.5	158.6	143.0	—
	SD	21.1	32.6	32.2	28.9	28.9	16.9	—
70+	$\bar{X}$	131.9	141.8	137.6	135.3	148.0	139.6	—
	SD	12.3	9.3	22.7	16.7	15.8	30.9	—
				Diastolic				
18-19	$\bar{X}$	75.4	74.7	86.0	74.1	75.5	73.9	—
	SD	4.1	6.5	10.4	10.9	8.7	9.1	—
20-29	$\bar{X}$	73.3	75.1	73.8	73.1	76.0	79.5	76.7 [a]
	SD	8.3	8.1	9.2	9.7	9.4	10.3	7.8
30-39	$\bar{X}$	75.4	73.2	80.3	81.5	82.6	80.9	86.0
	SD	8.7	8.5	10.6	11.7	12.2	10.0	11.0
40-49	$\bar{X}$	81.4	82.2	84.4	87.0	90.6	89.3	88.6
	SD	10.9	12.0	12.6	12.1	15.2	13.5	10.1
50-59	$\bar{X}$	84.6	82.5	84.4	87.8	89.2	84.0	90.0 [b]
	SD	13.3	15.0	14.3	17.9	12.7	12.3	11.4
60-69	$\bar{X}$	82.3	83.0	90.1	84.9	96.9	89.3	—
	SD	13.7	19.0	13.7	14.9	16.6	11.9	—
70+	$\bar{X}$	83.4	70.6	78.2	78.6	89.5	87.8	—
	SD	7.8	6.8	12.4	9.3	11.3	18.9	—

[a] Age group 18-29.
[b] Age group 50 and above.

Western Samoa is lowest, followed by the Manu'a, Tutuila, and Pago Pago groups. The two female migrant samples from Hawaii and California are not easily placed in this trend.

The higher means for systolic pressure in the 18- to 20-year-old women relative to the 20- to 29-year-olds from the most traditional Samoan samples has two possible explanations: (1) the higher weight and fatness of these young women relative to their older peers, and (2) anxiety among young adult women from traditional Samoa about researchers performing anthropometric and blood pressure measurements.

These results pertain to cross-sectional data and must be treated

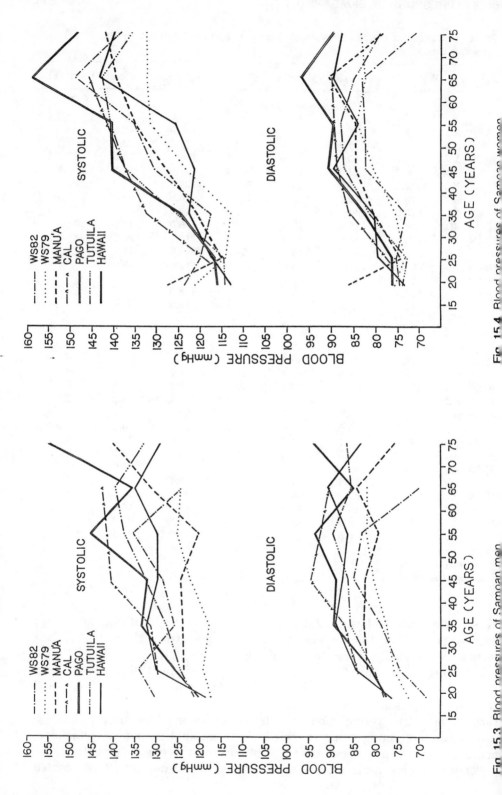

Fig. 15.3 Blood pressures of Samoan men.

Fig. 15.4 Blood pressures of Samoan women.

cautiously. Individual and group exposure to modern life in the Samoan archipelago differs by age in both duration and intensity. The decrease of mean blood pressure in old age may reflect differential mortality and/or different life histories, that is, cohort effects. Age/blood pressure curves of Samoans are obviously contingent on when and where those Samoans were born and what they experienced. Inferences and processes of biological aging or exposure to modern life require longitudinal data.

The blood pressure of Samoans is similar to other Polynesian groups, considering the general degree of exposure to modern life. Among most Polynesian groups there are sharp increases in blood pressure with adult age as they become exposed to modern life (Prior and Tasman-Jones 1981; Zimmet et al. 1980). Tokelauan nonmigrants and Pukapukans from the Cook Islands have similar or lower pressures than Western Samoans and Manu'ans. Tokelauan migrants in New Zealand are comparable to Samoans referred to here as the Tutuila group, but have lower pressures than the Pago Pago group. However, in none of the Samoan groups are the age increases in blood pressure as high as in Rarotonga (Prior et al. 1966a).

Blood pressures in rural and urban Western Samoa (Zimmet et al. 1980) are higher than values reported here for Western Samoans, perhaps due to differences in measuring diastolic pressure, and regional and village variation, such as that shown here. However, the urban Western Samoans do have significantly higher blood pressure even after adjustment for body mass.

Relative to other Pacific populations such as the Melanesians studied by the Harvard Solomon Islands Project (Page et al. 1977, 1974; Page and Friedlaender n.d.), or New Guineans (Sinnet and Whyte 1973), Samoans are much more modernized, fatter, and have higher blood pressures.

In comparison to Western societies (Roberts and Maurer 1977; Epstein and Eckhoff 1967) the age increases in blood pressure among all the Samoan groups is generally similar, except for the most traditional males. All of the Samoan groups show consistent increases through adulthood. Thus, Samoans are not comparable to groups relatively unexposed to modern life, such as hunting and gathering, or horticultural populations, for example, the San !Kung (Truswell et al. 1972) or Yanomamö (Oliver et al. 1975). These and other populations (Shaper 1972; Lowenstein 1961; Donnison 1929) are characterized by no age increase in adult blood pressure.

Hypertension in the samples

Table 15.13 presents the age-specific and overall percentages of hypertension in the Samoan samples. There is more hypertension with modernization among Samoans of both sexes. In particular, among the nonmigrant American Samoans, the Pago Pago men and women have more hyper-

Table 15.13 Percentage of Definite Hypertension in the Study Samples: Systolic Blood Pressure 160 mmHg or Above and Diastolic Blood Pressure 95 mmHg or Above

Age	18-34		35-54		55+		All Ages	
	# / N[a]	Percent	# / N	Percent	# / N	Percent	# / N	Percent
Men								
WeSam79	1/66	1.5	2/48	4.2	4/26	15.4	7/140	5.0
WeSam82	2/56	3.6	4/31	12.9	3/15	20.0	12/102	11.8
Manu'a	2/22	9.1	3/43	6.9	5/44	11.4	10/109	9.2
Tutuila	9/132	6.8	35/121	28.9	23/88	26.1	67/341	19.6
Pago Pago	16/113	14.2	30/101	29.7	19/56	33.9	65/270	24.1
Hawaii	20/115	17.4	27/94	28.7	11/46	23.9	58/255	22.7
California	6/36	16.7	16/38	42.0	6/25	24.0	28/87	32.8
Women								
WeSam79	1/86	1.2	6/56	10.7	5/42	11.9	12/184	6.5
WeSam82	0/45	0.0	2/34	5.9	4/16	25.0	6/95	6.3
Manu'a	1/40	2.5	19/103	18.4	15/53	28.3	35/196	17.9
Tutuila	6/169	3.6	38/167	22.8	21/83	25.3	65/419	15.5
Pago Pago	8/169	4.7	43/159	27.0	32/67	47.8	83/395	21.0
Hawaii	12/96	6.1	26/107	24.3	8/40	20.0	46/343	13.4
California	2/57	3.5	15/65	23.1	12/37	32.4	29/159	18.2

[a] Total persons with hypertension divided by total sample.

tension in the youngest, oldest, and overall age groups. The percentage of hypertension among men and women is rather similar in the middle adult ages (35-54) among the more modernized groups, suggesting that the pandemic adiposity characterizing these Samoans is more important in hypertension risk than other differences in exposure to modern life. Because of an increasing variance of blood pressure with age in this age group, there may be a stratum in each locale of Samoans more similar in hypertension but different in mean values. A similar trend has been reported in Tokelauans (Ward et al. 1979). Among the two migrant samples there is an overall similarity within the sexes. The strikingly high rates for older women in all groups, even the most traditional American Samoans, suggest considerable risk for cardiovascular disease mortality.

By comparison with other Polynesians, traditional Samoan rates of hypertension are higher than Pukapukans and are similar to Tokelauans. The Tutuila group has more hypertension than Tokelauans, but less than New Zealand Maoris. Samoan men in California have rates similar to those in Rarotonga, although the number of cases is small. Hypertension prevalence is higher in another study of Western Samoans (Zimmet et al. 1980) relative to those shown here, perhaps due to bias from sampling of hypertensives or those concerned about high blood pressure. It could also be due to village-specific factors associated with hypertension. Regardless of comparison to other studies, these percentages for Samoans strongly indicate that Samoans outside of traditional life are very susceptible to hypertension.

Blood pressure correlates

Blood pressure correlates are first considered bivariately, followed by their multivariable relationships within each sample. Different types of data and different analytic strategies were used by some investigators in some samples, and the focus is on the three American Samoan and the Hawaiian samples with similar data. Data are presented for the Western Samoan and California samples when available or similarly analyzed.

Age
Table 15.14 presents the age and blood pressure correlations for the three American Samoan and the Hawaiian groups. Age is significantly related to both systolic and diastolic blood pressure in all the Samoan groups except traditional American Samoan males. This reflects the increase in blood pressure with cross-sectional age noted above.

Adiposity: Association of single measures
Triceps skinfold thickness is often suggested as the best single indicator of adiposity in different groups. This may not be the case for Samoans, but intensive research has not yet been performed. In most of the American

Table 15.14 Bivariate Correlations of Age with Blood Pressure in American Samoans and Hawaiian Samoans

	Manu'a	Tutuila	Pago Pago	Hawaii
		Men		
Systolic	.22[a]	.28[b]	.34[b]	.14[a]
Diastolic	.03[c]	.21[b]	.28[b]	.18[a]
		Women		
Systolic	.38[b]	.46[b]	.53[b]	.43[b]
Diastolic	.24[b]	.33[b]	.45[b]	.33[b]

[a] $p = .01$.
[b] $p < .00001$.
[c] not significant.

Table 15.15 Bivariate Correlations of Triceps Skinfold Thickness with Blood Pressure among Samoan Adults

	WeSam	Manu'a	Tutuila	Pago Pago	Hawaii
			Men		
Systolic	.21[a]	.07[c]	.52[b]	.51[b]	.36[b]
Diastolic	.29[b]	-.01[c]	.52[b]	.53[b]	.37[b]
			Women		
Systolic	.35[b]	.40[b]	.53[b]	.45[b]	.33[b]
Diastolic	.40[b]	.44[b]	.52[b]	.47[b]	.33[b]

[a] $p < .001$.
[b] $p < .0001$.
[c] not significant.

Samoan and Hawaiian samples triceps skinfold is very highly correlated with weight. Table 15.15 shows the bivariate correlations of triceps skinfold thickness and blood pressure in the combined Western Samoan samples, the three American Samoan and the Hawaiian samples.

Adiposity measured by triceps skinfold thickness is strongly associated with blood pressure in all samples except the traditional American Samoan men, which is attributable to their leanness. Age-adjusted blood pressure correlations with triceps are also large and do not differ significantly from those in Table 15.15. For example, among Tutuila women the partial correlations of triceps with age-corrected systolic and diastolic pressure are .50 and .49, respectively. Among the California Samoans, the body mass index (BMI) is used due to unreliability of the very large skinfolds concordant with their massive adiposity. BMI is strongly and positively correlated with blood presure in women but not men (Pawson and Janes 1981). Either adiposity is truly unrelated to blood pressure in the men, which is doubtful, or BMI does not assess the adiposity of these men precisely enough.

Subcutaneous fat distribution

The ratio of subscapular skinfold to the sum of triceps and subscapular skinfolds is used as an index of relative subcutaneous fat distribution (Frisancho and Flegel 1982). High ratios indicate relatively more trunk fat or a centripetal distribution, and low ratios more limb fat or a centrifugal distribution. Fat distribution is distinct from fat patterning models based on specific locations of adipose tissue. Among the Samoan groups, females are characterized by lower ratios than males due to deposition of fat on the upper arm with age and obesity, and lower ratios are associated with modernization in males (McGarvey 1984a).

In the combined Western Samoan sample this ratio is positively associated with age-adjusted diastolic blood pressure in men ($r = .17$, $p < .01$). The analysis of fat distribution and blood pressure in the three American Samoan and the Hawaiian groups included the effects of age, and an age-squared term to better model age effects, the sum of triceps and subscapular skinfolds, and the ratio described above. Among Manu'a men, the ratio was positively related to both systolic ($F = 6.9$, $p < .01$; $N = 108$) and diastolic blood pressure ($F = 4.9$, $p < .03$; $N = 108$). The ratio effects are three times as influential as the sum of skinfolds on systolic pressure (standardized betas, ratio - .025 vs. sum - .008). The ratio was the only correlate to diastolic. Among women in the Tutuila group, there was a negative association of the ratio with systolic pressure ($F = 7.8$, $p < .01$; $N = 414$) and a similar trend with diastolic ($p = .07$). The relative effects of the ratio are one-fifth those of the sum of skinfolds. The negative association of the ratio with blood pressure among the Tutuila women means that relatively more fat on the arm is related to increased blood pressure. This may be accounted for by the overall massive adiposity of the Tutuila group women (McGarvey 1980) and the possibility for greater deposition of fat on the limbs in fat Samoan women (McGarvey 1984a; Greksa and Baker n.d.). In other words, with increasing adiposity the sum of skinfolds is related to blood pressure and the extra fat on the arms of the very fat produces the additional but small centrifugal ratio effect on blood pressure.

In summary, there is a centripetal fat distribution and blood pressure association in traditional lean men, a centrifugal distribution and blood pressure association in the very fat women, and no association of fat distribution and blood pressure among the other relatively fat Samoans. The large amounts of fat measured by the sum of skinfolds probably swamps the secondary and relatively weak ratio effects, except in the very fat who are able to continue to deposit fat on the limbs. Further work should include more skinfold sites, longitudinal analyses, and technical considerations of the measurement of skinfold thickness in a population with great adiposity.

Salt intake

There is a relationship between salt intake and blood pressure in several modernizing populations (Page et al. 1974; Prior et al. 1968), although this is not always found in Western societies. In the Samoan samples salt intake is determined via dietary questions concerning salt use and the amounts used. Two questions concern salt use in cooking and eating, and two concern amounts used. The salt quantity answers were categorized into three categories: small, moderate, and large.

Salt use is significantly associated with diastolic pressure among the Pago Pago men. Those who used salt in both cooking and eating had the highest pressure, those who used salt in only cooking had lower pressures, and those not using salt had the lowest pressure (cook and eat—89.5; cook—87.1; no salt—83.6; $F = 3.32$, $p = .04$). Amount of salt is also significantly related to diastolic blood pressure among Pago Pago men, due to the significantly different contrast of no salt users—83.6 mmHg with the small and moderate amount users—91.9 mmHg ($F = 6.15$, $p = .003$). Among Pago Pago women, those who use large amounts of salt have higher systolic pressure (143.2 mmHg) than moderate amount users (127.6 mmHg) or nonusers (133.7 mmHg) ($F = 4.61$, $p = .01$). In multivariable analysis with blood pressure adjusted for age and adiposity, salt amount remains directly and significantly related to systolic blood pressure among Pago Pago women ($F = 3.1$, $p < .05$). In Pago Pago men, salt interacted with adiposity such that only among the leaner men was salt amount directly related to diastolic blood pressure ($F = 3.2$, $p < .05$). The adjustment for adiposity altered the bivariate association of blood pressure and reported salt intake by removing it in Hawaiian women and attenuating it in Pago Pago men. Other studies found similar effects suggesting fatter people use more salt.

These results suggest, especially in women in modernized villages, that dietary intake of salt through cooking and table use, and probably through processed foods, may be an important factor in blood pressure variation.

Psychosocial factors related to blood pressure

Changes in social interaction patterns, norms, expectations for social and economic mobility, and achievement motivation follow changes in education and occupation associated with social and economic modernization. These are potentially stressful depending on domain of life and its salience, social supports, and personal coping styles, and may increase blood pressure.

The conceptual hypothesis that psychosocial factors of modernization affect Samoan blood pressure led to the following specific hypotheses

(McGarvey 1984b). (1) Status incongruity between occupation and education is potentially stressful (Smith 1967) and may be related to blood pressure. Those with low education and managerial jobs or those with high education with menial jobs are examples of status incongruity, although the former may be more likely in rapidly modernizing societies like Samoa. (2) Formal education is related to blood pressure and may interact with complaint behavior. Schooling is a potent exposure to modern life and may serve as a single measure of exposure to and participation in modern life (Inkeles and Smith 1974). Education may be directly related to blood pressure in early stages of sociocultural change or modernization through upward social mobility and high achievement motivation leading to stress, and may be inversely related to blood pressure through medical knowledge and health habits. (3) Expression of psychological complaints or emotions is related to blood pressure. Complaints are expressions of dissatisfaction resulting from the impingement of external events, the individual's ability to behaviorally adapt, and the propensity to complain (Jenkins 1971). Although sociocultural factors influence recognition of distress and willingness to complain, we may consider complaining and expressing strong emotions measures of coping with stress. Education may change traditional enculturated patterns of coping such as complaints and emotional expression, interacting with blood pressure and coping relationships. (4) Number of children is related to blood pressure among women participating in the money economy or exposed to modern life through residence (Scotch 1963). Desired family size may change among individual women due to opportunities for female education and occupation, and conflicts with family formation norms may create stress for some women.

The Cornell Medical Index (CMI) was used in several groups of Samoans to assess the patterns of complaints. The CMI is *a priori* classified into sections based on common content and theme according to standard medical practice at the time of its construction (Brodman et al. 1951, 1949). The translation of the CMI was done by acculturated Samoans in Hawaii and then back-translated in American Samoa by Samoan medical personnel. Conceptual equivalence (Kalimo et al. 1970) was used in the translation in order to produce confidence in the meaning of the terms and the intent of the items. The CMI seeks yes or no responses to a number of symptoms, including 51 items about emotions, moods, and feelings, which are referred to here as psychosocial complaints and as emotional expression. Total number of psychological complaints and factor-analysis-based scales are used in analyses with blood pressure.

Status incongruity

To detect stress effects of status incongruity on blood pressure a two-factor analysis of variance with interaction was performed with education and occupation and age and adiposity-adjusted blood pressure. In Pago Pago men the interaction of education and occupation was significantly related to both systolic ($F = 2.77$, $p < .03$) and diastolic ($F = 3.04$, $p < .02$) blood pressure. Men with less than 7 years of education and managerial jobs had higher adjusted systolic and diastolic blood pressure compared to wage and subsistence workers (169.2/109.8 mmHg vs. 137.0/89.2 mmHg vs. 136.0/89.1 mmHg; post hoc $p < .01$). In men with 7 to 11 years of education, wage laborers had significantly higher diastolic pressure than those not in the money economy (87.7 mmHg vs. 80.4 mmHg, post hoc $p < .05$).

The effect of complaint score on blood pressure among Pago Pago men in the education and occupation groups described above was analyzed to detect stress mediating or coping influences. Men with low education ($N = 44$) show a positive association between complaint score and blood pressure as main effect (systolic, $F = 7.2$, $p < .01$; diastolic $F = 11.0$, $p < .01$) and occupation. Blood pressure is higher in those who are in managerial jobs and have high numbers of psychological complaints. Inspection of the interactions, although they are both only marginally significant, shows the positive complaint and blood pressure association present primarily in those with managerial jobs.

In Pago Pago men with 7 to 11 years of education ($N = 39$), psychological complaints are inversely related to systolic ($F = 8.1$, $p < .01$) and diastolic ($F = 4.3$, $p < .05$) pressure, and complaint score and job status interact significantly with systolic pressure ($F = 4.0$, $p < .02$). Among the managerial occupations, adjusted systolic blood pressure is inversely related to complaint score. That is, those men with moderate education and in managerial jobs, who report no or few complaints have significantly higher systolic than those reporting more complaints (164.8 mmHg vs. 116.8 mmHg, post hoc $p < .02$). Regardless of occupation, men with moderate education who complain less have higher systolic and diastolic pressure than their counterparts who complain more (146.4/89.8 mmHg vs. 128.3/81.7 mmHg, post hoc systolic $p < .01$, diastolic $p < .05$). Although blood pressure differences are large and significant, the rather small sample sizes suggest caution.

American Samoan men most exposed to modern life through residence in Pago Pago villages may be in stressful situations due to the demands of managerial occupations, but have relatively low education. Many of the men in this occupational category are not only managers in the office or industrial sense but are village officials and leaders, such as mayors and clergy. As such they may be expected to do well for their villages, compared to other villages, in territorial political, economic, and religious affairs. It may be stressful for men raised traditionally, now in positions of leadership in the economically least traditional villages of a rapidly modernizing population. This result fits with the overall concept of status incongruity (Jenkins 1971; Smith 1967) between job and education, specifically a high status job with little education, and with that of sociocultural mobility (Syme et al. 1965). Both concepts are relevant in

Table 15.16 Nonparametric Correlations of Cornell Medical Index Psychosocial Complaints with Blood Pressure in American Samoa and Hawaii

Spearman's R	Manu'a		Tutuila		Pago Pago		Hawaii	
	Men	Women	Men	Women	Men	Women	Men	Women
Systolic	.01	-.05	.01	-.13[c]	.15[b]	-.12[b]	-.01	.07
Diastolic	.05	-.01	.07[a]	-.10[b]	.08	-.08[a]	-.12	.02

[a] $10 < p < .05$.
[b] $p < .05$.
[c] $p < .01$.

modernizing societies because they mean living an adult life different from that of one's childhood or from one's parents.

In Pago Pago men with moderate education and jobs in the cash economy, complaint score is inversely associated with adjusted blood pressure. Nontraditional wage labor, probably with supervision and evaluation of timed productivity, seems to be stressful for the noncomplaining males. Other studies link suppression of strong emotions to elevated blood pressure (Gentry et al. 1982; Osfeld and Shekelle 1967).

Blood pressure differences attributed to status incongruity are found only in modernized men and not at all in women. Males outside of the Pago Pago area reside in villages less exposed to commercial and industrial activities and also have more available agricultural land. Perhaps work has different purposes or intensity of purpose than among the more modernized men. Most women remain at home and among those who are employed, work may not yet be crucial to individual or family socioeconomic mobility. More data on achievement orientation (Kasl and Cobb 1970) and social support (House et al. 1982; Berkman and Syme 1979) are needed to understand stressors among adults entering the money economies of rapidly changing traditional societies.

Emotional complaints and education

Table 15.16 shows significant negative bivariate associations of psychological complaints with systolic and diastolic pressure among the Tutuila women and with systolic pressure among Pago Pago women. Among Pago Pago men there is a significant positive association between systolic pressure and complaint score psychosocial complaints.

Table 15.17 shows that education and age-adjusted blood pressure are inversely related, mostly in the more modernized American Samoan samples. There is a strong negative association between systolic blood pressure and education in all but the Manu'a men and a similar relationship for diastolic pressure in all except the Manu'a men and women. Those with more years of education have significantly lower blood pressures.

Table 15.17 Nonparametric Correlations of Education with Blood Pressure in American Samoa

Spearman's R	Manu'a		Tutuila		Pago Pago	
	Men	Women	Men	Women	Men	Women
Systolic	-.20[a]	-.19[b]	-.18[c]	-.34[c]	-.22[c]	-.28[c]
Diastolic	-.03	-.13	-.14[b]	-.24[c]	-.23[c]	-.24[c]

[a] .10 < p < .05.
[b] p < .05.
[c] p < .01.

Table 15.18 Nonparametric Correlations of Triceps Skinfold Thickness and Education with the CMI MR Score in American Samoa

Spearman's R CMI M-R Score	Men		Women	
	Triceps	Education	Triceps	Education
Manu'a	-.08	.15	.12	.20[a]
Tutuila	.01	.12[a]	-.16[b]	.13[a]
Pago Pago	.07	.07	-.07	.25[b]

[a] p < .05.
[b] p < .01.

Table 15.19 Nonparametric Correlations of Education with Triceps Skinfold Thickness in American Samoa

Spearman's R Triceps	Men Education	Women Education
Manu'a	-.01	.03
Tutuila	-.03	-.26[a]
Pago Pago	-.02	-.15[b]

[a] p < .01.
[b] p < .05.

Table 15.18 shows fatness is inversely related to complaint score only in Tutuila women. Among all women and Tutuila men education is directly related to complaint score. Those with more education complain more. Also, as Table 15.19 shows, fatness is inversely related to education in Tutuila women ($r = -.26$, $p < .01$) and Pago Pago women ($r = -.15, p < .05$).

The positive association of education and complaints among all women and Tutuila men may reflect a growing awareness of psychological concepts and terms with education or an increased proclivity to identify and express emotions and feelings. The inverse association between complaints and adiposity in Tutuila women may be due to the covariation of adiposity and complaints with education. That women with more education are less fat and complain more can be attributed to individual acculturation processes.

Age- and adiposity-adjusted blood pressure are not related to complaints or educational level. The earlier associations between complaints and blood pressure and education and blood pressure are due to the influence of age and adiposity, as well as the fatness and education association. Psychological awareness and tendency to complain may be related to fatness both independently and indirectly through education in females.

These results suggest that elevated blood pressure is one indirect but important outcome of this direct link, and support the prediction of Ostfeld and D'Ari (1977) that in many populations undergoing rapid sociocultural change, fatness may be the key in blood pressure variation and that psychosocial factors may be principally related to adiposity. Although the causal sequence among these variables is simultaneous in the analyses, it is probably not the case. For example, it is more likely that among women education influences both emotional complaint score and fatness and all three exert some influence on blood pressure. This suggests a need for causal modeling.

The lack of any simple relationship between education and blood pressure after control of age and adiposity in other groups is not surprising given the strong inverse relation of cross-sectional age and education in a rapidly modernizing society. Nonetheless, first results from a preliminary study focusing on education and blood pressure, show strong inverse associations within age groups. For example, among Pago Pago men 18 to 34 years old, education is negatively related to systolic blood pressure (Spearman's $r = -.33$, $p < .02$) but is unrelated to adiposity, and a similar relationship exists for Pago Pago men aged 50 years and over. The inverse association in modernized Samoans of education and blood pressure and the bivariate association of low education with adiposity and blood pressure among Samoan women is similar to findings from the United States (Hypertension Detection and Follow-up Program Cooperative Group 1977; Stunkard 1975; Syme et al. 1974). This suggests a risk of elevated blood pressure among those Samoans who are poorly educated and exposed to modern life.

The suggestive results linking low complaint scores and blood pressure among modernized Samoan women led to detailed analyses of their complaint pattern and its association with education and blood pressure (McGarvey and Weishaar 1984). A factor analysis of the CMI items determined whether any underlying common factors existed in the CMI responses, detected any unique Samoan patterns of emotional complaint, and determined which items explained most of the variation. Factor-based scales of emotional expression or complaint are constructed by

Table 15.20 Emotional Expression and Blood Pressure in Women from Modernizing Villages of American Samoa (*N* = 634)

| Education | Emotional Expression | Interaction of Education and Emotional Expression | |
		Systolic	Diastolic
0-6 yr	Low: 0-2 complaints	132.8[a]	84.1[b]
0-6 yr	High: 3+ complaints	126.6	80.6
7+ yr	Low: 0-2 complaints	130.3	82.5
7+ yr	High: 3+ complaints	128.9	82.7

[a] Systolic: with low education, difference between low vs. high emotional expression $p = .02$.
[b] Diastolic: with low education, difference between low vs. high emotional expression $p = .03$.

summing positive responses to high-loading items. The first factor explained most of the variation, indicating a general tendency to complain about strong emotions, and after orthogonal rotation contained items from the CMI inadequacy, depression, and anger sections (McGarvey 1984a).

Table 15.20 shows that among those with low education, low expressors have significantly higher systolic and diastolic blood pressure. Among those with high education there is no difference between low and high emotional expression or complaints.

The opportunities and demands of modern life may conflict with traditional female social roles and as a consequence, women are experiencing strong negative emotions. Those with low education may be both unfamiliar or unable to express emotion as suggested above and also be relatively unable to participate in the opportunities of economic modernization, resulting in stress and increased blood pressure.

There is a consistent trend in the Samoan data for low emotional complaint or expression associated with higher blood pressure both in univariate and multivariable analyses. Either Samoans are not expressing strong negative emotions because they do not experience them or they are suppressing them. The latter is more likely, following several lines of evidence indicating that emotional control is a key concept in Polynesian child rearing and adult interactions. Research among Samoans has noted the rigid social control of behaviors (Howard 1979), the strict child discipline (Shore 1976; Young 1974), punishment for expression of strong emotions, the control of anger (Gerber 1975), and the adult suppression of strong emotions, in the face of personal injury and pain (Ablon 1973). These learned, or enculturated, patterns may be psychosocial adaptations to small island and circumscribed village life (see Chapter 7). However, in the presence of novel economic, social role, and psychosocial events and stimuli concomitant with modernization, these patterns of emotional expression may prove maladaptive.

Number of children

The association between number of children and blood pressure is tested among married and divorced Samoans between 18 and 50 years of age. Among Pago Pago women there is a significant inverse association between adjusted systolic pressure and the number of children ($F = 2.9$, $p = .05$). Women with less than five children have higher blood pressure than those with six or more children. Because the Pago Pago villages are populated by migrants from Western Samoa, a further analysis is stratified by females both in American Samoa or Western Samoa. The age- and adiposity-adjusted systolic blood pressure for Pago Pago women born in Western Samoa with 0 to 5 children are significantly higher than for their counterparts with more than five children ($F = 3.92$, $p < .05$; 0-5 children, 128.3 mmHg vs. 6+ children, 116.9 mmHg).

Western Samoan migrant women living in Pago Pago villages of American Samoa are exposed to modern life. Having fewer children probably indicates attempts to control fertility, which may be stressful in a high fertility society with traditional female roles. The failure to find higher blood pressure in mothers with many children contrasts with Scotch's (1963) work among South African Zulus. However, American Samoa may not be an ideal society to detect stress effects on mothers with children. Economic modernization has raised the standard of living and thus, family income may be perceived as adequate without the participation in the labor force of adult women. In addition, women working outside the village often depend on relatives from extended, multigenerational families to provide child care.

Multivariable analyses

Multivariable analyses are presented with the focus on the relative contribution of adiposity and age to blood pressure variation, although other significant factors are discussed. This is restricted to the three American Samoan and the Hawaiian samples where similar data and methods were used. Table 15.21 presents the standardized betas and the amount of blood pressure variation explained by age and triceps skinfold thickness in the American Samoan and Hawaiian samples.

In women, age is slightly more influential than fatness on systolic blood pressure, whereas fatness is more related to diastolic pressure. This is probably due to the women's striking adiposity and the effect of being adipose over the span of adult years. In men fatness is much more important than age in explaining blood pressure variation. These findings in Samoans confirm the long-standing association between fatness and blood pressure (Tyroler et al. 1975; Chiang et al. 1969).

The amount of blood pressure variation explained by the variables ranges from 5 percent and 15 percent in Manu'a men to almost 40 percent in several samples of women. These are typical of epidemiologic

Table 15.21 Multivariable Analyses of Blood Pressure among Samoan Adults

	Manu'a	Tutuila	Pago Pago	Hawaii
Men				
Systolic β				
Age	—	.172	.233	—
Triceps	.008	.427	.379	.693
Ratio[a]	.025	—	—	—
Years in Hawaii	—	—	—	.160
R^2	.15	.29	.22	.16
Diastolic β				
Age	—	.144	.138	.092
Triceps	—	.361	.409	.500
Ratio[a]	.080	—	—	—
R^2	.05	.25	.21	.16
Women				
Systolic β				
Age	.423	.378	.730	.301
Triceps	.375	.527	.575	.272
Salt intake	—	—	.141	—
R^2	.37	.39	.39	.20
Diastolic β				
Age	.302	.245	.347	.264
Triceps	.397	.492	.408	.202
Years in Hawaii	—	—	—	.093
R^2	.25	.28	.39	.19

[a] The ratio is: subscapular skinfold / (triceps skinfold + subscapular skinfold).

blood pressure studies (Chakraborty et al. 1977), but it is noteworthy that a simple measure of adiposity explains so much in Samoans.

Modernization

We first assessed the effect of modernization by comparing three American Samoan groups based on residential ecological differences in general modernization: Manu'a versus Tutuila versus Pago Pago. Adult blood pressure is directly related to residential exposure to modern life. The absolute differences in blood pressure among the three American Samoan samples are significant, and age and triceps correction does not fully account for the differences (McGarvey and Baker 1979).

Table 15.22 shows the results from a more complete analysis restricted to subjects with all available information on age, fatness, salt intake, and complaint scores (McGarvey 1980).

Modernization differences among the samples are significantly and positively related to male systolic and diastolic pressure, along with age

Table 15.22 Mutivariable Analysis in American Samoan Groups of Blood Pressure; Modernization Term Is Three-level Index Corresponding to Residence Groups Manu'a, Tutuila, and Pago Pago

	Systolic			Diastolic		
	F	p	β	F	p	β
Men (N = 444)						
Main effects						
Triceps	84.7	.01	.783	80.2	.01	.460
Age	26.5	.01	.263	10.5	.01	.102
Modernization	6.0	.01	.151	4.9	.01	.140
R^2 for blood pressure	.26			.22		
Mean adjusted blood pressure:						
Manu'a (N = 53)		123.6			80.9	
Tutuila (N = 257)		130.2			84.8	
Pago Pago (N = 134)		133.3			86.4	
Women (N = 716)						
Main effects:						
Triceps	171.0	.01	.616	162.5	.01	.374
Age	182.2	.01	.632	75.8	.01	.254
Psychosocial complaints	3.6	.05	-.249	3.5	.06	-.153
Modernization	2.1	.12	—	8.7	.01	.130
R^2 for blood pressure	.41			.32		
Mean adjusted blood pressure:						
Manu'a (N = 105)		129.4			82.2	
Tutuila (N = 337)		129.2			81.2	
Pago Pago (N = 274)		132.2			85.0	

and triceps skinfold. However, triceps skinfold is much more influential on systolic (five-fold) and diastolic (three-fold) pressure variation than the residential modernization term. This confirms the bivariate and multivariable results shown above.

Among women, the modernization term is significantly related to diastolic pressure and approaches significance in systolic. Comparison of the betas shows that modernization has a relatively weak effect on diastolic variation compared to triceps skinfold, about one-third, and age, about one-half. It has slightly less of an effect than the psychosocial stress measure on diastolic pressure.

The results suggest that known factors of exposure to modern life, fatness, and emotional complaints have a substantial effect on the blood pressure of women, so that the general modernization differences are not significant. In men, however, there are still blood pressure differences attributable to unknown residential factors associated with exposure to modern life.

Blood pressure is also directly related to an additive index of modernization or acculturation based on education, occupation, English language ability, and residence history, similar to that used by Bindon (1981). The index varies significantly with residence and both are associated with blood pressure (McGarvey, unpublished data). Future work on indices of exposure to modern life should focus on single variables such as education (McGarvey 1984b), or years spent outside of Samoa and on more careful index construction (Patrick et al. 1983).

Migration

The migration of traditional peoples to modern settings is a special case of modernization and differs from *in situ* modernization. The effect of migration on blood pressure is one of the most often studied phenomena of modernization (Prior et al. 1977 Cassel 1974; Prior 1974; Florey and Cuadrado 1968; Cruz-Coke et al. 1964; Trulson et al. 1964).

In order to detect migration effects we took advantage of the presence of Samoans in Hawaii. Initially we sought migration effects on blood pressure by comparing Samoans in Hawaii with Samoans in Samoa. These initial contrasts showed no significant differences when Hawaiian Samoans are compared with all of American Samoans (McGarvey and Baker 1979).

Consequently, migration effects are analyzed by comparing the migrants in Hawaii with nonmigrants in American Samoa based on area of origin. Using age and triceps corrected blood pressures, migrant men from Western Samoa, the islands of Manu'a, and north shore Tutuila villages have higher diastolic pressure than their counterparts in American Samoa (87.5 mmHg vs. 82.1 mmHg; $F = 4.3$, $p = .05$). A similar nonsignificant trend was found for systolic pressure of men. Migrant men from the Pago Pago area of American Samoa and Apia, Western Samoa, have lower systolic and diastolic blood pressure than their counterparts in Pago Pago (125.8 mmHg vs. 134.0 mmHg; $F = 4.4$, $p = .05$; 81.4 mmHg vs. 86.7 mmHg, $F = 4.4$, $p = .05$). In a similar analysis with age and adiposity, migrant women from Tutuila had higher diastolic blood pressure than nonmigrants (McGarvey and Baker 1979).

The interpretation of these observations is not clear. First, with regard to higher pressures in migrants from traditional Samoa, these migrants have the least education, smallest proportion of English speakers, and the shortest length of residence in Hawaii (McGarvey 1980). The sudden greater exposure to modern life in Hawaii relative to a more traditional premigration life and the relative recency of residence may lead to the higher blood pressure in men. Second, with regard to the lower blood pressure in migrants from modernized Samoa, these migrants have resided the longest in Hawaii and have more education and greater English abilities than the nonmigrants (McGarvey 1980). These migrants most exposed to modern life may be changing and reducing blood pressure by dietary, physical activity, and health-care behaviors. Both of these observations and the inferences offered here suggest that increased time in Hawaii may lead to biobehavioral adjustments leading to lower blood pressures. Return migration to Samoa, further migration to California, mortality of those with high blood pressure, however, may also result in the observations presented here.

One aspect of migration affecting contrasts of migrants and nonmigrants is selective migration of migrants. Follow-up research in American Samoa in 1981 permitted a study to test for selective migration among those sampled in 1976 but reported as migrant in 1981 (McGarvey 1983). Twenty-seven percent of all men and 24 percent of women migrated and the migrants were significantly ($p < .001$) younger than nonmigrants: men—35.2 versus 44.2 years; and women—37.0 versus 42.5 years. Among men under age 40, migrants weighed less (81.3 vs. 89.4 kg, $p < .01$), had thinner subscapular skinfolds (19.4 vs. 25.5 mm, $p < .01$) and lower blood pressure (123.4/79.9 vs. 128.9/85/6 mmHg, $p < .01$) than nonmigrants. This is also noteworthy for men from ages 18 to 50 from the Pago Pago villages. For women under 30 years from Manu'a, the migrants had higher systolic and diastolic pressure than nonmigrants. Among older adults few differences were found and these were thicker skinfolds and higher blood pressures in the migrants, especially women.

The results from the study of premigration differences provide indirect support for the idea that migration increases blood pressure but do not support the idea that biobehavioral adaptations in Hawaii reduce blood pressure. No premigration differences were found among traditional American Samoan men. This suggests that the migration effect of higher pressures in the traditional men may be due to the new environment of Hawaii rather than selective migration. The selective migration results indicate lower premigration blood pressure in men less than 50 years old from Pago Pago. This strongly suggests that the lower blood pressure found in migrant men in Hawaii from the Pago Pago area compared to nonmigrants is due to selective migration. This is not conclusive because we are not dealing with the same individuals over time. There is still a possibility that certain new migrants to Hawaii from the more modernized areas of Samoa adapt more readily to life there and have lower blood pressures.

The selective migration differences in older subjects need to be understood in light of who first migrates from the village and who follows (Prior et al. 1977). Certainly further work must consider marital status, education, and family migration histories. In earlier work on the migration effects on blood pressure Hanna and Baker (1979) showed that residence in Hawaii was related to blood pressure. The association among origin in Samoa, destination in Hawaii, and selective migration needs to be analyzed.

Familial aggregation

We attempted to assess the effect on blood pressure of shared genes and shared family environments in Samoans undergoing modernization. We studied nuclear families and attempted to measure both intra- and extrafamilial changes due to modern life. There is familial aggregation of blood pressure among Samoans, with sibling and parent/offspring correlations of age- and weight-corrected blood pressure similar in magnitude to other populations, ranging from 0 to .30, with most from .10 to .30 (McGarvey et al. 1980). Juvenile sibling correlations increase with degree of modernization, defined by location, and adult sibling correlations decrease. Assuming similar distribution of additive genes among samples, the results may be due to environmental factors and their interaction with genes. Specifically we suggest greater interfamilial and lesser intra-familial environmental variation with greater exposure to modern life. For example, adult siblings in Hawaii are likely to share less of a common environment due to nucleated residence patterns. Likewise, juvenile siblings in Hawaii will share a common family environment more than in Samoa. The assumption of similar genes for blood pressure among the areas needs to be tested considering the evidence for selective migration among modernized males.

The preliminary estimate of the additive genetic effects derived from Western Samoa is .25 (McGarvey et al. 1980), and is similar to other populations (Tyroler 1977). However, this is specific to this Samoan groups and not the other Samoans. Both genetic and cultural heritability estimates await more sophisticated methodologies of family aggregation study.

There is an increased level of spouse concordance for blood pressure associated with modernization in American Samoa (James et al. 1983). In addition those couples married less than 15 years have significant positive concordance for blood pressure. Sociodemographic data on the couples show that concordance for education increases and that for area of birth decreases. We suggest that educational strata may be forming, directing mate choice away from kin and village influences and toward individual choice. These forces may be resulting in similarity in spouse blood pressure with modernization.

They also suggest that familial aggregation of blood pressure with modernization may have both genetic and sociocultural bases. The series of studies by Ward and colleagues among the Tokelauans (Ward 1983; Ward and Prior 1980; Ward et al. 1979) are suggestive of a scenario for Samoans. The results in Samoans indicate that there may be family clusters of high blood pressure due to the genetic segregation stemming from spouse concordance and the gene/environment interactions.

From a population and evolutionary view none of the work discussed above indicates an increased Samoan susceptibility to high blood pressure, per se. However, family studies of adiposity could suggest susceptibility to the major risk factor for hypertension. There are suggestions that Polynesians may be prone to adiposity (see Chapter 17).

SUMMARY

The multivariable analyses in the American Samoan and Hawaiian groups and the comparisons of the age/sex means from all the study locales provide consistent, unambiguous support for the primary hypothesis. Exposure to modern life is directly related to elevated blood pressure in the Samoan population. The results suggest that better support could be offered if an appropriate measure of exposure to modern life could be devised for individuals.

The secondary hypotheses about specific biological and sociocultural effects on blood pressure are supported by results showing how adiposity, age, salt intake, and socioecological and psychosocial factors are all related to blood pressure in several samples. The cross-sectional design of the research makes it impossible to establish precedence among these variables, but it is likely that causal modeling would improve our understanding of the processes of exposure to modern life and their direct and indirect blood pressure effects.

Figure 15.5 shows how some of the data discussed here may be used in such a causal model. Structural variables such as birthplace and migration largely determine sociodemographic factors, such as education and occupation, which affect stress and coping. Biology and behavior are to a limited extent affected by these preceding factors, but combine and interact with sociodemographic and stress and coping factors to determine blood pressure levels and variation. Blood pressure is related to all of the above through both direct and indirect pathways. Causality is implied by

Fig. 15.5 Causal model for blood pressure in modernizing Samoans.

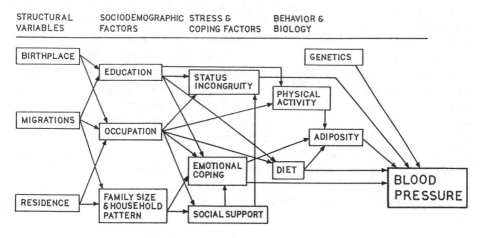

the ordering of variables and their paths. For example, migration affects occupation, which affects physical activity, adiposity, and finally blood pressure. Some indirect relations to blood pressure are also shown. Along this simple pathway migration, occupation, and physical activity may influence blood pressure independently of their association with adiposity. Of course, there are probably many different potential pathways, and the overall model shown in Fig. 15.5 may be too simplistic. The model suggests future analyses of extant data, such as social support and blood pressure, and indicates where new research should point, for example longitudinal investigations of these interrelated factors to explicitly test hypothetical causal relations.

Regardless of future directions, the present results show that the most consistent and powerful correlates in the multivariable analyses are adiposity and age. Clearly, in this modernizing population adiposity is powerfully related to blood pressure. Ostfeld and D'Atri (1977) hypothesize that obesity is the primary factor in elevated blood pressure among populations experiencing rapid sociocultural change. The present work supports that hypothesis, but suggests salt intake and psychosocial factors also play a role.

On a more general point, the characteristic relationship of factors with blood pressure described here will probably change with the secular changes in the population's exposure to modernization. This has been demonstrated for coronary heart disease in a rural U.S. population (Morgenstern 1980). The cohort effects of intensity, duration, and types of exposure to modern life may produce different concomitants of blood pressure and modernization, and different estimates of their effects. This is certainly good reason for further study with longitudinal samples.

In conclusion, the Samoan research project has shown consistent, strong, and biologically plausible evidence in favor of the proposition that exposure to modern life leads to elevations in blood pressure. The major factors in this modernization-related increase are adiposity, salt intake, and socioecologic and psychosocial stress.

We do not know how past traditional Samoan biocultural adaptations are involved in the present and future risk to health posed by their blood pressure patterns. However, the striking adiposity of Samoans exposed to modern life and its effects on blood pressure suggest that the thrifty genotype (Neel 1982, 1962) can be applied to Samoans (Baker 1981) and other Polynesians (Prior 1977). Similarly, cultural patterns of emotional expression formerly adaptive (Howard 1979; Gerber 1975) may place some Samoans at risk (McGarvey 1984b). One last speculation concerns the differences in renal- and ionic-balance physiology in some populations (Gillum 1979), and the question whether the Samoan history of tropical residence and ocean-voyaging discovery and settlement of islands has produced adaptations in water balance and control.

Future human biological research on blood pressure should focus on individual and population interactions with the environment, using the method of human ecology, and search for the suite of biological and sociocultural factors affecting blood pressure and their evolutionary and adaptive antecedents.

Chapter 16

Samoan Coping Behavior

ALAN HOWARD

SAMOAN COPING STRATEGIES

The main purpose of this chapter is to identify significant problems commonly shared by Samoans as they confront increasingly complex environments, and to describe their coping responses to these problems. Some of the problems are endemic to Samoan culture; that is, they are rooted in the traditional life-style, while others are a consequence of changes in economic and social conditions that are affecting the Samoan populace. Still other problems are associated with dislocations such as migrations to New Zealand, Hawaii, the U.S. mainland, and other destinations where Samoans are ethnic minorities. As might be expected, the coping strategies used are composed from a mixture of practical considerations—recognizable to anyone having to deal with the daily contingencies of life in the modern world—and some distinctly Samoan, or at least Polynesian, modes of problem solving. In Chapter 6 Ala'ilima and Stover presented data on coping from the subjective standpoint of individual Samoans. Their presentation demonstrates the considerable variability both in problems confronted and in coping responses. In this chapter we look for commonalities, although the reader should keep in mind that an increase in variability is one of the major consequences of the macroprocesses affecting the Samoan population.

CONCEPTUAL FRAMEWORK

One of the major problems for comparative anthropologists lies in the variability of conceptual usage from study to study, and it seems that the more widely used the term the greater its range of usage. Concepts like stress, coping, modernization, and acculturation, like the concept of culture itself, have been defined in innumerable ways (and more often left implicit), depending on the purposes of the ethnographer or comparative analyst. While it would be presumptuous to propose definitive conceptualizations here, it is important that we define key concepts as we use them, if for no other reason than to clarify the bases of our judgments.

The concept of *stress* has proved so thorny that many recent commentators have refused even to attempt a definition (Mason 1975:29). There are, however, a number of identifiable models of stress (Scott and Howard 1970), and it will be useful for our discussion to specify the one we are using here. This might be termed the "social stress/generalized susceptibility model" as developed by Cassel (1976) and Syme and Berkman (1976). From their perspective, stress is viewed as the result of a disjunction between demands of the social environment and an individual's coping resources or ability to solve problems. Coping resources span social and personal factors and include, for example, the emotional and instrumental assistance provided by others, as well as personal attributes. Although coping resources can buffer the effects of environmental stress, if demands are not matched by resources, and if recurrent or acute problems are not solved, the stress process is seen as promoting a susceptibility to disequilibrium that contributes to the onset of disease, psychopathology, and/or social deviance.

The concept of *modernization* has also encountered difficulties, and has been justly criticized for implying a unilineal progression from a traditional base to a monolithic notion of contemporary, urbanized, Western society. Its usage frequently has been in the intellectual tradition of classical evolutionism, with "traditional" substituted for "savagery" and/or "barbarism," and "modern society" for "civilization," and it is subject to all the criticisms that brought on the demise of that paradigm. Earlier generations of anthropologists used the term *acculturation* to refer to the cultural (as distinguished from the material) changes experienced by a population encountering the various agencies of modern Western societies, but that concept, too, with its implications of lineal transformation from one single base line culture toward another, has proved overly simplistic. Indeed, one of the lessons we have learned by extensive examination of populations undergoing change is that responses are highly diversified within as well as between cultures. Cultural contact almost invariably produces, over a period of time, a myriad of new possibilities for most of the individuals it engages, just as it generates new problems. For this reason I prefer the concept of *cultural diversification* to refer to the processes of relevance to Samoan coping behavior. I originally used this concept to deal with responses to changes affecting Hawaiian-Americans (Howard 1974), and the same rationale holds for contemporary Samoans. To paraphrase the argument:

The viewpoint is based on an assumption that contemporary Samoa represent a collectivity of individuals attempting to cope with an increasing complex milieu. Their world is no simple mix of things Samoan and Wester At one level their world is dominated by legal, political, occupational, a educational institutions imposed by the prevailing American (or in the case Western Samoa, New Zealand) social system, albeit with modifications of particularly Samoan nature. At another level Samoans are affected by t cultural prescriptions of *fa'aSamoa*, and by the norms generated within loc communities, within work groups, and extended families. These relate ways of speaking, using time and space, and behaving toward vario categories of people, and they influence the organization of daily routin Local norms differ from community to community, and may or may not compatible with Western expectations. Some are derivative from earli forms of Samoan culture, some are amalgamations of several traditions, wh. still others are innovative adaptations to new circumstances, including t problems posed by change itself. The point is that Samoans face a wid range of problems than did their ancestors, and they have at their disposal greater range of cultural options to deal with them. In short, cultur diversification has the advantage of including all those processes by whi cultural repertoires are expanded, whether or not they are associated wi particular historical traditions. (Adapted from Howard 1974:83-84)

Cultural diversification entails, for Samoans as well as for Hawaiia Americans, an expansion of both problems and problem-solving tec niques, or *coping strategies*, as we shall refer to them. The bas framework for examining "problems" within the context of stress researc derives from studies of stressful life events (Kaplan et al. 1983). Th underlying premise is that a life change produces a disequilibrium tha imposes a period of readjustment, during which the individual is mo vulnerable to stress and its consequences (Kessler et al. 1985:533). Wor on stressful life events has been facilitated by the Social Adjustme Rating Scale (see Chapter 8 for a discussion of Samoan responses Although it was initially presumed that any life change, whether positi or negative, increased the probability of illness, and that effects wer general rather than specific [i.e., not predictive of a particular illne (Homes and Rahe 1967)], recent research suggests that such features a undesirability, magnitude, time clustering, and uncontrollability ar crucial to the relationship between life events and pathology (Thoits 198 Dohrenwend and Dohrenwend 1981). This framework is of theoretic import for coming to grips with the health implications of stress amon Samoans, particularly insofar as it involves migration. A move to a ne environment, and especially to New Zealand, Hawaii, or the U.S mainland, generally involves several significant life changes, either simu taneously or in quick succession. It ordinarily involves changes i household composition, work, finances, eating habits, and recreation activities, to name a few. But these rather distinct "events" by no mean exhaust the potential stress involved in relocation. Chronic stress, in the form of "daily hassles" (Kanner et al. 1981) may take an even greater toll. For Samoans, as with other struggling immigrants, solving problems on a daily basis in a strange linguistic and cultural environment can take a great deal of energy and many resources. At the very least these daily hassles can be expected to exacerbate the effects of life events (Pearlin et al. 1981; Brown and Harris 1978).

Another line of stress research has emphasized the psychological characteristics of individuals as opposed to external conditions. Interest in this approach has been greatly enhanced by the attention given to the so-called *type A*, or *coronary-prone, syndrome*. The characteristics at the center of the syndrome are generally considered to be an exaggerated sense of time compression, competitive achievement striving, aggressiveness, and hostility. Research by Glass (1977) strongly suggests that two response factors are of particular importance in linking these characteristics to physiological indicators of stress, namely personal control and anger arousal. Thus Glass found that it was not simply that type A persons were inherently more aggressive than type B persons, but that they became so in response to a specific set of circumstances that threatened their sense of mastery (1977:70). This implies that coronary-prone individuals are more subject to anger arousal in response to frustration. Although one would not expect Samoans—particularly rural Samoans—to show type A personality characteristics [although they do show up with modest frequency in questionnaire data[1] (Graves et al. 1983)], there is good reason to believe that they experience heightened emotional arousal and threats to their sense of control in new environments. Indeed, violent outbursts are well documented among Samoans and have become a central part of their ethnic stereotype.

It is generally agreed that a variety of additional psychosocial factors implicate vulnerability to stress. Resources ranging from financial means to communication skills have been documented as mitigating variables. Commonly included are such assets as cognitive flexibility, social support, and effective coping strategies (Haan 1982).

Cognitive flexibility is considered to be an asset because it allows individuals to consider a range of alternative solutions to a problem rather than relying strictly on past procedures. It is associated with novel problem-solving behavior as well as the capacity to readjust one's mental program to fit new circumstances. Individuals lacking in cognitive flexibility are presumed to be effective problem solvers as long as conditions are congruent with those in which they were initially socialized, but they are expected to experience more acute difficulties when conditions change.

Perhaps the variable that has received the most attention over the past few years has been social support, "a term that has been widely used to refer to the mechanisms by which interpersonal relationships presumably protect people from the deleterious effects of stress" (Kessler et al.

1985:541). Among the benefits attributed to social support have been access to the expression of positive affect or emotional support, to expressions of agreement with a person's beliefs and feelings, to advice and information, to sympathetic listeners and to material aid in times of hardship (House 1981). The initial enthusiasm that permeated the literature on social support in the 1970s has been dampened somewhat by a growing awareness of the complexities involved. For example, recent evidence suggests that low-density networks characterized by weak ties can aid adaptive strivings more than high-density networks under some circumstances (Hirsch 1980, 1979). There is also accumulating evidence that different types of support have different effects and are associated with differential outcomes (Cohen and McKay 1984; House and Kahn 1984). Other complications include the possibility of being enmeshed in networks of relationships that are partially supportive but also involve a good deal of negative affect, which may add to the strains of adaptation. Furthermore, received support is generally offset by the obligation to provide it, and for some the burden of obligations may outweigh the benefits accrued. As we shall see, the evidence for Samoans is mixed. Group support clearly plays an important role in the adaptation of Samoan immigrants to the United States and New Zealand, but there is just as clearly a price to be paid.

The concept of coping has also received a good deal of attention in recent years. The earlier emphasis on coping behavior has yielded to a greater concern for cognition, although most definitions retain both a behavioral and cognitive aspect. Thus, Kessler et al. define coping as "the cognitive and behavioral effort made to master, tolerate, or reduce demands that tax or exceed a person's resources" (1985:550; Cohen and Lazarus 1979; Pearlin and Schooler 1978). The emphasis is on the active role individuals play in structuring the world around them and in managing resources when responding to adaptive problems.

Coping strategies can be categorized in a wide variety of ways, and although no typology is generally agreed upon, Pearlin and Schooler (1978) list three dimensions common to most: altering the problem directly, changing one's way of viewing the problem, and managing emotional stress aroused by the problem. A wide variety of coping strategies have been identified, including assertive problem solving, tension reduction through exercise, humor, crying, drinking, drugs, etc, engaging in diversions; aggressively attacking frustrating persons or obstacles, and so on. Of special concern for our purposes is the degree to which individuals attempt to cope in a self-reliant versus an other-reliant manner. As we shall see, Samoan culture places a heavy emphasis on familial and communal obligations, and encourages group efforts at problem identification and resolution. Since this is often seen as being at odds with the demands of life in the modern world, the tension between self- and other-reliant coping strategies has been a focal point for research on Samoan adaptation to new environments.

It is my intention to summarize the findings on Samoan coping strategies from a variety of sources, including the studies being reported on in this volume and previously published literature. The discussion is divided into two major segments. The first deals with aspects of Samoan culture that have a bearing on coping styles and factors associated with successes and failures of coping. The second segment concerns Samoan responses to specific kinds of life events, particularly those associated with urbanization and emigration. My focus will be on adaptive problems associated with work and economic demands; crises, including illness; and the management of anger.

COPING SAMOAN STYLE

To begin our discussion of Samoan coping style let us examine the literature dealing with Samoan personality. Of particular concern will be cognitive styles, emotional patterning, self-esteem, and sense of control, since these aspects of personality directly implicate coping behavior.

Cognitive styles

Given the Samoan emphasis in child-rearing on obedience to adult authority (Chapter 7), the heavy emphasis on ritualization and the overall weight of tradition on social life, one might expect individuals to develop an alloplastic cognitive structure, that is, one that cannot easily be altered to accommodate unusual experiences or observations contrary to those previously made and incorporated. This in turn might be seen as an impediment to learning new coping skills in new environments (Howard 1966).[2]

Although I know of no data sets that directly measure Samoan cognitive styles, the personality data that are available suggest that the learning milieu in traditional Samoan settings in fact favors an alloplastic cognitive style. Thus Torrence (1962), using both verbal and nonverbal tests with a sample of 1000 Western Samoan school children, found them to rank lowest in original thought when compared with similar samples for Australia, Germany, India, and southern blacks. He attributed low creativity among Samoans to four specific cultural values: (1) an emphasis on remembering well, (2) an acceptance of authority hierarchies, (3) submission, and (4) doing nothing until told. The inference one might draw from these data is that Samoans can be expected to experience a good deal of adaptive strain when attempting to meet the demands of a new environment. We must be cautious about making such an inference, however, since the primary strategy Samoans use to cope with adaptive requirements, and the stresses and strains they induce, is social rather than psychological.

Social support

In concluding their review of personality data on Samoans, Holmes et a
(1978) present the following image:

> The picture that all these studies present is essentially similar, and there
> consistency through time. It reveals the Samoans as a people valuing ord
> in things familial and ceremonial, with a great tendency toward mutual a
> and support, especially when directed toward family or other members of tl
> in-group. There is a strong tendency to conform to the will of the group and
> reject opportunities for leadership or personal gain. Samoans are tenacio
> and conservative. They do not value personal autonomy, personal achiev
> ment, or being the center of attention. They are not an aggressive people, n
> are they particularly sensual or creative, and, just to throw in one mar
> opinion [Maxwell 1969], they are extremely extroverted. (Holmes et a
> 1978:470)

This composite image is based on the work of some 15 students
personality utilizing a broad range of methods, including folklore ar
literature analysis; projective, verbal and nonverbal testing; observatic
and interviewing; controlled laboratory observation; personality invei
tories and value schedule analysis, among others (Holmes et a
1978:453). The overwhelming emphasis on social relations in th
"personality" portrait has been affirmed by ethnographers and soci
logists. Mead, for example, wrote that: "The dominant note in Samoa
society is its prevailingly social emphasis. All of a Samoan's interest, a
of his emotion, is centered upon his relationship to his fellows within a
elaborate and cherished social pattern" (1969:80).

Social strategies based on kinship rights and obligations and reciproc:
exchange are the primary means by which Samoans cope with problem
Furthermore, in many instances it is the family group rather than tl
individual that formulates strategies. This is often the case when
comes to decisions to migrate, particularly for women (Shankman 1976
Thus Graves et al. (1983) report that only half of the men and 16 percei
of the women in their New Zealand sample claimed to have made tl
decision to migrate mainly by themselves. Their informants asserted th:
it is particularly common for families to send their single daughters t
New Zealand because they are considered more likely than sons to sen
remittances back home. The financial exigencies involved in migratin
make it very difficult for an individual to go it alone, and over three
fourths of the Samoans in Graves' sample reported that their fares wer
paid mainly by their families (Graves et al. 1983:14).

Family-oriented decision making may be seen as a reflection of tl
Samoan preoccupation with status in the traditional mold, in which hon
is a familial responsibility. Parents and family elders typically scrap
enough money together to send promising youngsters abroad so that the

can get a well-paying job and send remittances back home. Younger ones might be sent abroad in order to get better quality educations. When remittances are sent home they are likely to be invested in conspicuous generosity, especially in lavish gifts to the local church and to community projects, which enhances the family's status. Concerning the centrality of the family to Samoan social life Fay and Vaiao Ala'ilima (1968) have written:

> It is difficult for someone steeped in Western individualism to grasp the Samoan idea that the smallest political unit is a family group. The family is regarded not as a plurality of individual opinions, but as a single political organism. True, it may have internal parts: its old people providing experience, its young people acting as arms and legs, and its chief being the central brain for formulating decisions. The strength of such a body depends, however, not on individual rights but on how effectively these organs perform their different functions collectively. Any glory gained by the family is shared equally by all.
>
> . . . the organic family concept means that decisions about community affairs are left to family chiefs with little resentment by other members. The arms and legs simply assume that the brain will operate on behalf of them all. A Samoan adolescent was recently asked, "What do you want to be when you grow up?" He answered without hesitation, "My chief has decided I shall be a pastor." Probing for signs of adolescent rebellion, the interrogator continued, "Yes, but what would you like to be yourself?" The boy acted surprised and confused. He couldn't say, not because he felt suppressed by his chief but because he had not thought of this as his individual decision before. Unlike a Western adolescent he did not sharply dissociate his personal opinion from that of his family on the matter. In some ways a Samoan boy may be freer than a Western boy. Chiefs are notably tolerant of the dress, social activities, and personal habits of their young men; but only so long as these do not affect the strength or public image of the group. When it comes to defending family reputation, position, and interests, traditional chiefs are expected to direct and their families to obey. (1968:13-14)

Self-esteem, an important variable frequently mentioned in the coping literature, is for Samoans therefore closely tied to their position in their family and their family's position in the broader Samoan community. To become the matai of a well-respected family, to hold a prestigious title, are nearly universal goals for Samoan men. Women contribute to their family's honor by supporting their husband's and son's ambitions, and by demonstrating social competence in their own right. Thus it is important to recognize that the way in which a Samoan is located socially, quite beyond the usual indicators of socioeconomic status (occupation, education, residence, etc.), is of considerable consequence for understanding his or her vulnerabilities and coping potential. Families, and to a lesser extent neighbors and church congregations, not only provide material resources

for coping, they are a continuing source of psychosocial resources
though there are costs involved as well, as we shall see).

The management of anger

This now brings us to the issues of emotional arousal and control. 1
Samoan cultural paradigm places a considerable emphasis on the c
straint of impulse. Indeed, the social control of personal action is at
core of the Samoan notion of culture:

> When Samoans speak or sing about *aganu'u* or "culture" they stress th
> aspects of social life associated with dignity and respectful deference. In
> Samoan sense of the term, then, culture excludes aggression, competition, a
> the unrestrained expression of personal impulses. Those aspects of ex
> ience, part of the natural order of things, are not, in themselves, cultural fa
> for Samoans. In its Samoan sense culture is control and the so
> institutions and understandings which are associated with control. (Sh
> 1977:410)

Yet most observers have commented on the extent of violence
Samoa, in contrast to the social norms. Freeman (1983), in his refutat
of Mead's ethnography, documents high rates of assault, rape, murd
and other forms of aggression in the Samoas, but he is by no means
first anthropologist to challenge Mead's idyllic portrayal. For examp
Robert Maxwell reported that during his relatively short period of fi
work a total of 30 of the 52 men in his village sample were involved
one or more fights (Maxwell 1969:223). Furthermore, it is the intens
of aggression that has often caught the attention of observers (see,
example, Lemert 1964:371).

In answer to the question, Why are Samoans aggressive? Kee
(1978) looks to both child-rearing practices and frustrations generated
social patterns. He notes that in Samoa the expression of aggression
not regulated by family rules but rather is heavily punished. If a chil
behavior is irritating to an adult, the child will be slapped or spanked.
explanation is given to the child, so any rules that underlie the puni
ment are obscure and implicit. In addition to frequent use of corpo
punishment, Samoan parents often follow it with displays of affecti
presumably to communicate to the child that he or she is loved a
wanted in spite of the punishment. The overall effect, according to Kee
is that parents provide aggressive models. They generate high levels
anger through frequent and severe punishment, and they link pain a
love, violence and pleasure in the child's mind [for an excellent descript
of parent/child relations in Samoa, see Gerber (1975)].

In addition to frustrations imposed by the shift from an indulg
infancy to a punitive childhood, Keene maintains that the requirement
sharing is a continuing source of frustration to Samoans. Thus, there a

many times when an individual is forced by custom to give up things he may want for himself or his family. It is also likely, he suggests, that the cultural requirement that aggressive feelings must not be expressed serves to intensify the hostile emotions engendered by the system of child rearing. The strict rules discouraging displays of hostility are such that there is little opportunity to vent hostility short of fighting. It therefore seems that in contrast to the Tahitian case, where redundant controls in conjunction with a gentle child-rearing strategy serve to produce a gentle character (Levy 1978), in Samoa powerful external controls have a pressure-cooker effect.

Shore (1982) reports that Samoans themselves attribute controlled behavior to external social constraints, while uncontrolled behavior is associated with the failure of self-control over personal desire. They tend to see their own anger as leading to tantrums and going wild.

In order to understand the character of Samoan aggression one must also take into account social structural principles. As has been pointed out, Samoa is highly politicalized in the sense that striving for status is a continuous preoccupation. Challenges to status are threats to self-esteem and to the esteem of one's family members, and it is expected that these challenges will be vigorously met. What is significant here is that the cultural apparatus oriented toward controlling anger and aggression requires clarity of status differences, on the unambiguous complementarity of junior and senior. In his incisive analysis of conflict in Samoa, Shore (1982, Chapter 11) shows that the flash points involve relationships that include structural ambiguities or contradiction, such that status differentials are uncertain. It is not simply a matter of young men, among whom most of the violence occurs, being less in control of their emotions. They are also in a position of considerable status ambiguity, so that the effects of social controls are seriously dampened.

As we shall see, the relationship between social and personal controls is of considerable significance for understanding behavioral and psychological responses to urbanization and migration to new environments, particularly where traditional social controls are inoperative or severely limited in their effectiveness.

PROBLEMS AND RESPONSES

As Harbison points out in Chapter 4, migration has played an important role as an adaptive strategy for Samoans in a context of high fertility, low mortality, and limited natural resources. As an aggregate strategy it has the advantage of drawing off surplus labor and increasing foreign exchange through remittances, although it has the disadvantage of increasing the dependency ratio, thus creating a greater burden for the productive individuals who remain. As a family strategy—and it is clear that emigration is, for the most part, a family rather than individual

decision—migration is a means of expanding the group's access to vari
resources, including new labor markets.

In order to highlight Samoan coping strategies in the context of t
resulting diversification, I will focus on studies of Samoans abroad,
New Zealand and the United States. This is not to imply that life with
the Samoas is without stress. In fact there are good indications, includi
an alarming increase in suicides (discussed below), that it may
stressful indeed (see Chapters 9 and 15). But whatever adaptive pro
lems confront Samoans in the villages, or in Apia and Pago Pago, th
are present in more extreme forms abroad, and hence provide more acu
insights into coping strategies.

Samoan migrants to New Zealand and the United States face all of tl
disadvantages of immigrants from nonindustrialized countries, and th
some. Language is often a problem, making it difficult to conduct tl
normal business of daily life outside the household;[3] discrimination bas
on negative ethnic stereotypes can be a problem that undermines se
esteem; the scarcity of familiar food requires changes in eating habits;
New Zealand and on the U.S. mainland cold weather requires adapti
adjustments; and given the large size of most Samoan families, tl
burden of dependent children often adds to financial problems. On a mo
general level, those Samoans who come directly from rural villages ha
simply had little training for, or experience with, urban life. What th
do have, however, in almost every instance, are relatives to rely on.

Housing

The initial problem most migrants confront is housing, and the domina
coping strategy is simple—stay with a relative until one can establi
one's own residence. In the United States an exceptionally hi
proportion of Samoans reside in public and low-income housing. T
demand for such housing remains strong and results in long waiti
periods; it also contributes to crowding. According to the 1980 censu
Samoans are reported as having the highest median number of persons
any ethnic group in the United States. For renter-occupied housing uni
the U.S. median is 1.99 persons per dwelling; for Samoans the median
4.23 persons (U.S. Dept. of Labor 1984:81).

In their New Zealand sample, Graves et al. found that the averaₚ
migrant lives with relatives for between 2 and 3 years, with only a fe
(5-10%) able to find their own place in less than two months.[4] Tl
situation in other overseas communities is essentially the same, althou
the time frames may vary according to local circumstances.[5] Th
arrangement has, of course, both advantages and disadvantages for ne
migrants and their host households. For the new immigrant the accor
modating household generally provides a supportive group of relative
who know the ropes and can assist in solving the basic problems
adjustment. They can steer the new arrival through appropria

channels, can assist in finding employment, and can provide material support until the person is able to pull his or her own weight. For the accommodating household providing aid is a means of expanding the local network of kinsmen that can be relied upon when needs arise, since the provision of a place to stay imposes strong obligations on the recipient to comply with future requests for assistance. If the new immigrant is able to find gainful employment shortly after arrival, the returns to the household may be more immediate, in the form of contributions to family income. On the debit side, households are under pressure to accept a broad range of relatives, including those who are likely to be troublesome and unproductive, resulting in an additional burden on household resources. For the ambitious newcomer the obligation of submitting completely to the authority of the household head, and turning over one's paycheck, may be a heavy price to pay.

Employment and economic problems

Kinsmen also play a dominant role in finding employment for new migrants. It is expected that all able-bodied men will get a job as soon as possible so that they can begin to contribute to family expenses, to engage in reciprocal exchanges, and send remittances back home. The situation for women is somewhat more variable, since some are expected to contribute in domestic rather than wage-earning roles.[6] Since most immigrants arrive with minimal skills, and without prior experience in the kinds of jobs available, they are likely to be limited to a somewhat small segment of the labor market. They are also less likely to have job-hunting skills, and so must rely on relatives who have already established themselves. In formal situations such as job interviews Samoan etiquette calls for lowering one's head and avoiding eye contact. Initiative is to be left in the hands of the interviewer, and questions answered softly and deferentially. Unfortunately, these displays of respect are often perceived as signs of apathy by prospective employers [see Shore and Platt (1984) for a discussion of communication problems related to Samoan employment].

The job markets vary from place to place, as do the regulations governing immigration and employment. Whereas immigration to the United States is unrestricted for American Samoans, Western Samoans going to New Zealand are required not only to arrange housing in advance, they must also present proof of guaranteed employment. Unemployment is therefore not much of a problem for Samoans in New Zealand, whereas it can be in the United States.

The U.S. census data for 1980 (U.S. Bureau of the Census 1983a) shows 9.7 percent of all Samoans in the labor force to be unemployed, with rates slightly higher for females than males. The unemployment rates are 81 percent higher for Samoan males, and 51 percent higher for Samoan females than for the U.S. population as a whole (U.S. Dept. of

Labor 1984:59). In Hawaii the percentage of unemployed Samoans in the civilian labor force is by far the highest of any ethnic group (Fran 1984:23). The unemployment rates for youths ages 16 to 19 are partic larly severe, reaching over twice the state average in California (U. Dept. of Labor 1984:60). Furthermore, as a consequence of both uner ployment and the low-paying jobs that are available, Samoan per capi income is among the lowest in the United States, trailed only by the Vietnamese. Using standardized indices of poverty, the census data sho 29.5 percent of Samoans to be below the poverty line compared to 12 percent of the population as a whole, with 19.6 percent living in "extren poverty" (below 75% of poverty level) compared with 8.3 percent of the total U.S. population (U.S. Dept. of Labor 1984:42).

Research by Pitt and Macpherson (1974), and by Graves et al. (198 1983, 1981, 1977) in New Zealand provides a good indication of the ir portance of the 'aiga in economic matters, from obtaining a job establishing economic viability. The Graves distinguished between thr types of adaptive strategies, which they labeled kin-reliance, pee reliance, and self-reliance, and developed questionnaires to measure ther They then interviewed substantial numbers of Samoans, Cook Islander Maoris, and Pakehas (European New Zealanders). They found the Pol nesians, and particularly the Samoans, to be higher than the Europear on 13 of 15 measures of kin-reliance. The results were significant at th .001 level (Graves et al. 1983:44).

The kin-reliant strategy shows up clearly in obtaining employmen and particularly first employment. Thus, 66 percent of the Pacific islar immigrants (including Samoans) in the Graves sample obtained their fir job through a relative, compared with 47 percent of New Zealan educated Polynesians (also including Samoans) and 43 percent European New Zealanders. When "present job" was the subject of inquir the figure fell to 35 percent, which was comparable with New Zealan educated Polynesians, but still well above the 20 percent of Europear who were kin-reliant. Pitt and Macpherson (1974), using a different dat base, found that 44.8 percent of Samoan immigrants who responded their inquiry located their first job through relatives, 11.7 percent wer through a friend, and 2.3 percent used the church. The remaining 41. percent either answered advertisements (19.7%), went through th Samoan or New Zealand governments (13.7%), or looked for jobs on the own (7.7%).

Pitt and Macpherson (1974) note that Samoan job stability has bee high in New Zealand, and that this is related to adaptive strategies. The found that the majority of migrants in their sample had had only one jc since arriving in New Zealand, and that the pattern was one of sho periods in various jobs until one found a "good" job. The notion of a goo job begins with relatively high take-home pay, with less concern fc working hours, job satisfaction, or status. According to these authors:

This attitude toward jobs has also been noticed by employers and the Chief Employment Officer in the Auckland office of their Department of Labour, who said that cutbacks in overtime lead to job changes and the restoration of overtime to the return of those who had left.

This emphasis on immediate earnings may be partly explained by the support patterns in which migrants are involved. A migrant's earnings must support his nuclear family, contribute to the support of the kin group in Samoa, and also to other expenses incurred within the family group—such as the fares of other migrants from Samoa. To do all this a migrant needs a good deal of money. Migrants are well aware of the variation in hourly rates of pay but they reason that industries with high hourly rates and relatively frequent rises are often prone to industrial unrest and stoppages. Thus small hourly raises may be achieved at the expense of major losses of immediate income.

This emphasis on take-home earnings also explains why few migrants, even those with the necessary educational qualifications, take clerical and other white-collar positions. It appears that, initially at least, they prefer jobs that offer higher pay even at the expense of a possible gain in status. This preference explains why only 4.4 percent of the total work force is employed in commercial and retail-clerical occupations. (Pitt and MacPherson 1974:82)

Other criteria that affect a job's desirability are the opportunities it presents for obtaining work guarantees for intending migrants and for providing employment for Samoans already in New Zealand who want to change their jobs. As Pitt and Macpherson point out, the best chances exist in industries with high labor turnover, both because job opportunities frequently occur and because the stability of Samoan labor is appreciated. An important result of this process is that work units in selected industries come to consist of groups of relatives and friends. The opportunity to work with other Samoans acts as an additional buffer for new migrants, especially if they have initial language difficulties. Although problems are encountered in new work situations, particularly with regard to adjusting to a quite different mode of giving and receiving instructions, asking questions, and other information processing procedures, Samoans seem to adapt well to initial confusions. They tend to avoid confrontations whenever possible, and to passively accept authority in most circumstances (Graves and Graves 1977; Pitt and Macpherson 1974:88-98).

Franco (1985a) found somewhat different attitudes prevailing among younger Samoan immigrants to Hawaii. Whereas older Samoans were described by his informants as willing to take any kind of employment, younger ones were described as desiring white collar jobs. Informants also made a distinction between Western Samoans, whom they perceived as more committed to "making it" in the Hawaiian job market, and

American Samoans, whom they perceived as more casual about employ
ment. They ascribed this difference to the economic backwardness c
Western Samoa and the lack of career opportunities there. Franco als
found that Samoans in Hawaii respond enthusiastically to jobs that offe
periodic advancement and other forms of recognition, while they becom
discouraged and indifferent to jobs that do not. Significantly, they se
work under the former conditions in much the same terms (*tautua*, o
"service") they see work for a matai. In the traditional system tautua i
the path to chieftainship. In contrast, jobs where mobility is blocked an
recognition not forthcoming are considered to be merely work (*galue*); the
elicit little commitment and are readily dropped.

Although circumstances differ in Hawaii and in various U.S. mainlan
communities, the general reliance on kin networks, and to a lesser exten
on friends, is pronounced everywhere among new Samoan immigrants
With time, however, increasing proportions of the immigrant populatio
move toward a self-reliant adaptive strategy. Thus in the Graves sampl
of Pacific island immigrants, whereas only 27 percent obtained their firs
job by themselves, 51 percent got their present job without help fror
kinsmen or friends (Graves et al. 1980:203). Among Polynesian mer
whom they consider to be the major determiners of what strategy will b
followed by their families, the Graves found a strong positive correlatio
($r = .27$) between length of time in New Zealand and self-reliant score
(Graves et al. 1983:44).

COPING WITH ILLNESS AND CRISES

The ways in which Samoans deal with illness have been a cause o
consternation among Western health practitioners, who typically perceiv
them as only seeking modern medical treatment as a final resort
Clinicians express particular consternation when children are involved
whom they see "as helpless victims of parental unconcern and who ofte
suffer from diseases whose course responds well to early intervention wit
appropriate therapy" (Cook 1983:138).

There appear to be two sets of reasons for this behavior. One focuse
on Samoans' experience with Western medical personnel, which is ofte
negative from the Samoan viewpoint. The other set of reasons is base
on the fact that there are alternatives to Western medicine that are insti
tutionalized within Samoan culture.

For an immigrant Samoan, going to a medical clinic, hospital emer
gency room, or comparable facility can be an exercise in frustration
Aside from language problems that may severely impair communication
the impersonality of the procedures, combined with intrusive questionin
about medical histories, is a disquieting occurrence. Often nurses an
doctors themselves are under pressure and reveal an impatience tha
aggravates communication problems, and it is not uncommon for parent

who bring children with advanced conditions to be directly rebuked. The characteristic Polynesian response to these experiences is avoidance. Avoidance both reduces the embarrassment and shame experienced in such encounters and minimizes anger toward authority figures that cannot be adequately expressed.[7]

Equally important is the alternative system of health care that has been retained with considerable vigor, even among overseas migrants. Aspects of Samoan practices have been described by McCuddin (1974), Kinloch (1980) and Cook (1983), and I will not dwell on them here. For our purposes what is important is that all agree that Samoans retain confidence in traditional practices and include them, along with Western medicine, in their repertoire of responses to illness. The question is thus what determines how much priority different individuals give to "traditional" or "modern" practices in their hierarchy of resort. Variations in response patterns depend upon a number of factors, including the family diagnosis of the problem (what kind of illness it is, how severe the symptoms are, etc.), past histories of such ailments within the family, the advice of respected relatives and friends, the relative availability, and expense of various kinds of healers and facilities, and so on.

Not surprisingly, in view of everything else we have learned about Samoan culture, illnesses are regarded as family problems first and foremost. For this reason Samoans are often reluctant to bring problems to the attention of outsiders until they are convinced they cannot be solved within the family (Kinloch 1980:25). Reinforcing the tendency to rely on family remedies is the Samoan (and widely shared Polynesian) theory of illness, which stresses behavioral factors as causal. The central notion is that one's "life essence" (to'oala) has been displaced from its normal location in the upper abdomen to various parts of the body where it may induce pain and other symptoms. What causes the to'oala to move are occurrences, and especially behaviors, that disrupt the normal order of things. They may include working too much, getting too little sleep, eating the wrong kinds of foods, and most seriously of all, acting immorally or in ways to disrupt interpersonal relations within the family. Among the more provocative offenses are failing to carry out one's family responsibilities properly, disobeying an authority figure or God's laws, and disrespectful behavior toward kinsmen and ancestors (Cook 1983:140).

The general healing principle this notion of causation entails is that the patient can be brought back to normal by the application of behavioral or physical opposites. Thus, if overwork is seen as the cause of an illness, rest is prescribed; a fever is treated by rubbing the patient with leaves dipped in cool water; and a chest cold attributed to exposure is dealt with by using Vick's Vaporub and wrapping the victim's chest in cloth. The most prevalent treatment, however, and the one that usually initiates the healing process, is massage, usually performed by an older family member. The idea behind this is that massage functions to put back into place, by directional stroking, the to'oala of the patient (Cook 1983:140;

McCuddin 1974:7). If an illness persists, and does not respond to home treatments, the patient may be brought to a clinic, hospital, or physician, or alternatively, to a Samoan specialist.

The literature suggests a pragmatic, but somewhat impatient approach when it comes to healers. If the condition does not respond to treatment quickly the prescribed regime may be dropped and alternative treatments sought. If the condition continues for a long period, or grows worse, this is often taken as a sign that something is wrong within the family and a family meeting is held to determine the probable cause. Anyone bearing grudges, or having a grievance against other family members, is expected to air their feelings and amidst confessions, apologies, and prayers to work toward reestablishing family harmony.[8]

Quite aside from the relative merits or hazards of Samoan views toward illness and treatment (and there are cases to be made in both directions), their beliefs provide them with an alternative coping strategy in the face of an impersonal and anxiety-provoking Western medical system. Their beliefs not only give them a sense of control over most ailments, by making them comprehensible and manageable, but they also act to strengthen family solidarity, and thus reinforce the dominant coping mechanism of kin-reliance.

The importance of kin-reliance as a coping strategy for dealing with acute crises is clearly manifest in Ablon's reports of Samoan reactions to a disastrous fire (Ablon 1973, 1971a). The fire occurred in a church social hall in San Francisco in 1964, killing 17 Samoans and severely burning many others. Attending physicians in hospitals remarked about the stoicism with which both the burn victims and their relatives accepted the event, in marked contrast to typical American responses. Those who provided long-term care described their patients as "stalwart and uncomplaining, no matter the seriousness of their burns" (Ablon 1973:170). The only complaint was that some patients did not follow instructions for self-care after leaving the hospital.

The Samoans Ablon interviewed 5 years later explained their ability to absorb such disasters as a result of the Samoan proclivity for hard work and their deep, fatalistic religious beliefs. They spoke of themselves as strong and able to take the inevitable hardships of life without complaint. Ablon, however, emphasizes the importance of family and community support, which, she comments, "may be more apparent to the non-Samoan observer than to the Samoan who takes for granted the extraordinary emotional, social, and financial support offered by the Samoan extended family and the larger Samoan community" (Ablon 1973:177). The evidence for such support in her interview material is overwhelming.

COPING WITH ANGER

We have already commented on the contradictions that appear in the literature on Samoa concerning aggression. On the one hand is Mead's idyllic image of Samoans as gentle and emotionally bland; on the other is Freeman's portrayal of them as among the most aggressive people in the world. As we pointed out, social context is a critical factor in understanding the nature of aggression in Samoa. In this section we are concerned with the implications of cultural diversification for the management of anger and aggression. If Shore (1983) and others are correct, and controls depend upon well-defined cultural principles and the clarity of social contexts, then one would expect the management of anger and aggression to become a more acute problem as the Samoans are drawn more fully into the modern world system. The same should hold true for migrants to New Zealand and the United States, and in fact the available information suggests that this is indeed the case.

Data on mortality and suicide within Samoa (Chapter 5; Bowles 1985; Macpherson and Macpherson 1985; Oliver 1985) indicate that violence to others and self has increased in recent years, presumably in response to "modernizing" influences. In American Samoa the rate of recorded homicides and suicides, while variable from year to year, appear to show a marked upswing since the 1970s, with both rates at 25 per 100 000 during 1980 (Chapter 5, Table 5.10). Age-specific rates for young men would be considerably higher, since women and older men are involved to a much lesser extent in such events.

There are, of course, problems with the official vital statistics records. Careful inquiry on a case-by-case basis suggests that violent deaths are considerably underreported. For example, in Western Samoa the official statistical abstracts list only four suicides for the years 1980-1982, whereas Bowles, working from inquest records, determined that 122 cases could reasonably be classified as suicide over the same period (Bowles 1985; Macpherson and Macpherson 1985:70). Bowles' data show a steady increase from 1970 (6 suicides) to 1981 (49 suicides) (see Table 16.1). While the suicide rate for the total population, based on Bowles's information, stands at 22.6 per 100 000 for the years 1981-1983, for males 14-24 they are 71.0 and for males 25-34 they are 75.6 (Bowles

Table 16.1 Suicide Rates for Western Samoa, 1981-1983 Average per 100 000

	Population	Deaths	Rate
Total population	156 349	106/3	22.6
Total males	81 027	76/3	31.3
Males 15-24	18 787	40/3	71.0
Males 25-34	8380	19/3	75.6
Total females	75 322	30/3	13.3
Females 15-24	19 570	21/3	35.8
Females 24-34	8155	5/3	20.4

Source: From Bowles 1985:23.

1985:23). During the peak year 1981, the suicide rate for males in th 25-34 age group reached 167 per 100 000. Furthermore, for even completed suicide there was evidence of an unsuccessful attempt (Bowl 1985:17). Clearly, such figures are of epidemic proportions.[9]

While comparable figures for suicide are not available for Samoan abroad, data on imprisonment in New Zealand and Hawaii indicate tha Samoans are considerably overrepresented in prison populations (see, fi example, Pitt and Macpherson 1974:108-112; State of Hawaii 1985). is also the case that where they interact with other groups in urba environments Samoans are perceived as highly aggressive, and ai known as quick-tempered brawlers (Gerber 1985). We must not k misled into generalizing about Samoans as crime-prone, since clearly th vast majority of Samoans are law-abiding. Furthermore, the likelihoc that selective law enforcement, prosecution, conviction, and incarceratic operates to the detriment of Samoans must be considered. Nevertheles such statistics and stereotypes do raise questions that ought to k approached in a forthright manner, and since a high proportion of Samoa arrests appear to involve physical aggression, the ways in which Samoar manage anger would seem to be a place to look for explanations.

The first point that needs to be made is that when Samoans talk abow emotions their attention is directed to interactions in which those emotior occur rather than to internal sensations (Gerber 1985, 1975). In othe words, it is the social context that commands interest. With regard ' anger, the ideal is expressed in the concept of *lotomama* (to be withow anger), particularly in the face of a situation that might provoke (Gerber 1985). When individuals do get angry, however, it is soci context that plays the determining role in how it is, or is not, expresse As pointed out earlier, within Samoa the flash points of aggressic generally occur in situations where status differences are minimal or ai ambiguous. Where clear hierarchy exists, expressions of anger ai inhibited to extremely muted. The prototype of hierarchical relations ai between parent and child, and as many observers have reported, ther are no socially acceptable ways of directly expressing one's anger towar parents. There are, however, several terms that can be used to indicat covert angry responses to parental demands, including *augata* (laziness *o'ono* (suppression of anger), *fiu* (fed up), and *musu* (reluctance) (Gerbe 1985). Of these the most interesting is musu, since it has specificall been linked with suicide.

Pratt, a missionary who lived in Samoa between 1839 and 1879, note that suicide is "mostly caused by anger within the family" (quoted i Macpherson and Macpherson 1985:36; see also Freeman 1983:220, 346 It is important to recognize, of course, that the term *family* refers to th extended family ('aiga), which for youths involves subordination to man adults. Concerning the situation of youth in Samoan society, Macpherso and Macpherson have written:

Samoan culture prescribes for adolescents a period in which they are expected to serve (*tautua*), not challenge, those who hold power over them. Adolescents are told that service is the path to power ... [Samoan] culture allows youth to raise sources of dissatisfaction in the family provided that appropriate deference is shown to the person with whom the matter is raised. A young person must make it clear that he or she is grateful for the opportunity to raise a matter which it is not their right to do. By implication he or she accepts that any outcome is final since the opportunity to raise the matter is a privilege accorded them and not a matter of right. . . .

A person who wishes to express continuing dissatisfaction with an outcome may become *musu*, in which state he or she becomes sullen and withdrawn; says very little to those around them; does no more than what they are told; and shows little interest in social life. In most cases one who is *musu* will treat a particular person with special disinterest to underscore the supposed source of their discontent. The Samoan concern with relationships and their maintenance leads those around the person concerned to attend to the source of the discontent. Where the matter is soluble gentle pressure is applied to both parties to move toward a compromise. Where an adult makes concessions care is taken to ensure that this is portrayed as generosity and not retreat. If a "reasonable" compromise is negotiated, but is not accepted by the young person, the sympathy for him or her is likely to wane quickly and is likely to be replaced with accusations of childishness (*fia pepe*), and immaturity (*le mafaufau*). The difficulty is that what mediators consider a "reasonable" compromise may not meet the expectations of the young person. In this situation the young person has three options, and their choice will be determined, at least to some extent, by their sense of injustice.

Where the matter involved is not a source of major annoyance the person may simply accept the suggestion that he or she forget the matter and be patient in the knowledge that his/her turn will come. . . .

Where a matter is of more significance a young person may demonstrate his/her intensity of feeling by running away to another village. This is a symbolic rejection of the legitimacy of the authority of those in power. . . .

If a person feels that a matter is of major importance and experiences an intense sense of dissatisfaction and injustice, he or she may be moved to an intense rage which both Gerber (1985) and Shore (1982) highlight in their accounts of Samoan emotion. The rage is said to "leap up" inside the person and take control. In that state people typically lash out, usually at inanimate objects with fists, knives, paddles and so on. . . . A number of cases of suicide which we documented occurred during or shortly after a display of rage. (Macpherson and Macpherson 1985:56-58)

Indeed, after reviewing the inquest records on suicide Oliver determined that "The precipitating event in half of the cases was a scolding or rebuke; and in 55 percent of the cases the triggering agent was one of the parents of the victim" (Oliver 1985:76).

Oliver also found that the ratio of matai to commoners was partic
larly high in the four villages with the highest incidents of suicide (Oliv
1985:76), further pointing to the significance of authority relations f
self-directed rage.

Macpherson and Macpherson hypothesize that the dramatic increase
the incidence of suicide within Western Samoa is the result of block
social mobility.[10] The combination of rapid population growth, an increa
in the dependency ratio, and a stagnant economy, along with high
aspirations and a decline in the possibilities for emigration, wage emplo
ment, or advancement within the traditional status system has led, th
conclude, to a stronger sense of relative deprivation and deepened sense
frustration (Macpherson and Macpherson 1985).

Although aggressive behavior among Samoan immigrants to tl
United States and New Zealand is almost legendary, there have been fe
careful studies of its nature. The typical speculation, exemplified l
Gerber in the conclusion to her recent paper on rage and obligation,
that when Samoans migrate the structure of supports for authorit
including parental authority, is eroded, leaving the burden of controls
the individual. But to the extent that the individual has been trained
rely on external authority and social submission, the appropriate cha
neling of underlying anger may not occur (Gerber 1985). The result is a
increased variability in ways of handling anger (see Chapter 8 f
examples) and a greater frequency of socially inappropriate outbursts
hostility. Other explanations stress the sensitivity of Samoans to sligh
to their social status, and their comparatively enormous body size (s
Chapter 11), which presumably makes reliance on physical force a tem
tation when settling differences with non-Samoans.

While all this may be so, such explanations are not very satisfyin
One would like to know more about the circumstances in which violenc
does occur, for again, it must be stressed, the vast majority of Samoar
do not engage in such behavior. The study of Graves et al. (1981, 198:
on drinking and violence in New Zealand provides some valuable clues
the contextual nature of Samoan aggression in overseas environment
They found that even though relatively few Samoans admitted to bein
regular drinkers (41% of the men and 4% of the women, compared to 90
and 86% of European New Zealanders), the Samoans account for a dispr
portionate amount of pub violence. Graves et al. relate this to the stron
group orientation of Samoan workers, since this leads them on the on
hand to stay in pubs longer and consume more alcohol, and on the othe
to join in to help a friend or relative when he gets into a fight. They als
found Samoans to be particularly sensitive to verbal assaults, and t
move quickly from a verbal to a physical level of conflict. The fact tha
Samoan women do not drink means that male drinking takes place i
settings, such as all-male pubs (rather than mixed lounge bars), wher
they are most likely to become engaged in fights.

The cumulative evidence makes it clear that the chief coping strateg for Samoans is interpersonal. When problems arise it is rarely considered à matter for individuals to resolve for themselves, but involves the extended family at a minimum and often more distant kin and community. The question we now will address is what are the health implications of such a coping strategy, especially for Samoans abroad.

THE HEALTH IMPLICATIONS OF SAMOAN COPING STRATEGIES

The fact is that perpetual involvement with kin networks can become enormously burdensome, especially for an upwardly mobile couple. The obligations of committing income to a broad range of relationships, or supporting the church, of sending remittances back home, make it difficult to save for investments in future socioeconomic advancements. In comparing the situational stressors to which Pacific island immigrants and European New Zealanders were exposed, the Graves found that although the levels of stress were comparable, they derived from different sources:

> Polynesian subjects . . . reported more stress in the area of money matters and kinship relationships: they are more likely to run out of money and receive a notice from a debt collector, and are more likely to have had relatives living with them or experience the death of someone close to them or the birth of a child. These situational stressors all follow from their large families and obligations to kinsmen, both financially and through hospitality, which put a strain on their resources even though their incomes are roughly comparable to their European neighbors. (Graves et al. 1983:37)

The implications of social involvement for health status are particularly interesting, but the results from existent studies are inconclusive. Thus, although Pawson and Janes (n.d.) found their measure of social support to correlate strongly and significantly with blood pressure in the expected direction among a sample of Samoans in California, other studies raise questions about the impact of social involvement on health. For example, among the Samoans in their New Zealand sample, the Graves found that for both men and women, kin-reliant and peer-reliant strategies are both associated with higher rates of reported health symptoms than those who emphasize self-reliant strategies. Furthermore, they found that the main factor responsible for this outcome is the commitment to "mutual aid." These results held only for the Samoan segment of their sample:

> Only among Samoans is . . . mutual aid strongly associated with more reported health symptoms ($r = .45$ and $.47$ for men and women, respectively). And only among Samoans is the number of friends and relatives within walking distance, the amount of money given to relatives, and the number of visits during the last two weeks (mainly with relatives and co-ethnics)

consistently associated with *more* reported health symptoms. In fact, among Cook Islanders and Europeans these relationships usually go in the *opposite* direction, though the magnitude of these correlations tends to be small (Graves et al. 1983:45-46)

These findings are congruent with those of Hanna and Baker (1979 see also this volume Chapter 15) that Samoans residing in rural commu nities on Oahu in the State of Hawaii suffer from higher blood pressure than those residing in urban Honolulu. The rural residents are much more involved in a tightly knit Samoan community and are likely to be burdened with heavy social obligations. Martz et al. (1984) found that i their American Samoan sample high levels of community involvement ar associated with intermediate stress levels as measured by overnight urinary hormone excretions. Those showing the most stress are the one who are most self-reliant, while those showing the least stress hav relatively modest levels of involvement with kin and community, but rel heavily on friendships for social support. This makes sense sinc friendships are voluntary relationships involving far less custom-lade obligations (see also Chapter 9, this volume). The evidence seems clear then, that the predominant Samoan coping strategy—relying on kins men—is one that has costs as well as benefits.

CONCLUSIONS

Like all formerly rural populations who are in transition, Samoans ar confronted with a myriad of adaptive problems. Foremost is the proble of making a living, of gaining access to resources so that they ca maintain a life-style that they value. It is also important for Samoans keep channels of mobility open, so that the interests of the family can l advanced. For Samoans social status, and particularly the prestige of th family, is a perpetual problem.

Overseas, Samoan immigrants are generally at a disadvantage as result of language problems, inferior schooling, and lack of experien with cosmopolitan culture. Even within the Samoas there are indicatio that social mobility is increasingly blocked. As a result, Samoans are population under stress.

Samoan coping styles are shaped to a considerable extent by th traditional culture. The resources they rely on to deal with problems a primarily social. They are not an entrepreneurial people, and their s cialization emphasizes conformity to social convention rather than in vidual enterprise. As a result Samoans do not seem to rely on an arr of personal qualities to solve problems or to deal with stress. There little training for independent decision making, for imaginative proble solving, or for internalizing controls over behavior and emotion. Cogniti flexibility is not a particular strength.

It is apparent that, for the most part, the dominant strategy Samoans do use, reliance upon others in their extensive social networks, has served them well thus far. It has allowed them to explore a variety of possibilities at minimal cost, and has diffused the adaptive burdens they would have to bear as individuals going it alone. But there have also been costs, and it remains to be seen how well it will hold up in the long run.

It is certain that one consequence of cultural diversification will be that the next generation of Samoans will have more choice than their parents and grandparents had. They will have been less thoroughly socialized in the traditional Samoan style, and will have learned a greater variety of coping strategies. It will be interesting to compare the health of those who choose traditional family-oriented strategies, or modifications of them, with those who choose to adopt more "modern," individualized coping mechanisms.

NOTES

1. For example, in the Seven Village Study (discussed in Chapter 8) 47 percent of the men and 64 percent of the women interviewed reported often being bothered by a lack of time.

2. Alloplastic contrasts with autoplastic, which refers to a cognitive style based on altering one's cognitive organization in the face of new information that does not fit into existing patterns.

3. Graves et al. (1983:14) indicate that 41 percent of their sample of Samoan immigrants to New Zealand report "a substantial strain to carry on a conversation in English." They claim that most Samoans are unable to communicate freely in English when they arrive, and are therefore dependent on relatives to help translate for them. The period of linguistic dependency lasts for approximately 2 years.

4. Accommodation presents special problems in New Zealand, since entry permits are only issued after an intending migrant can prove he or she has a suitable accommodation waiting. Whatever housing is designated must then be inspected to make sure it complies with overcrowding regulations. Furthermore, Samoans do not become eligible for housing assistance until they have been in New Zealand for 5 years (see Pitt and Macpherson 1974:31-37).

5. Douthit and Lung (1974:1) for example write that in Hawaii, "Housing is a problem singled out by Samoan immigrants to be an immediate concern. A substantial number of Samoans live in substandard housing. Severely overcrowded conditions are given as the primary undesirable factor."

6. In Hawaii Samoan women have the lowest percentage of participation in the labor force of any documented ethnic group (37.7%); the next lowest group is Vietnamese women (44.6%). All other groups are over 50 percent, with Japanese women leading at 63.4 percent (U.S. Bureau of the Census 1983:66).

7. Kinloch (1980:24) points out that two additional factors may influence medical personnel's impressions that Samoans, and Samoan parents in

particular, behave inappropriately. One is that they are reluctant to use
telephone to talk with medical personnel; hence, they make appearances wh
others would seek advice by phone. The other is that Samoan parents tend no
tell doctors from whom they are seeking a second opinion of previous med
attention, thus leading doctors to imply prolonged negligence.

8. The basic causal notion here, if I can extrapolate from other Polynes
cultures, is that intrafamily conflict irritates ancestral spirits, who show th
displeasure by bringing illness or other forms of misfortune.

9. According to Murphy, "The international statistics of suicides during
1970s do not show any other country to have a suicide rate in males 15-24 wh
is as high as the Western Samoan one" (reported in Oliver 1985:74). While t
statement must be taken against the background of notorious underreporting
suicides in almost all official statistics, it nevertheless underscores the magnit
of the problem in Western Samoa.

10. This echoes the comments of Franco's informants concerning the imp
tance of job mobility in Hawaii. In fact the same term *musu* (sullenness) v
used to describe reactions in both instances.

Chapter 17

Perspectives on Health and Behavior of Samoans

PAUL T. BAKER
JOEL M. HANNA

As we discussed in the introductory chapter of this volume, our research was formulated to investigate changes in the behavior and health of Samoans associated with the processes of modernization. Other investigations have shown that a predictable pattern of change in health and behavior commonly occurs during modernization, but there is considerable between-population variation in specific responses. Thus, at the outset it was not clear in which ways Samoans would be unique and in which ways they would follow the patterns of other populations.

The theoretical framework that we employed was that best known to anthropologists, an evolutionary approach. Accordingly, we assumed that the population was probably best adapted to the natural environment and traditional customs of Samoa. We then postulated that a change in either natural or social environment would have potentially adverse health effects. These could include the following: (1) an increased longevity with modernization, but also many manifestations of ill health to accompany it; (2) in slowly changing situations such as *in situ* modernization there would be fewer adverse health effects than during a period of rapid change such as that produced by migration; (3) there would be fewer health problems for the young than for the elderly; and (4) the most serious effects would be seen early in the change process. Our research has generally supported these premises, but they are obviously very broad and lack explanatory power. Therefore, in this final chapter we take a somewhat different approach in trying to reach some generalizations and hypotheses based upon the integrated results of the various studies.

While the health and behavioral responses of Samoans exposed to

modern environments are in general similar to those reported for other
populations, there are some differences in degree, and in some specifics
commonly observed changes do not appear. Furthermore, in some of the
specific deviations, Samoan findings often run counter to the conventional
explanations for certain diseases and behaviors in modern societies. For
these reasons in this chapter we emphasize the unusual changes and
explore the results for possible explanations. Finally, we suggest what we
consider would be useful lines of future research.

SOME PRELIMINARY CONSIDERATIONS

Reliance on cross-sectional data to describe a longitudinal process clearly
presents a number of problems: for example, to make an interpopulation
comparison between Western Samoan villages, American Samoan, and
Hawaiian Samoan communities, and to use them as examples of how
increasing levels of exposure to modern society affects Samoans, require
certain assumptions. First, it requires the assumption that Samoan
populations in Western Samoa and American Samoa initially shared a
common natural environment, gene pool, and cultural heritage. As Gag
described it in Chapter 2, the natural environment does vary significantly
among the Samoan islands, and as detailed in Chapter 3 social structure
and community status varied from island to island even before the
beginning of colonial control. With the imposition of colonial rule in 1900
Western Samoa and American Samoa became influenced by somewhat
differing cultural systems. Even though all of the colonial powers involved derived from a northern European cultural heritage, the types of
educational and political institutions to which the two Samoan island
groups were exposed differed in a number of ways that certainly affected
the later changes in Samoan behavior and health.

A second major assumption that must be made is that migration per se
does not affect the population and that migrants are representative of
nonmigrants. In the initial phases of the research program we felt that
migration might have a significant effect on many of the biological
characteristics of Samoans, since obviously it was a different experience
to be a minority migrant in Hawaii than it was to remain in American
Samoa, even though both resulted in involvement with the occupation,
foods, and communication systems of the United States. The migrant
experienced a very different social and work environment when they
migrated to the San Francisco area than did those who migrated to Oahu
in Hawaii. As discussed in Chapter 1, Parsons (1982) found that the
Samoan migrants to Hawaii were somewhat different in allelic frequencies and morphology from those who did not migrate. However, it
important to note that neither she nor other investigators found substantial differences in body size, blood pressures, or fertility between Tutuila
residents and the migrants to Hawaii. The migrants to San Francisco

area do appear to differ in some of these characteristics from both Hawaii and Tutuila, but it is not clear how much of this may be related to sampling-technique variation.

We suggest that both the differences between Western and American Samoa and the effects of migration, because of the many intrapopulation comparisons, may not be as significant as they first seem in detecting the effects of modernization. Thus, the village-to-Apia in Western Samoa, Manu'a-to-Tutuila in American Samoa, and village-to-urban Oahu comparisons reinforced the finding from the interpopulation analyses, suggesting that the violations of the assumptions are not important.

While the various inter- and intrapopulation comparisons all suggested that modernization exposure results in comparable biological changes irrespective of migration, this may not be equally valid for behavioral changes. As the life histories in Chapter 6 illustrate, the migrant finds acceptance of new U.S. social values rather difficult. Samoans attempted to maintain many traditional childhood socialization practices after migration, but as discussed in Chapter 7, the value systems reflected in the schools and other institutions of Hawaii often result in different social pressures than are found in the modernized areas of Samoa. Indeed, in Chapter 16 Howard views the coping behavior of the Samoans as primarily specific to an environmental context rather than as part of an overall change-related continuum. However, he also notes in Chapter 8 that it was not possible to measure these behaviors in a quantitative manner; thus, the nature of many behavioral changes could not be specified in the study. As a final problem of interpretation, we also lack substantial comparative information on the behavior of migrants to California and New Zealand. In our research, therefore, health and behavioral changes resulting from migration cannot be separated from those resulting from modernization. If anything, we have shown that the two processes have comparable and similar influences upon the population.

EXPECTED VS. OBSERVED CHANGES

In general, the cross-sectional data and the longitudinal demographic data conform to a pattern found in other traditional groups affected by modern societies. After a period of stability in the late 1800s, the Samoan populations began to increase, first in the early 1900s in American Samoa, then some two decades later in Western Samoa. The differences in the onset of growth was primarily related to the 1918 influenza epidemic in Western Samoa, which caused high mortality. This contrasted to American Samoa, where a strict quarantine excluded the epidemic (McArthur 1967).

After 1950 the death records from American Samoa permit a better perspective on changes in mortality. They show that although the American Samoans already had a relatively low death rate from infectious disease and a relatively high rate of death from cardiovascular disease, this was still in a very active period of mortality transition. By the 1970s the ratios of deaths from infectious disease versus degenerative and chronic disease generally resembled those of modern countries. The limited data available from Hawaii and California reveals a similar pattern of mortality in the migrant communities. As discussed in Chapter 5 the life-expectancy and cause-of-death patterns for Western Samoa in the 1980s still suggest a transition and resemble those from American Samoa in the 1960s.

While these longitudinal demographic data imply that the pattern of health-related biological characteristics of Western Samoans should be very similar to those found in American Samoa, the cross-sectional data from the villages of Western Samoa fail to support this supposition. It is likely that the worldwide suppression of high mortality infectious diseases such as smallpox, cholera, plague, syphilis, and filariasis, probably increased life expectancy more than did local and specific medical practices. Nevertheless, it is also important to note that other studies suggest that this is often balanced by nutritional, gastrointestinal, and degenerative diseases that did not occur in Western Samoa. Thus life expectancies have risen to near the levels of modern countries even in Western Samoa.

The cross-sectional data on the behavioral and biological characteristics of the Samoan subpopulations provide more detailed evidence as to how the change has altered health indices. It is in these detailed comparisons that we also found the greatest discrepancies between the responses of Samoans and the modal responses of other modernizing populations. Table 17.1 is a summary of how some of the Samoan responses compare to those found for other modernizing societies. In general, the table includes only characteristics where sufficient data are available to make a reasonable estimate of the expected response. Emotional stress has been included even though Howard (see Chapter 8) notes it is very difficult to quantify and may be impossible to describe as an expected modal response. Despite this, it has been included because of potential importance as a health indicator.

ATYPICAL RESPONSES

In this section we consider several areas in which the Samoan population seems atypical in its responses to modernization and migration and we explore some possible causes.

Fertility

Numerous studies have shown that changes in fertility as well as mortality accompany exposure to modernization. Thus, many traditional peoples exhibit a decline in mortality, initially accompanied by high

Table 17.1 Some Modal Behavior and Biological Responses of Traditional Populations to Modernization Compared to Samoan Responses

Response category	Modal response	Samoan response
Fertility	Decline, especially for educated women and women in labor force	Less decline than expected; little if any effect of educational level or work status
Body weight	Modest increase	Massive increase
Maturation rates and adult height	More rapid infant and child growth; earlier puberty; increased height of adults	Little evidence of more rapid growth; no evidence of earlier puberty; ambiguous evidence of increased adult height
Work capacity	High initial VO$_2$ max; moderate decline with influence of modernization	Moderate initial VO$_2$ max; substantial decline
Fat and protein	Substantially increased absolutely and relative to total calories	Little if any change relative to total caloric intake
Emotional stress	Believed to rise	Evidence for initial rise; later changes variable
Blood lipids	Rise	Rise, but does not reach U. S. standard levels
Blood pressures	Moderate rise	Rise to U. S. levels
Cardiovascular disease	Rises, but exact rates unknown for most populations	Rises, but does not reach U. S. levels as middle-aged mortality cause

fertility, which gradually falls as the attitudes and behaviors of modernization appear (Preston 1976; Teitlebaum 1975). Since studies of many modern societies have also shown an inverse relationship between fertility and the percentage of women in the work force and between fertility and educational attainment by women, it has been assumed that the changing status of women was related to subsequent decline in fertility.

The longitudinal and cross-sectional data described in Chapter 4 indicate that the Samoan populations only partially fit this pattern and that the education and increased employment of Samoan women does not appear to reduce fertility. While the Samoans clearly showed high

fertility during the nineteenth century, the decline during the twentie
century has been irregular and confined to specific locations. Even whe
agriculture is not an important aspect of economic behavior, as in Ame
ican Samoa and the Hawaiian immigrant populations, the total complet
fertility remains at about five or more children.

Harbison notes in Chapter 5 that this is still compatible with econom
demographic theory if we assume the individual maximizes well-beir
within the larger cultural, economic, or household *context* (emphasis our:
Thus high fertility benefits both parents and children so long as there
the option of outmigration for employment and remittances are return
to the 'aiga. She further suggests that educated Samoan women as w
as those engaged in the labor force are also likely to maintain hig
fertility as long as the 'aiga favors high fertility and removes the ec
nomic penalty to the woman by providing household or child care.

Body weight

A number of studies have shown that traditional peoples tend to ga
weight as they become modernized (Coyne et al. 1984; Kasl and Berkm:
1983). This is often attributed to disease reduction and better nutritio
which result in improved health. The Samoan response is a massi
weight gain. Indeed Pawson and Janes (1981) and Bindon (1984b) clai
that the adult Samoans living in modern societies are the heaviest adul
in the world. Even Samoan infants and children are among the heavie
if compared to other adequately sampled populations.

The reasons for this extreme weight gain are not immediately eviden
Bindon and Zansky's review in Chapter 10 of this volume shows that th
children in Western Samoa are similar in height and weight to the chil
ren in modern societies. In Chapter 11 Pawson notes that even as adul
the Western Samoan village men remain near the U.S. means of weigl
for height until middle age. However, based upon data from America
Samoa and one Western Samoan village Bindon and Zansky suggest tha
all Samoan infants have a tendency toward rapid weight gain during th
first semester of life. Some as yet unidentified factor in the Wester
Samoan village environment seems to prevent a continuation of th
massive weight gain beyond 6 months of age, because in American Samo
and Hawaii the massive gain continues through infancy. Our data ar
not adequate to identify the causes, but this age corresponds to the pos
weaning period of malnutrition that was reported in earlier descriptions c
traditional Samoan villages (Chapter 12).

The diet and activity data described in Chapter 12 lend little credenc
to weight gain as a simple function of increased caloric intake or as
function of a change in dietary compostion. Instead, Bindon's (1982) an
Pelletier's (1984) studies suggested that among possible factors contribu
ting to weight gain, a reduction in daily energy expenditure may be mos
important.

The potential for enormous weight gain among Polynesians is a well-documented phenomenon dating to the early European explorers who noted that individuals of high social status tended to be corpulent (Beaglehole 1961). Today there seems to be an inverse relationship between the degree of non-Polynesian admixture and average weight, with those populations showing the greatest admixture also showing the lowest average weights (Baker 1984a). This has led Prior (1981) and Baker (1984a) to suggest that the Polynesians were subject to selective pressures, which predisposes them to rapid weight gain under favorable environmental conditions. The present findings that even in traditional Western Samoa there is a rapid weight gain during early infancy appears to support the hypothesis. If this proves correct, it does not explain why body weights for Polynesians in such places as Western Samoa, Puka-Puka, and other traditional settings (Baker 1984a) are near the weight for height norms for modern society. Wood and Gans (1984, 1981) have suggested that there are heavy parasitic loads in Western Samoa. Although this might inhibit weight gain, there is now no evidence supporting this speculation.

Maturation rates and adult height

As populations become modernized, increases in growth rates and stature are usually observed. There is a decline in the age at puberty (Eveleth and Tanner 1976), and adults who have developed in the new environment tend to be taller than those living in a traditional society (Tobias 1972). These changes are frequently attributed to improved nutrition, reduced exposure to parasites, and the control of infectious diseases. Temporal increases in growth rates and adult stature have also occurred in modern society, where they are called secular trends. The secular trends have also been attributed to improved nutrition and disease control. This view may be challenged in some instances, because secular trends have also been observed in upper social classes whose nutritional status presumably did not appreciably improve (Damon 1968).

The research into the growth and adult morphology of Samoans does not suggest that growth rates for stature or overall adult stature itself are affected by the changing environment. The stature of contemporary adult Samoans approximates that reported for Samoans in the early 1900s (Kraemer 1902). From the comparisons discussed in Chapter 11 it can also be observed that there is little variation according to place of residence. The adult sample from California tends to be taller than the samples from other areas, but the maximal overall variation among the other adult samples is only 1.2 cm for men and 1.6 cm for women. A comparison of younger with older adult Samoans shows that the younger are somewhat taller in most samples, but using cross-sectional data we cannot determine whether this represents a modest secular trend or is the result of aging and differential mortality.

The sample comparisons of height by age in children also suggests that there were some slight differences, with the Hawaiian immigrant children taller at a given age than others, but Bindon and Zansky consider these differences minor. While no specific puberty data were collected, the growth-velocity data resemble those describing U.S. white children and is probably indicative of a relatively young age of sexual maturation.

The lack of population differences in growth rates, age at puberty, and adult stature may be explained by the fact that the diets of all groups appeared to be adequate and similar in terms of overall content. However, given major differences in weight by age among the groups and the probability of a differential exposure to both infectious and parasitic diseases, the results are of considerable importance in understanding physical growth and development.

Work capacity

In many traditional societies the work capacity of men, as measured by aerobic capacity (VO_2/unit body mass) is relatively high in comparison to a cross-section of U.S. men (Shephard et al. 1978). Exceptions to this are not uncommon and generally appear in groups suffering chronic malnutrition or a high prevalence of disease (Bouchard et al. 1981 Astrand and Rodahl 1977). Among the healthy and fit populations a decline in work capacity frequently occurs with the process of modernization Because habitual physical activity and aerobic capacity are strongly correlated, it is generally assumed that this reduction in fitness is primarily a result of declining levels of physical activity (Spurr 1983; Edholm 1967). For groups with relatively low capacity, some improvement may occur as modernization brings improved nutrition and health status.

The Samoan men on Tutuila and Hawaii appear to have remarkably low aerobic capacities. Greksa et al. in Chapter 13 of this volume note that these capacities are probably below those found for Easter Islanders, a group cited as having one of the lowest measured values. Samples of Samoan men from Western Samoa measured by a less reliable technique appear to have substantially higher aerobic capacities among the village and manual laborers. Nevertheless, none of these Samoan groups appear to reach mean values as high as those found in the fitter, traditional populations or even the fitter, modernized populations (Greksa 1980).

A primary contributor to the low aerobic capacity among Samoan men in Tutuila and Hawaii appears to be degree of body fatness. From regression analysis on data from these studies, Greksa et al. calculated that if the effect of the body fat content was statistically removed, the men's aerobic capacity would be quite high. They also estimated the influence of habitual physical activity as predicted from occupation and found that this was less important than fatness in determining aerobic capacity, although a clear separation of these two variables was not possible. The data from Western Samoa suggest that changes in physical activity may have made a greater impact than is typically described in other studies, but we still have the impression that the aerobic capacity of Samoan villagers is lower than other traditional peoples with adequate nutrition. This impression is in agreement with data in Chapters 2 and 13 that suggest that there probably did not exist a high sustained work requirement in the traditional Samoan environment.

These findings also have somewhat unusual behavioral implications for Samoans as they become more fully involved in modern society. As Greksa et al. demonstrate, their relatively low average aerobic capacity would not have handicapped the population's ability to meet the work requirements found in their traditional society. Even in contemporary American Samoa the available employment does not require substantial numbers of men with high aerobic capacity. However, as migrant Samoans become part of modern society in Hawaii and California, a significant number of the men seeking jobs may find that work requirements exceed their current capacity. The consequences of this are difficult to predict, but the low work capacity may exclude them from some positions open to members of other migrant populations.

Food intake

While many traditional populations had nutritionally adequate diets, those practicing subsistence agriculture were frequently low in fat and protein intake. Modernization of these groups has in cases resulted in diets containing more fats and protein. An exception to this trend may be the Pacific Islanders, whose traditional consumption of coconut and fish provided high fat intake and a good supply of high-quality protein. As a result, they probably were less influenced by dietary change than other modernizing groups.

The Samoans traditionally used coconut in combination with reef resources such that the traditional diet was probably high in both protein and fat. All of the Samoan subpopulations studied have diets that have been altered by imported foods, but as noted in Chapters 2 and 12, the more traditional villages still obtain most of their calories from traditional sources. As Hanna et al. note (Chapter 12), the relative caloric intake from fat, protein, and carbohydrates is very much the same for all groups from traditional to modern. Even saturated fat intake did not increase because of the high consumption of coconut in traditional villages. These results do not suggest an unchanging diet with modernization, rather it makes a dietary explanation less attractive in accounting for other changes following modernization.

Emotional Stress

Studies of peoples undergoing rapid social change or of peoples migrating from traditional to modern societies provide evidence that social change can be emotionally stressful. While such observations are suggestive, the quantification has proved to be difficult, making comparisons between different peoples and situations impractical. We cannot simply assume

that such changes are stressful. As Howard cautions in Chapter 8, som
of the changes in attitudes and social conditions may actually reduce th
degree of emotional stress.

The life histories in Chapter 6 suggest that at least some Samoans fin
the change difficult. T. Baker in Chapter 7 points out that the tradition
Samoan socialization practices place most Samoans in direct conflict wit
the values manifest in U.S. legal and educational institutions. Thu
children often face conflicting school and family demands, while thei
parents' early training in child-rearing and work-scheduling also place
them in direct conflict with the expectations of the larger society. Despit
this evidence that the changing environment may produce emotiona
stress, attempts to examine the nature and extent of the stress throug
questionnaires has proved unreliable. In Chapter 8 Howard suggests tha
while under some conditions questionnaires can provide useful data, th
problem of the question's *meaning* and the *context* in which it is inter
preted make answers to standard stress questionnaires very difficult t
interpret.

The studies of hormone excretion in urine probably represent th
strongest evidence that many of the Samoans are stressed by th
changing environment. Since the samples are select and small and th
situations are limited, no general description of changes in stress over
continuum of change is possible. The results do suggest that in Western
Samoa villagers may experience less stress than those who are mor
involved in the institutions of modern society. The American Samoan
study has suggested that certain aspects of Samoan culture may reduce
stress levels for those who adhere to group principles, but how this socia
support may mediate stress in migrant communities remains to be
studied.

Blood lipids

As noted by Pelletier and Hornick in Chapter 14, the plasma cholestero
and triglyceride levels of traditional populations tend to be low. This is
true even when the traditional diet was high in fat, including saturated
fat. As populations modernize *in situ* or as they migrate to urban
settings, blood lipids tend to increase, with the exception of the HDL seg
ment, which often falls. These changes in blood lipids are usually linked
to changes in diet and physical activity.

The levels of plasma cholesterol of the Samoans in the villages of
Western Samoa and on the island of Ta'u in American Samoa tend to be
low as compared to contemporary U.S. and European values. The values
obtained in Apia remain at a traditional level, but on the main island of
American Samoa and among Hawaiian migrants cholesterol is somewhat
higher. Even so the averages remain substantially below expectations for
modernized populations. The HDL-cholesterol averages tend to show an
inverse relationship to modernization, with the highest average values

found in the villages and on Ta'u. Samoans living in Hawaii and on Tutuila had lower average values. However, for this cholesterol fraction, the Apia men seem to resemble the Tutuila and Hawaiian groups rather than rural villagers.

The blood lipid studies indicate that the higher activity levels and lower body fatness of Samoan villagers may have contributed to their lower total-cholesterol and higher HDL fractions. Of particular significance is the relatively low plasma cholesterol level found for men in Hawaii and Tutuila. Values were below U.S. expectations despite the U.S.-type diet, high levels of body fat, and relatively low physical fitness of these Samoan men. This represents a major deviation from the blood lipid profile, which is typical of modern societies but resembles the findings by Prior et al. (1981) for other Polynesian groups. It may have profound implications for the subsequent development of coronary heart disease.

Blood pressure

Many surveys of blood pressure in traditional populations have shown that adults have lower blood pressures at any age than adults in modern society. In some cases there seem to be no increase in blood pressure with age, which is in sharp contrast to modernized society where age-related increases are expected. While this may be subject to question due to the cross-sectional nature of most survey data, there are groups that seem totally free of hypertension as defined by WHO established criteria (Hornabrook et al. 1974; Huizinga 1972). Further, longitudinal and cross-sectional studies have found that these populations manifest higher blood pressures when exposed to modernization.

The Samoan studies show a similar pattern with relatively low adult blood pressures in Western Samoan villages and in Manu'a, and higher blood pressures in parts of Tutuila. It is particularly significant that the California, Hawaii, and Pago Pago harbor area samples do not substantially differ in blood pressure despite the considerable differences in their respective environments. McGarvey and Schendel note in the conclusion to Chapter 15 that measures of body fatness show significant associations with blood pressure and that they think emotional stress and salt intake also influence individual values. The relative fatness of the various samples may, therefore, help to explain which blood pressures vary between Samoan samples, but it fails to explain why, in Western Samoan villages where adult height, weight, and skinfolds approximate U.S. norms, the rates of hypertension and blood pressures by age are much lower than those reported from the United States.

McGarvey and Schendel in reviewing blood pressure measurements on Samoan children found that they were generally comparable to other Polynesian children and were within the ranges reported for U.S. mainland populations. They also report that a child's weight has a higher correlation to blood pressure than does his or her age, a finding similar to

that of other investigations. One aspect of the results that seems unusu.
is that the heavier Samoan children from the Hawaiian sample have bloc
pressures by age category that are not significantly different from th
children measured in Western Samoan villages. Indeed, if regressions ·
blood pressure by age are calculated, holding the effects of weigl
constant, the Western Samoan children have significantly higher bloc
pressures than any of the other groups. Since the parents of thes
children formed a significant portion of the adult Western Samoa
sample—which was the adult sample with lowest blood pressure—there
clearly a discordance between children and adults.

Cardiovascular disease

Deaths from cardiovascular disease are rare in most traditional popula
tions even though this is the leading cause of death in modern societ
The rarity of CVD in traditional populations is explained in part by th
relative youth of most such populations and by the high rates of dea·
from infectious disease. Nevertheless, in a number of traditional societi·
where the data seems reliable, the rates for CVD appear to be very lo
among middle-aged and elderly adults (Trowell 1981; Zapata and Mart
corena 1968). But, given the low frequency of CVD risk factors-
hypertension, smoking, elevated cholesterol—a low rate of middle-ag
CVD can be anticipated.
    The initial analysis of American Samoan death records from the yea·
1963 to 1973 (Crews and McKeen 1982) led us to believe that the CV
death rate was low in Samoa but was on the rise. The subseque·
analysis of American Samoan records from 1950 to 1981 (Crews 198·
did not support the hypothesis of consistently rising rates despite th
1963 to 1973 rise. If, as suggested in Chapter 5, some of the CVD ra·
in the 1950s was the result of residual heart-valve damage from rhe·
matic heart disease, then the data could be interpreted as showing som
rise in middle-aged cardiovascular death rates resulting from ather·
sclerosis and hypertension in the 1950s. It must, however, be emph·
sized that such a cause for the CVD death rates in the 1950s is pure·
speculative. Even if CVD rates have risen, the rate remains significant·
below that for the U.S. population. The death rates calculated fro·
Hawaii and from California are also low.
    The high frequencies of hypertension and obesity among Samoans ·
Tutuila, Hawaii, and California should contribute to a high CVD rate ·
middle age. Similarly, the accumulated evidence for rather high em·
tional stress in these groups also suggests an increased risk. Th
cumulative effects of such risk factors are presumed to be delayed; thu·
the observed low levels of CVD deaths may represent a time lag in the·
manifestation. Unless the causal precursors for CVD develop in chil·
hood, however, the time frame for change in American Samoa shou·

have resulted in increasing rates by the 1970s. Available evidence does not suggest such an increase.

The short-term prospective mortality study based upon American Samoa data also yielded some unexpected results. As found in studies on other populations, hypertension was a significant predictor for both total mortality and CVD-related mortality, but body weight and body weight by height were not predictors. Given the extreme variance in the weights of the Samoans studied, this finding was particularly interesting, for if taken in conjunction with the other risk factors, it suggests that body weight and fatness in Samoans may have a different influence on CVD death risk than is observed in U.S. black and white populations.

A separate analysis of the small prospective sample of American Samoans, which included some additional social and life experience data, suggested that the death of a spouse increased the probability of the death of the surviving partner. This is in agreement with life events research in the United States (Holmes and Rahe 1967). This analysis also showed that the probability of death increased for men in proportion to the number of years of education. This contrasts with U.S. studies where the probability of death declines with the years of education. These associations are statistically significant only for overall risk of death and not specifically for death from cardiovascular-related causes. Nevertheless, they are suggestive that psychosocial stress may contribute to CVD, since both death of a spouse and a greater involvement in modern society through education can be interpreted as indicators of stress. The supporting evidence for a direct linkage between emotional stress and death from CVD is admittedly weak since the association may have occurred because of intervening variables.

Figure 17.1 is an attempt to summarize the results of our studies in relation to cardiovascular disease and its precursors. The study of linkages is obviously incomplete and the fact that the linkage determinations were not all made on the same individuals in the same environment further weakens our conclusions. Yet, such an overview reemphasizes that the Samoan tendency toward extreme obesity does not appear to produce the high level of cardiovascular disease that might be anticipated. One factor that may reduce risk is the low level of total-cholesterol found in all Samoan populations studied—including those from Hawaii and Tutuila. Given the low levels of physical fitness and apparently limited physical activity of these populations, an elevated level of cholesterol should be expected. We have no immediate explanation for why the Samoan levels are lower than anticipated. And the investigations of HDL-cholesterol do not clarify the problem, since these levels are relatively low in all populations.

From the composite evidence we suggest that there are two reasonable hypotheses to explain the Samoan pattern with respect to CVD. First, it may be that the physiological responses of many Samoans to high body

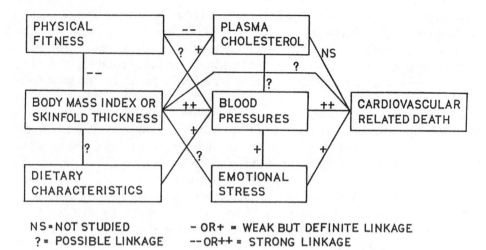

Fig. 17.1 The linkages between potential cardiovascular risk factors and cardiovascular-related deaths, as reported in the Samoan Project studies.

weights, a high fat diet, and low levels of physical activity are different from responses of other groups, such as U.S. whites and blacks. Second, the effects of these variables may be developmental in nature and depend upon experiences during the growth process. Given the composite information on other Polynesian populations and on other Pacific Islanders, we are inclined toward the first hypothesis, but because of the nature of evidence we must consider the second equally valid. Future longitudinal work will help resolve the possibilities.

OVERALL GENERALIZATIONS

In many respects the study of the Samoans has revealed both biological and behavioral changes that parallel those found for other traditional groups exposed to the influences of modern societies. Although their physiological work capacity has declined, blood pressures and total plasma cholesterol have risen. Death from infectious disease has decreased, while type II diabetes and perhaps cardiovascular disease has risen. Behaviorally, the Samoans have readily accepted the material culture of modern society but have found many aspects of the change both distressing and psychologically stressful. Social control of behavior has lessened, but the level of interpersonal conflict has risen. Yet, when the specifics of change are considered, the Samoans show many differences from other groups in both the degree and intensity of response. As described in the preceding chapters and summarized in this one, the deviations from the expected are numerous.

While all the causes for the unusual biological and health changes cannot be deduced, it may be that the Samoan tendency toward obesity could be a key. The effect of this massive weight gain demonstrably reduces work capacity and increases blood pressure. It probably also contributes to the frequency of type II diabetes. Increased fat with modernization is common, but the pervasiveness and magnitude of the change among Polynesians suggest that it may result from a process not normally found in most other populations. The unusual nature of the process is evidenced by (1) the magnitude of the gain, (2) its appearance in the infants of all the Samoan group studied, (3) lower average weights as a function of non-Polynesian admixture, (4) increased weights without increased statures, (5) the lack of clear association between relative weight and CVD, and (6) the traditional association between status and obesity.

The tendency of the Samoans, along with other Polynesians, to become obese may have both genetic and cultural components, but the history and prehistory of the Polynesians indicate a very favorable set of conditions for the development of a population with the genetic predisposition to gain weight rapidly. As first suggested by Prior (1971), and later elaborated by Baker (1984a), the migratory process by which the Polynesians reached their islands and the environments they encountered would have greatly favored individuals capable of rapid weight gain. It would have also provided optimal conditions for the fixations of such a gene or genes in the populations.

If the weight gain response of the Samoans is a key to most of their unusual biological responses to modernization, it leaves unexplained why, with other high CVD risk factors, they have relatively low plasma cholesterol levels. Certainly this is not because of low cholesterol intakes; even the triglyceride levels show high fat intakes. The study of the Masai by Biss et al. (1971) may be pertinent. This population have high intakes of saturated fat and even cholesterols, yet are free of atherosclerosis and heart disease. Metabolic studies have shown that under certain conditions endogenous cholesterol production may be reduced by high dietary fat. We cannot judge how pertinent these findings are for the Samoans.

The unusual behavioral responses of Samoans to participation in modern society include the retention of high fertility and perhaps a more physically and emotionally aggressive response to the new environment. It is far from clear to us how unusual these responses are given the lack of well-quantified data, but if these suggested responses are correct, then it may be that the social control mechanism represented by the 'aiga is a significant factor. This social structure, which is gradually losing behavioral control power in the more modernized Samoan communities, was in the past pivotal in decisions relating to reproduction, economic activity, and social interaction.

As we have frequently noted throughout this volume, many of our findings are tentative and the explanations for the results even more subject to future modification. Nevertheless, the conclusions raise a number of questions both of considerable significance about what is happening and what will happen to populations in transition from traditional to modern societal settings, and of importance for the existing populations that we classify as modern.

While comparable studies on other populations undergoing similar transitions should obviously be a primary focus of future research, the considerable data base collected on the Samoans offers many opportunities for both comparative survey research and more specific problem analysis. In survey research a more detailed examination of the behavior and health of Samoans in California would help resolve many of the questions raised in the present analysis. For example, in a variety of measures such as weight, type II diabetes, and CVD the question remains whether the American Samoan and Hawaiian Samoan values are maximums, or does life in California lead to even higher responses. Studies of Samoans in New Zealand would be equally valuable even though the existing studies suggest that behavioral responses may be similar to migrants in Hawaii. Given the very different health-care system and work situation in New Zealand compared to Hawaii and California, it might improve the determination of causation if it was known how the health characteristics of Samoans in New Zealand differ.

Among the specific questions raised by the existing studies, three seem particularly valuable for future efforts. First, what are the reasons for the massive body weights of the modernized Samoans? And, as a corollary, what are the reasons for the very unusual infant growth pattern in Western Samoa? Both should be studied in detail, and the investigative methods needed appear to be both those of quantitative genetics and biochemistry. A second question is why the high relative weight of Samoans in American Samoa, Hawaii, and California does not seem to result in high death rates from cardiovascular disease. Of particular potential importance are the apparently low plasma cholesterol levels. A number of approaches, including the further examination of natural experiments with more attention paid to insulin production, might prove helpful.

Finally, the study of stress in Samoans using increasingly sophisticated designs for examining how the corticosteroid hormones relate to change and stress would be particularly valuable in understanding the process of change. In all of these studies a greater emphasis on the study of women is needed, since in a variety of the detailed physiological studies undertaken in the present program limited resources and personnel kept the samples restricted to men.

While the list of useful research that could be generated for the future is obviously much greater than has been suggested, we would like to conclude by noting that a base line of data has now been developed from which future efforts can develop analyses of longitudinal change in behavior and health.

References

Abell, L. L., B. B. Levy, B. B. Brodie, and F. E. Kendall. 1952. A simplified method for the estimation of total cholesterol in serum and demonstration of its specificity. J. Biol. Chem. 195:357-66.

Ablon, J. 1970. The Samoan funeral in urban America. Ethnology 9:209-27.

Ablon, J. 1971a. Bereavement in a Samoan community. Br. J. Med. Psychol. 44:329-337.

Ablon, J. 1971b. Retention of cultural values and differential urban adaptation. Social Forces 49:385-93.

Ablon, J. 1971c. The social organization of an urban Samoan community. Southwest J. Anthrop. 27:75-96.

Ablon, J. 1973. Reactions of Samoan burn patients and families to severe burns. Soc. Sci. Med. 7:167-78.

Abraham, S., G. Collins, and M. Nordsieck. 1971. Relationships of childhood weight status to obesity in adults. HSMHA Health Rep. 86:273-82.

Abraham, S., C. L. Johnson, and M. F. Najjar. 1979. Weight by height and age for adults 18-76 years, United States, 1971-1974. Vital and Health Statistics Series II DHEW Publication No. (PHS) 79-1656. National Center for Health Statistics, Hyattsville, Md.

Adams, C. F. and M. Richardson. 1981. Nutritive Value of Foods. Washington, D.C.: G.P.O.

Ala'ilima, F. C. 1982. The Samoans of Hawaii. Social Process in Hawaii 29:105-12.

Ala'ilima, F. C. and V. Ala'ilima. 1968. Fa'a Samoa: Traditional Samoan perceptions.

Al-Issa, I. 1982. Gender and adult psychopathology. In Gender and Psychopathology, ed. I. AlIssa. New York: Academic Press.

Allen, T. H., M. T. Peng, K. P. Chen, T. F. Huang, C. Chang, and H. S. Fang. 1956. Prediction of total adiposity from skinfolds and the curvilinear relationship between internal and external adiposity. Metabolism 5:346-52.

American Council on Science and Health. 1980. Diet Modification: Can It Reduce the Risk of Heart Disease? New York.

American Samoa Dept. of Education. 1976. Think Children. Pago Pago, American Samoa.

American Samoa Dept. of Education. 1977. Think Children. Pago Pago, American Samoa.

American Samoa Dept. of Education. 1980. Think Children. Pago Pago, American Samoa.

American Samoa Development Planning Office. 1976. Report on the 1974 Census of American Samoa. Part 1. Basic Information. Prepared by M. J. Levin and P. N. D. Pirie with the assistance of E. S. Marcus and E. Gebauer. Honolulu: East-West Population Institute.

American Samoa Development Planning Office. 1980. Statistical Bulletin: Annual Report on Economic Indicators. Pago Pago: Development Planning Office.

American Samoa Govt. 1952. Annual Report of the Governor of American Samoa. Pago Pago: Govt of American Samoa Printing Office.

American Samoa Govt. 1956a. Census of American Samoa: September 25, 1956. Pago Pago: Govt of American Samoa Printing Office.

American Samoa Govt. 1956b. Report of the Governor of American Samoa. Pago Pago: Govt of American Samoa Printing Office.

American Samoa Govt. 1957. Report of the Governor of American Samoa. Pago Pago: Govt. of American Samoa Printing Office.

American Samoa Govt. 1959. Report of the Governor of American Samoa. Pago Pago: Govt. of American Samoa Printing Office.

American Samoa Govt. 1960. Report of the Governor of American Samoa. Pago Pago: Govt. of American Samoa Printing Office.

American Samoa Govt. 1962. Report of the Governor of American Samoa. Pago Pago: Govt. of American Samoa Printing Office.

American Samoa Govt. 1968. American Samoa 1968 Annual Report. Pago Pago: Govt. of American Samoa Printing Office.

American Samoa Govt. 1974. American Samoa 1974 Annual Report. Pago Pago: Govt. of American Samoa Printing Office.

American Samoa Govt. 1976a. "American Samoa 1976 Annual Report. Pago Pago: Govt. of American Samoa Printing Office.

American Samoa Govt. 1976b. State Plan for American Samoa: For Comprehensive Health Planning 1976. Pago Pago: Govt. of American Samoa Printing Office.

American Samoa Govt. 1978. American Samoa 1978 Annual Report. Pago Pago: Govt. of American Samoa Printing Office.

American Samoa Govt. 1979. American Samoa 1979 Annual Report. Pago Pago: Govt. of American Samoa Printing Office.

American Samoa Govt. 1980. American Samoa 1980 Annual Report. Pago Pago: Govt. of American Samoa Printing Office.

Amerson, A. B., W. A. Whistler, and T. D. Schwaner. 1982. Wildlife and Wildlife Habitat of American Samoa. 1. Environment and Ecology. U.S. Dept. of the Interior, Fish and Wildlife Service, Washington, D.C.

Andres, R. 1980. Effect of obesity on total mortality. Intl. J. Obesity 4:381-86.

Andrews, G., C. Tennant, D. M. Hewson, and G. E. Vaillant. 1978. Life event stress, social support, coping style, and risk of psychological impairment. J. Nerv. Ment. Dis. 166:307-16.

Arky, R. A. 1982. The role of diet and exercise in the care of patients with diabetes mellitus. In Genetic Environmental Interactions in Diabetes Mellitus, ed. J. Melish, J. Hanna, and S. Baba. Amsterdam: Excerpta Medica.

Astrand, I. 1960. Aerobic work capacity in men and women with special reference to age. Acta. Physiol. Scand. 49(suppl. 169):1-92.

Astrand, I. 1967. Degree of strain during building work as related to individual work capacity. Ergonomics. 10:293-303.

Astrand, P.-O. and K. Rodahl. 1977. Textbook of Work Physiology. 2d. ed. New York: McGraw-Hill.

Axelrod, J. and T. D. Reisine. 1984. Stress hormones: Their interaction and regulation. Science 224:452-59.

Baker, P. T. 1965. Multidisciplinary studies of human adaptability; Theoretical justification and method. Yrbk. of Phys. Anthrop. 13:2-12.

Baker, P. T. 1977a. Problems and strategies. In Human Population Problems in the Biosphere: Some Research Strategies and Designs, ed. P. T. Baker. MAB Technical Notes 3. Paris: UNESCO.

Baker, P. T. 1977b. Environment and migration on the small islands of the South Pacific. In Human Population Problems in the Biosphere: Some Research Strategies and Designs, ed. P. T. Baker. MAB Technical Series Notes 3. Paris: UNESCO.

Baker, P. T. 1981. Migration and human adaptation. In Migration, Adaptation and Health in the Pacific, ed. C. Fleming and I. Prior. Epidemiology Unit Wellington Hospital, Wellington, New Zealand.

Baker, P. T. 1984a. Migration, genetics and the degenerative diseases of South Pacific Islanders. In Migration and Mobility, ed. A. Boyce. London: Taylor & Francis.

Baker, P. T. 1984b. The adaptive limits of human populations. Man, n.s. 19:1-14.

Baker, P. T. In press. Modernization, migration and health: A methodological puzzle with examples from the Samoans. Indian Journal of Anthropology.

Baker, P. T. and J. S. Dutt. 1972. Demographic variables as measures of biological adaptation: A case study of high altitude human populations. In The Structure of Human Populations, ed. G. A. Harrison and A. J. Boyce. Oxford: Clarendon Press.

Baker, P. T. and J. M. Hanna. 1981. Modernization and the biological fitness of Samoans: A progress report on a research program. In Migration, Adaptation, and Health in the Pacific, ed. C. Fleming and I. Prior.. Epidemiology Unit Wellington Hospital, Wellington, New Zealand.

Baker, T. S. 1976. Child care, child training and environment. *In* Man in the Andes, ed. P. T. Baker and M. A. Little. Stroudsburg, Pennsylvania: Dowden, Hutchison & Ross.

Baldauf, R. B., Jr. 1981. Educational television, enculturation and acculturation: A study of change in American Samoa. Intl. Rev. Educ. 27:227-45.

Bang, H. O., J. Dyerberg, and A. B. Nielson. 1971. Plasma lipid and lipoprotein patterns in Greenlandic west coast Eskimos. Lancet 1:1143-85.

Bannister, J., S. Stanley, and M. Levin. 1978. Application of new and revised demographic techniques to Western Samoan data. Honolulu: East-West Population Institute.

Barclay, G. 1958. Techniques of Population Analysis. New York: Wiley.

Barrau, J. 1965. L'humide et le sec: An essay on ethnobiological adaptation to contrastive environments in the Indo-Pacific area. J. Polyn. Soc. 74:329.

Beaglehole, E. 1947. Trusteeship and New Zealand's Pacific Dependencies. J. Polyn. Soc. 56:128-57.

Beaglehole, J. D., ed. 1961. The Journals of Captain James Cook on His Voyages of Discovery. Cambridge Univ. Press for Hakluyt Society.

Beaglehole, R., E. F. Eyles, C. Salmond, and I. A. M. Prior. 1978. Blood Pressure in Tokelauan children in two contrasting environments. Amer. J. Epid. 108:283-88.

Beaglehole, R., I. A. M. Prior, E. Eyles, and V. Sampson. 1979. High density lipoprotein cholesterol and other serum lipids in New Zealand Maoris. New Zeal. Med. J. 90:139.

Beaglehole, R., C. Salmond, and E. Eyles. 1977a. A longitudinal study of blood pressure in Polynesian children. Amer. J. Epid. 105:87-89.

Beaglehole, R., C. Salmond, A. Hooper, J. Huntsman, J. M. Stanhope, J. C. Cassel, and I. A. M. Prior. 1977b. Blood pressure and social interaction in Tokelaun migrants in New Zealand. J. Chron. Dis. 30:803-12.

Beaglehole, R., C. Salmond, and I. A. M. Prior. 1975. Blood pressure studies in Polynesion children. *In* Epidemiology and Control of Hypertension, ed. O. Paul. Miami: Symposia Specialists.

Beaton, G. H. 1985. Uses and limits of the use of RDA for evaluating dietary intake data. Amer. J. Clin. Nutr. 41:155-64.

Beckerman, S. 1977. Protein and population in tropical Polynesia. J. Polyn. Studies. 86:73-79.

Bellwood, P. 1978. The Polynesians: The Prehistory of an Island People. London: Thames & Hudson, Ltd.

Bellwood, P. 1980. The peopling of the Pacific. Scientific American 243:138-48.

Bennett, P. H. and W. C. Knowler. 1979. Increasing prevalence of diabetes in Pima (American) Indians over a ten year period. *In* Diabetes 1979: Proceedings of the 10th Congress of the International Diabetes Federation, ed. W. K. Waldhausal. Oxford: Excerpta Medica.

Bennett, P. H., N. B. Rushforth, M. Miller, and P. M. Lecompte. 1976. Epidemiological studies of diabetes in the Pima Indians. Recent Progress in Hormonal Research, 32:333-76.

Berchtold, P., M. Berger, E. Greiser, M. Dohse, K. Irmscher, F. Gries, and H.

Zimmerman. 1977. Cardiovascular risk factors in gross obesity. Intl. J. Obesity 1:219.

Berkman, L., and S. Syme. 1979. Social networks, host resistance, and mortality: A nine-year follow-up study of Alameda County residents. Amer. J. Epid. 109:186-204.

Bindon, J. R. 1981. Genetic and Environmental Influences on the Morphology of Samoan Adults. Ph.D. diss., The Pennsylvania State Univ., University Park.

Bindon, J. R. 1982. Breadfruit, banana, beef, and beer: Modernization of the Samoan diet. Ecol. Food Nutr. 12:49-60.

Bindon, J. R. 1984a. The body build and composition of Samoan children: Relationships to infant feeding patterns and infant weight-for-length status. Amer. J. Phys. Anthrop. 63:379-388.

Bindon, J. R. 1984b. An evaluation of the diet of three groups of Samoan adults: Modernization and dietary adequacy. Ecol. Food Nutr. 14:105-115.

Bindon, J. R. In press. The influence of infant feeding patterns on growth in American Samoa. Med. Anthrop. 1986

Bindon, J. R., and P. T. Baker. 1985. Modernization, migration, and obesity among Samoan adults. Annals of Hum. Biol. 12:67-76.

Bindon, J. R. and D. L. Pelletier 1986. Patterns of growth in weight among infants in a rural western Samoan village: A semi-longitudinal study. Ecol. Food and Nutr. 18:135-43.

Bink, B. 1962. The physical work capacity in relation to working time and age. Ergonomics 5:25-28.

Bishop, J. 1977. American Samoa: Which road ahead? Pacific Studies 1:47-53.

Biss, K., K.-J. Ho, B. Mikkelson, L. Lewis, and C. B. Taylor. 1971. Some unique biological characteristics of the Masai of East Africa. New Engl. J. Med. 284:694-99.

Blackburn, H. and R. W. Parlin. 1966. Antecedents of disease: Insurance mortality experience. Annals of the New York Academy of Sciences 134:965-1017.

Blair, D., J.-P. Habicht, E. Sims, E. A. H. Sims, D. Sylvester, and S. Abraham. 1984. Evidence for an increased risk for hypertension with centrally located body fat and the effect of race and sex on this risk. Amer. J. Epid. 119:526-40.

Block, W. D. and K. J. Jarett. 1969. An automated technique for the quantitative determination of serum total triglycerides. Amer. J. Med. Techn. 35:93-102.

Bonjer, F. H. 1962. Actual energy expenditure in relation to the physical work capacity. Ergonomics 5:29-31.

Bonjer, F. H. 1968. Relationship between working time, physical working capability and allowable caloric expenditure. In Muskelarbeit and Muskeltraining, ed. W. Rohmert. Stuttgart: Verlag.

Bonjer, F. H. 1971. Temporal factors and physiological load. In Measurement of Man at Work, ed. W. T. Singleton, J. G. Fox, and D. Whitfield. London: Taylor & Francis, Ltd.

Born, T. J. 1970. American and Western Samoa: A Comparative Study of Population Growth and Migration.

Bouchard, C., M.-C. Thibault, and J. Jobin. 1981. Advances in selected areas of human work physiology. Yrbk. of Phys. Anthrop. 24:1-36.

Bowles, J. R. 1985. Suicide and attempted suicide in contemporary Western Samoa. *In* Culture, Youth and Suicide in the Pacific: Papers from an East-West Center Conference, ed. F. X. Hezel, D. H. Rubenstein, and G. M. White. Honolulu: East-West Center.

Boyden, S. V. 1970. Cultural adaptation to biological maladjustment. *In* The Impact of Civilization on the Biology of Man, ed. S. V. Boyden. Toronto: Univ. of Toronto Press.

Bray, G. 1979. Obesity. Disease-A-Month 26:1-85.

Brady. I., ed. 1983. Speaking in the name of the real: Freeman and Mead on Samoa. Amer. Anthrop. 85:908-47.

Brand, R. J., R. H. Rosenman, R. I. Sholtz, and M. Friedman. 1976. Multivariate prediction of coronary heart disease in the Western Collaborative Group Study compared to findings of the Framingham Study. Circulation 53:348-55.

Brennan, E. R. 1982. Secular change in age-specific cause of death in Sanday, Orkney Islands. Soc. Sci. Med. 16:155-64.

Broadhead, W. E., B. H. Kaplan, S. James, E. H. Wagner, V. J. Schoenbach, R. Grimson, S. Heyden, G. Tibblin, and S. H. Gehlbach. 1981. The epidemiological evidence for a relationship between social support and health. Amer. J. Epid. 117:521-537.

Brodman, K., A. J. Erdman, I. Lorge, and H. G. Wolff. 1949. The Cornell Medical Index: An adjunct to medical interview. JAMA 140:530-34.

Brodman, K., A. Erdman, I. Lorge, and H. G. Wolff. 1951. The CMI questionnaire 2. As a diagnostic interview. JAMA 145:152-57.

Brown, D. E. 1981. General stress in anthropological fieldwork. Amer. Anthrop. 83:74-92.

Brown, G. W. and T. O. Harris. 1978. Social Origins of Depression: A Study of Psychiatric Disorder in Women. New York: Free Press.

Brown, V. J., J. Hanna, and G. Severson. 1984. A quantitative dietary study of native and migrant Samoans. Amer. J. Phys. Anthrop. 63:142. (Abstr.).

Buskirk, E. R. and H. L. Taylor. 1957. Maximum oxygen intake and its relation to body composition, with special reference to chronic physical activity and obesity. J. Appl. Physiol. 11:72-78.

Cameron, S. S. 1962. Vegetation and forest resources. *In* Western Samoa: Land, Life and Agriculture in Tropical Polynesia, ed. J. W. Fox and K. B. Cumberland. Christchurch, New Zealand: Whitcombe & Tombs, Ltd.

Campbell, A., P. E. Converse, and W. L. Rogers. 1976. The Quality of American Life. New York: Russell Sage Foundation.

Carnoy, M. 1974. Education as Cultural Imperialism. New York: David McKay Col.

Carroll, M. D., S. Abraham, and C. M. Dresser. 1983. Dietary intake source data. Vital and Health Statistics Series 11, No. 231. Public Health Service, Washington, D.C.

Cassel, J. C. 1974. Hypertension and cardiovascular disease in migrants: A potential source for clues. Intl. J. Epid. 3:204-06.

Cassel, J. C. 1976. The contribution of the social environment to host resistance. Amer. J. Epid. 104:107-23.

Castelli, W. P., J. T. Doyle, T. Gordon, C. G. Hames, M. C. Hjortland, S. B. Hulley, A. Kagan, and W. J. Zukel. 1977. HDL cholesterol and other lipids in CHD - Cooperative Lipoprotein Phenotyping Study. Circulation 55:767-72.

Cavalli-Sforza, L. L. and W. F. Bodmer. 1971. The Genetics of Human Populations. San Francisco: W.H. Freeman.

Chakraborty, R., W. Schull, E. Harburg, M. A. Schork, and P. Roeper. 1977. Heredity, stress and blood pressure. A Family set method — 5. Heretability estimates. J. Chron. Dis. 30:683-99.

Charney, E., H. C. Goodman, M. McBride, B. Lyon, and R. Pratt. 1976. Childhood antecedents to adult obesity. New Engl. J. Med. 295:6-9.

Chavez, A., C. Martinez, R. Galvan, M. Coronado, H. Bourges, D. Diaz, S. Basta, M. Nieves, and O. Garciduenas. 1973. Nutrition and development of infants from poor rural areas. 4. Differences attributable to sex in the utilization of mother's milk. Nutr. Rep. Intl. 7:603-9.

Chen, P. 1973. Samoans in California. Social Work 18:41-48.

Chiang, B. W., L. V. Perlman, and F. H. Epstein. 1969. Overweight and hypertension: A review. Circulation 39:403-21.

Clement, D. 1982. Samoan folk knowledge of mental disorders. In Cultural Conceptions of Mental Health and Therapy, ed. A. Marsella and G. White. Boston: Reidel.

Clements, F. W. 1970. Some effects of different diets. In The Impact of Civilization on the Biology of Man, ed. S. V. Boyden. Toronto: Univ. of Toronto Press.

Clogg, C. C. 1978. Adjustment of rates using multiplicative models. Demography 15:523-39.

Cohen, F. and R. S. Lazarus. 1979. Coping with the stresses of illness. In Health Psychology, ed. G. C. Sone, F. Cohen, and N. E. Adler. San Francisco: Jossey-Bass.

Cohen, S. and G. McKay. 1984. Social support, stress and the buffering hypothesis: A theoretical analysis. In Handbook of Psychology and Health, ed. A. Baum, J. E. Singer, and S. E. Taylor. Hillsdale, N.J.: Erlbaum.

Colson, E. 1971. The Social Consequences of Resettlement: The Impact of the Kariba Resettlement Upon the Gwembe Tonga. Manchester: Manchester Univ. Press.

Connelly, D. M. and J. M. Hanna. 1978. Social and health associations of obesity among a Samoan migrant population: A preliminary report. Paper presented at 77th Annual Meeting of the American Anthropological Association, Los Angeles.

Connor, W. E., M. T. Cerqueira, M. S. Rodney, R. B. Wallace, M. R. Malinaw, and H. R. Casdorph. 1978. The plasma lipids and lipoproteins and diet of the Tarahumara Indians of Mexico. Amer. J. Clin. Nutr. 31:1131-42.

Cook, J. M. 1983. Samoan patterns in seeking health services -- Hawaii, 1979-81. Hawaii Med. J. 42:138-42.

Cook, S. F. and W. Borah. 1971. Essays in Population History: Mexico and the Caribbean. Berkeley:Univ. of California Press.

Cornoni-Huntley, J., W. Harlan, and P. Leaverton. 1979. Blood pressure in adolescence. The U.S. Health Examination Survey. Hypertension 1:566-71.

Coulter, J. W. 1941. Land Utilization in American Samoa. Bernice Bishop Museum Bulletin 170. Honolulu: Bernice Bishop Museum.

Cowgill,D. O. and L. D. Holmes, eds. 1972. Aging and Modernization. New York: Appleton-Century-Croft.

Coyne, T., (and J. Badcock and R. Taylor, eds.) 1984. The Effect of Urbanization and Western Diet on the Health of Pacific Island Populations. Technical Paper No. 186. Noumea, New Caledonia: South Pacific Commission.

Crawley, M. M. B. 1984. Vital Statistics Sample Survey Report. Western Samoa Dept. of Statistics. Apia: Western Samoa.

Crews, D. E. 1982. Modernization and changing patterns of infant mortality in American Samoa. Amer. J. Phys. Anthrop. 57:178. (Abstr.).

Crews, D. E. 1985. Mortality, survivorship and longevity in American Samoa 1950 to 1981. Ph.D. diss. The Pennsylvania State Univ., University Park.

Crews, D. E. and T. S. Baker. 1982. Aging and cultural change in Samoa: A comparison of traditional and modern life-styles. The Gerontologist 22:260. (Abstr.).

Crews, D. E. and P. C. MacKeen. 1982. Mortality related to cardiovascular disease and diabetes mellitus in a modernizing population. Soc. Sci. Med. 16:175-81.

Cruz-Coke, A., R. Etcheverry, and R. Nagel. 1964. Influence of migration on blood pressure of Easter Islanders. Lancet 1:697-99.

Currey, L. 1962. Weather and climate. In Western Samoa: Land, Life and Agriculture in Tropical Polynesia, ed. J. W. Fox and K. B. Cumberland. Christchurch, New Zealand: Whitcombe & Tombs, Ltd.

Dahl, L. K. 1960. Possible role of salt intake in the development of essential hypertension. In Essential Hypertension, ed. K. D. Bock and P. T. Cottie. Berlin: Springer-Verlag.

Daly, M. B. and H. A. Tyroler. 1972. Cornell Medical Index response as a predictor of mortality. Brit. J. Prev. Soc. Med. 26:159-64.

Damon, A. 1968. Secular trend in height and weight within old American families at Harvard 1870-1965. 1. Within twelve four-generation families. Amer. J. Phys. Anthrop. 29:45-50.

Darlu, P., M. F. Couilliot, and F. Drupt. 1984. Ecological and cultural differences in the relationships between diet, obesity and serum lipid concentrations in a Polynesian population. Ecol. Food Nutr. 14:169-83.

Davidson, J. W. 1967. Samoa Mo Samoa: The Emergence of the Independent State of Western Samoa. Melbourne: Oxford Univ. Press.

Davidson, J. W. 1976. Lauaki Namulau'ulu Mamoe: A traditionalist in Samoa politics. In Pacific Islands Portraits, ed. J. W. Davidson and D. Scarr. Canberra: Australian National Univ. Press.

Dawber, T. R., W. B. Kannel, A. Kagen, R. K. Donabedian, P. M. McNamara, and

G. Pearson. 1967. Environmental factors in hypertension. *In* The Epidemiology of Hypertension, ed. J. Stamler, R. Stamler, and T. N. Pullman. New York: Grune & Stratton.

Day, J., M. Carruthers, A. Bailey, and D. Robinson. 1976. Anthropometric, physiological and biochemical differences between urban and rural Masai. Atherosclerosis 23:357-61.

Dembroski, T. M., S. M. Weiss, J. L. Shields, S. G. Haynes, and M. Feinleib, eds. 1978. Coronary-Prone Behavior. New York: Springer-Verlag.

Dimsdale, J. E. and J. Moss. 1980. Plasma catecholemines in stress and exercise. JAMA 243:340-42.

Dobbert, M. L. 1975. Another route to a general theory of cultural transmission: A systems model. Anthrop. and Educ. Quarterly 6:22-26.

Dohrenwend, B. S. and B. P. Dohrenwend. 1981. Life stress and illness: Formulation of the issues. *In* Stressful Life Events and Their Contexts, ed. B. S. Dohrenwend and B. P. Dohrenwend. New York: Prodist.

Donnison, C. P. 1929. Blood pressure in the African native. Lancet 1:6-7.

Douthit, D. and J. Lung. 1974. Samoan Demographic Project Final Report. Honolulu: Univ. Of Hawaii.

Dufaux, B., G. Assman, and W. Hollman. 1982. Plasma lipoprotins and physical activity: A review. Intl. J. Sports Med. 3:123-36.

Durnin, J. V. G. A. 1976. Nutrition. *In* Man in Urban Environments, ed. G. A. Harrison and J. B. Gibson. Oxford: Oxford Univ Press.

Durnin, J. V. G. A. and R. Passmore. 1967. Energy, Work and Leisure. London: Heineman Educational Books.

Dutt, J. S. and P. T. Baker. 1978. Environment, migration and health in southern Peru. Soc. Sci. Med. 12:29-38.

Dutt, J. S. and P. T. Baker. 1981. An analysis of adult mortality causes among migrants from altitude and sedentes in coastal Peru. *In* Health in the Andes, ed. J. Bastien and J. Donahue. American Anthropological Association Special Publication No. 12. Washington, D.C.: American Anthropological Association.

Easterlin, R. A. 1975. Does economic growth improve the human lot? Some empirical evidence. *In* Nations and Households in Economic Growth: Essays in Honor of Moses Abramovitz, ed. P. A. David and M. W. Reder. Palo Alto: Stanford Univ. Press.

Eaton, S. and M. Konner. 1985. Paleolithic nutrition: A consideration of its name and current implications. New Engl. J. Med. 312:283-89.

Edholm, O. G. 1967. The Biology of Work. New York:McGraw-Hill.

Eid, E. E. 1970. Follow-up study of physical growth of children who had excessive weight gain in the first six months of life. Brit. Med. J. 2:74-76.

Ekblom, B. and E. Gjessing. 1968. Maximal oxygen uptake of the Easter Island population. J. Appl. Physiol. 25:124-29.

Ember, M. 1964. Commercialization and political change in American Samoa. *In* Explorations in Cultural Anthropology: Essays in Honor of George Peter Murdock, ed. W. H. Goodenough. New York: McGraw-Hill.

Ember, M. 1971. Political authority in the structure of kinship in aboriginal

Samoa. *In* Polynesia: Readings in a Culture Area, ed. A. Howard. Scranton: Chandler Publishing Co.

Embree, E. R. 1934. New school for chief's sons in Samoa. Political Science Quarterly 5:228-31.

Embree, J. F. 1941. Acculturation among the Japanese of Kono, Hawaii. Mem. Amer. Anth. Asso. No. 59.

Epstein, F. and R. Eckhoff. 1967. The epidemiology of high blood pressure — Geographic distributions and etiological factors. *In* Epidemiology of Hypertension, ed. J. Stamler, R. Stamler, and T. Pullman. New York: Grune & Stratton.

Erhardt, C. L. and J. E. Berlin, eds. 1974. Mortality and Morbidity in the United States. Cambridge, Mass.: Harvard Univ. Press.

Ernst, N. and R. I. Levy. 1980. Diet, hyperlipidemia and atherosclerosis. *In* Modern Nutrition in Health and Disease, ed. R. S. Goodhart and M. E. Shils. Philadelphia: Lea & Febiger.

Escudero, J. 1980. Hypertension, ischemic heart disease and mortality in Mexico: An epidemiological review. *In* Advances in the Management of Cardiovascular Disease: Vol. 1, ed. W. T. Foley. Chicago: Year Book Medical Publishers, Inc.

Evans, J. G. and I. Prior. 1969. Indices of obesity derived from height and weight in two Polynesian populations. Brit. J. Prev. Soc. Med. 32:56-59.

Eveleth, P. B. and J. M. Tanner. 1976. Worldwide Variation in Human Growth. New York: Cambridge Univ. Press.

Fairbairn, I. J. 1973. The National Income of Western Samoa. Melbourne: Oxford Univ. Press.

Farrell, B. H. and R. G. Ward. 1962. The village and its agriculture. *In* Western Samoa: Land , Life and Agriculture in Tropical Polynesia, ed. J. W. Fox and K. B. Cumberland. Christchurch, New Zealand: Whitcombe & Tombs, Ltd.

Ferguson, R. A. 1934. A dental survey of school children in American Samoa. J. Amer. Dent. Assoc. 21:534-39.

Filer, L. 1978. Early nutrition: Its long term role. Hospital Practitioner 13:89-95.

Fisk, E. K. 1962. Planning in a primitive economy: Special problems of Papua and New Guinea. Economic Record 38:462-78.

Fixler, D. E., J. Kautz and K. Dana. 1980. Systolic blood pressure differences among pediatric epidemiological studies. Hypertension 2(suppl. 1): 3-7.

Fleming, R., A. Baum, M. M. Gisriel, and R. J. Gatchel. 1982. Mediating influences of social support on stress at Three Mile Island. J. Hum. Stres 8:14-21.

Fleming, C. and I. A. M. Prior, eds. 1981. Migration, Adaptation, and Health in the Pacific. Epidemiology Unit Wellington Hospital, Wellington, New Zealand.

Florey, C. du V. and R. Cuadrado. 1968. Blood pressure in native Cape Verdeans and in immigrants and their descendants living in New England. Hum. Biol. 40:189-211.

Fomon, S. J. 1974. Infant Nutrition. Philadelphia: W.B. Saunders.

Fomon, S. J. 1980. Factors influencing food consumption in the human infant. Intl. J. Obesity 4:348-50.

Foresman, L. 1982. Consistency in catecholemine excretion in laboratory and natural settings: Correlational and variance components analysis. Scand. J. Pychol. 23:99-106.

Fortes, M. 1970. Social and psychological aspects of education in Taleland. *In* From Child to Adult, ed. J. Middleton. Garden City, N.Y.: The Natural History Press.

Fox, J. W. and K. B. Cumberland. 1962. Western Samoa: Land, Life, and Agriculture in Tropical Polynesia. Christchurch, New Zealand: Whitcombe & Tombs, Ltd.

Franco, R. W. 1984. A Demographic Assessment of the Samoan Employment Situation in Hawaii. Honolulu: East-West Population Institute.

Franco, R. 1985a. Samoan Perception of Work: Moving Up and Moving Around. Ph.D. diss., Univ. of Hawaii, Honolulu.

Franco, R. 1985b. Socialization and Samoan International Movement. Univ. of Hawaii, Dept. of Anthropology, Honolulu.

Frankenhauser, M. 1975. Sympathetic-adrenomedullary activity, behavior and the psychosocial environment. *In* Research in Psychophysiology, ed. P. H. Venables and J. Christie. New York: Wiley.

Frankenhauser, M. 1983. The sympathetic-adrenal and pituitary-adrenal response to challenge: comparison between sexes. *In* Biobehavioral Bases of Coronary Heart Disease, ed. T. M. Dembrowski, T. H. Schnidt, and G. Blumchen. Basel, Switzerland: Karger.

Freeman, D. 1983. Margaret Mead and Samoa: The Making and Unmaking of an Anthropological Myth. Cambridge, Mass.: Harvard Univ. Press.

Freis, E. D. 1976. Salt, volume, and the prevention of hypertension. Circulation 53:589-95.

Friedlaender, J. S. 1975. Patterns of Human Variation: The Demography, Genetics, and Phenetics of Bougainville Islanders. Cambridge: Harvard Univ. Press.

Friedlaender, J. S., W. W. Howells, and J. G. Rhoads. In press. The Solomon Islands Project: A Long-Term Study of Health, Human Biology, and Cultural Change. Oxford: Oxford Univ. Press.

Frisancho, A. R. 1975. Functional adaptation to high altitude hypoxia. Science 187:313-19.

Frisancho, A. R. and P. Flegel. 1982. Advanced maturation associated with centripetal fat patterning. Hum. Biol. 54:717-27.

Fulwood, R., S. Abraham, and C. Joahanson. 1981. Height and weight of adults, ages 18-74. Vital and Health Statistics Series 11, Washington, D.C.: Public Health Service.

Gad, M. T. and J. H. Johnson. 1980. Correlates of adolescent life stress as related to race, SES and levels of perceived social support. J. Clin. Child Psychol. 9:13.

Gage, T. B. 1982. Ecological theories of diet and food production: A case study of Samoan subsistence agriculture. Ph.D. diss., The Pennsylvania State Univ., University Park.

Gage, T. B. 1984. Subsistence agriculture in Samoa: Seasonality of productio and work. Paper prepared for the Samoan Studies Conference, April 198 The Pennsylvania State Univ., University Park.

Garcia-Palmieri, M. R., R. Costas, J. Schiffman, A. A. Colon, R. Torres, and I Nazario. 1972. Interrelationship of serum lipids with relative weigh blood glucose and physical activity. Circulation 45:829-36.

Garn, S. 1983. Fatness and mortality in the west of Scotland. Amer. J. Clir Nutr. 38:313-19.

Gearing, F. 1975. Overview: A cultural theory of education. Anthrop. and Edu Quarterly 6:1-9.

Gearing, F., A. Tindall, A. Smith, and T. Carroll. 1975. Structures c censorship, usually inadvertant: Studies in a cultural theory of educatio Anthrop. and Educ. Quarterly 6:1-22.

Gentry, W. D., A. Chesney, H. Gary, R. Hall, and E. Harburg. 1982. Habitua anger-coping styles. 1. Effect on mean blood pressure and risk for essentia hypertension. Psychosom. Med. 44:195-202.

Gerber, E. R. 1975. The cultural patterning of emotions in Samoa. Ph.D. diss Univ. of California, San Diego.

Gerber, E. R. 1985. Rage and obligation: Samoan emotion in conflict. In Person Self and Experience: Exploring Pacific Ethnopsychologies, ed. G. Whit and J. Kirkpatrick. Berkeley: Univ. of California Press.

Gillum, R. F. 1979. Pathophysiology of hypertension in blacks and whites. / review of the basis of racial blood pressure differences. Hypertensio 1:468.

Gilson, R. P. 1963. Samoan descent groups: A structural outline. J. Polyn. Soc 72:372-77.

Gilson, R. P. 1970. Samoa 1830 to 1900; The Politics of a Multicultura Community. New York: Oxford Univ Press.

Glass, D. C. 1977. Behavior Patterns, Stress and Coronary Disease. New York Wiley.

Glueck, C. J., J. L. Taylor, D. Jacobs, H. A. Morrison, R. Beaglehole, and O. D Williams. 1980. Plasma high density lipoprotein cholesterol: Association with measurements of body mass. Circulation 62 (suppl. 4): 62-69.

Goldfarb, A. I. 1971. Predictors of mortality in the institutionalized aged. Ir Prediction of Life Span: Recent Findings, ed. E. Palmore and F. C. Jeffers 79-94. Lexington, Massachusetts: Heath Lexington Books.

Gordon, T., W. P. Castelli, M. C. Hjortland, W. B. Kannel, and T. R. Dawber 1977. Predicting coronary heart disease in middle-aged and older persons The Framingham Study. JAMA 238:497-99.

Gordon, T., M. Fisher, N. Ernst, and B. M. Rifkind. 1982. Relation of diet t LDL-cholesterol, VLDL-cholesterol and plasma total cholesterol an triglycerides in white adults. Arteriosclerosis 2:502-12.

Graham, M. A. 1983. Acculturative stress among Polynesian, Asian, an American students on the Brigham Young Univ. — Hawaii campus. Intl J. of Intercultural Relations 7:79-103.

Grattan, F. J. H. 1948. An Introduction to Samoan Custom. Apia, Wester Samoa: Samoa Printing and Publishing Co.

Graves, N. B. 1978. Growing up Polynesian: Implications for Western Education. *In* New Neighbors: Islanders in Adaptation, ed. C. Macpherson, B. Shore, and R. Franco. Santa Cruz: Center for South Pacific Studies, Univ. of California.

Graves, N. B. and T. D. Graves. 1974. Adaptive strategies in urban migration. Ann. Rev. Anthrop. 3:117-51.

Graves, N. B. and T. D. Graves. 1977. Preferred adaptive strategies: An approach to understanding New Zealand's multicultural workforce. New Zealand J. of Indust. Relations. 2:81-90.

Graves, N. B. and T. D. Graves. 1978a. The impact of modernization on the personality of a Polynesian people. Human Organization 37:115-35.

Graves, T. D. and N. B. Graves. 1978b. Evolving strategies in the study of cultural change. *In* The Making of Psychological Anthropology, ed. G. Spindler and L. Spindler. Berkeley: Univ. of California Press.

Graves, T. D. and N. B. Graves. 1980. Kinship ties and the preferred adaptive strategies of urban migrants. *In* The Versatility of Kinship, ed. L. Cordell and S. Beckerman. New York: Academic Press.

Graves, T. D. and N. B. Graves. 1985. Stress and health among Polynesian migrants to New Zealand. J. Behav. Med.

Graves, T. D., N. B. Graves, V. N. Semu, and I. Ah Sam. 1981. The social context of drinking and violence in New Zealand's multi-ethnic pub settings. *In* Social Drinking Contexts, ed. T. C. Harford and L. S. Gaines. NIAAA Research Monograph 7, Rockville, Md.

Graves, T. D., N. B. Graves, V. N. Semu, and I. Ah Sam. 1983. The priceof ethnic identity: Maintaining kin ties among Pacific immigrants to New Zealand. Paper presented at the 15th Pacific Science Congress, February, Dunedin, New Zealand.

Greksa, L. P. 1980. Work requirements and work capabilities in a modernizing Samoan population. Ph.D. diss., The Pennsylvania State Univ., University Park.

Greksa, L. P. In press. Evaluation of work capability: An example from American Samoa. J. of Human Ergology.

Greksa, L. P. and P. T. Baker. 1982. Aerobic capacity of modernizing Samoan men. Hum. Biol. 54:777-88.

Greksa, L. P. and P. T. Baker. N.d. Age changes in the patterning of subcutaneous fat among adult Samoans. Dept. of Anthropology, The Pennsylvania State Univ., University Park.

Greville, T. N. E. 1951. Short methods of constructing abridged life tables. *In* Handbook of Statistical Methods for Demographers: Selected Problems in the Analysis of Census Data, ed. A. J. Jaffe. Washington, D.C.: U.S. Bureau of the Census, U.S. Govt. Printing Office.

Guthrie, H. A. 1979. Introductory Nutrition. 2d ed. Saint Louis: Mosby.

Haan, N. 1982. The assessment of coping, defense and stress. *In* Handbook of Stress: Theoretical and Clinical Aspects, ed. L. Goldberger and S. Brenitz. New York: Free Press.

Hamill, P. V. V., T. A. Drizd, C. L. Johnson, R. B. Reed, and A. F. Roche. 1977. NCHS Growth Curves for Children Birth-18 Years. U.S. Vital and Health

Statistics Series 11, No. 165. Washington, D.C.: DHEW Publication No. 78-1650.

Hanna, J. M. and P. T. Baker. 1979. Biocultural correlates to blood pressure of Samoan migrants to Hawaii. Hum. Biol. 51:481-97.

Hanna, J. M. and S. T. McGarvey. In press. The biomedical characteristics of the Samoan community in Hawaii.

Hansen, J. E. 1979. Sociocultural Perspectives on Human Learning. Englewood Cliffs, N.J.: Prentice-Hall.

Harbison, S. F. 1979. Preliminary report on the fertility of American Samoans. Dept. of Anthropology, The Pennsylvania State Univ.

Harbison, S. F. 1981. Family structure and family strategy in migration decision making. In Migration Decision Making, ed. G. DeJong and R. Gardner. New York: Pergamon Press.

Harbison, S. F. and T. S. Baker. 1980 The comparison of educational systems in developing populations: Western Samoa and American Samoa. Dept. of Anthropology, The Pennsylvania State Univ., University Park.

Harbison, S. F., T. S. Baker, and M. Levin. 1981. Education and fertility: A Samoan example. Paper presented at the Annual Meeting of the Population Association of America, March 29, Washington, D.C.

Harbison, S. F. and M. Weishaar. 1981. Samoan migrant fertility: Adaptation and selection. Human Organization, Fall:268-73.

Harburg, E., E. Blackelock and P. Roeper. 1979. Resentful and reflective coping with arbitrary authority and blood pressure: Detroit. Psychosom. Med. 41:189.

Harburg, E., J. C. Erfurt, L. S. Hauenstein, et al. 1973. Socio-ecologic stress, suppressed hostility, skin colour, and black-white male blod pressure: Detroit. Psychosom. Med. 35:276.

Harlan, W., J. Cornoni-Huntley, and P. Leaverton. 1979. Blood pressure in childhood. The National Health Examination Survey. Hypertension 1:559-65.

Harrison, G. A. 1973. The effects of modern living. J. Biosocial Sciences 5:217-28.

Harrison, G. A., ed. 1977. Population Structure and Human Variation. New York: Cambridge Univ. Press.

Harrison, G. A. and A. J. Boyce., eds. 1972. The Structure of Human Populations. Oxford: Clarendon Press.

Hawthorne, H. B. 1944. The Maori: A study in acculturation. Amer. Anthrop. New Series 46:Part 2.

Hayes, G. T. and M. J. Levin. 1983. How Many Samoans? An Evaluation of the 1980 Census Count of Samoans. Honolulu: East-West Population Institute.

Heiss, G., N. J. Johnson, S. Reiland, C. E. Davis, and H. A. Tyroler. 1980. The epidemiology of plasma high density lipoprotein cholesterol levels: The Lipid Research Clinics Program Prevalence Study Summary. Circulation 62 (suppl. 4):116-36.

Henkin, J. H. and L. Dickinson. 1972. Urbanization, diet and potential health effects in Palau. Amer. J. Clin. Nutr. 25:348-53.

Henkin, J. H., D. Reed, D. LeBarthe, M. Nichaman, and R. Stallones. 1970.

Dietary and disease patterns among Micronesians. Amer. J. Clin. Nutr. 23:346-57.

Henry, J. P. and J. C. Cassell. 1969. Psychosocial factors in essential hypertension: Recent epidemiological and animal experimental evidence. Amer. J. Epid. 90: 171-200.

Herd, J. A. 1978. Psychological correlates of coronary-prone behavior. *In* Coronary-Prone Behavior, ed. T. M. Dembroski, S. Weiss, J. Shields, S. Haynes, and M. Feinleib. New York: Springer-Verlag.

Hill, H. B. 1977. The use of nearshore marine life as a food resource by American Samoans. Master's thesis, Univ. of Hawaii, Honolulu.

Himms-Hagen, J. 1984. Thermogenesis in brown adipose tissue as an energy buffer. New Eng. J. Med. 311:1549-58.

Hirsch, B. J. 1979. Social networks and the coping process. *In* Social Networks and Social Support, ed. B. Gottlieb. Beverly Hills: Sage.

Hirsch, B. J. 1980. Natural support systems and coping with major life change. Amer. J. Community Psychol. 8:159-72.

Hirsch, S. 1958. The social organization of an urban village in Samoa. J. Polyn. Soc. 67:266-301.

Hodgson, J. L. and E. R. Buskirk. 1977. Physical fitness and age, with emphasis on cardiovascular function in the elderly. J. Amer. Geriatr. Soc. 25:385-92.

Holmes, L. D. 1958. Ta'u: Stability and change in a Samoan village. Wellington, N.Z.: Polynesian Society Reprints No. 7.

Holmes, L. D. 1972. The role and status of the aged in a changing Samoa. *In* Aging and Modernization, ed. D. O. Cowgill and L. D. Holmes. New York: Appleton-Century-Crofts.

Holmes, L. D. 1974. Samoan Village. New York: Holt, Rinehart & Winston.

Holmes, L. D. 1978. Aging and modernization: The Samoan aged of San Francisco. *In* New Neighbors: Islanders in Adaptation, ed. C. Macpherson, B. Shore, and R. Franco. Santa Cruz: Center for South Pacific Studies, Univ. of California, Santa Cruz.

Holmes, L. D. 1980. Factors contributing to the cultural stability of Samoa. Anthropological Quarterly 53:188-97.

Holmes, L. D. 1984. Samoan Islands Bibliography. Witchita, Kans.: Poly Concepts Publishing Co.

Holmes, L. D. and E. Rhoads. 1983. Aging and change in Samoa. *In* Growing Old in Different Societies, ed. J. Sokolovsky. Belmont, Calif.: Wadsworth Press.

Holmes, L. D., G. Tallman, and V. Jantz. 1978. Samoan personality. J. Psychol. Anthrop. 1:453-72.

Holmes, S. 1954. A quantitative study of family meals in Western Samoa with special reference to child nutrition. Brit. J. Nutr. 8:223-39.

Holmes, T. and R. Rahe. 1967. The social readjustment rating scale. J. Psychosom. Med. 11:213-18.

Holtzman, E., U. Goldbourt, T. Rosenthal, S. Yarrai and H. N. Newfeld. 1983. Hypertension in middle-aged men: Associated factors and mortality experience. Israel J. of Med. Sci. 19:25-33.

Hornabrook, R. W. 1977. Human ecology and biomedical research: A critica review of the International Biological Programme in New Guinea. I* Subsistence and Survival: Rural Ecology in the Pacific, ed. T. P. Bayliss Smith and R. G. Feachem. London: Academic Press.

Hornabrook, R. W., G. G. Crane, and J. M. Stanhope. 1974. Karkar and Lufa An epidemiological and health background to the human adaptabilit studies of the International Biological Programme. Phil. Trans. Royal Soc London Series B 268:293-308.

Hornick, C. A. 1979. Heart disease in a migrating population. Ph.D. diss., Th* Univ. of Hawaii, Honolulu.

Hornick, C. A. and B. Fellmeth. 1981. High density lipoprotein, insulin an* obesity in Samoans. Atherosclerosis 39:321-28.

Hornick, C. A. and J. Hanna. 1982. Indicators of coronary risk in a migratin Samoan Population. Med. Anthrop. 6:71-79.

House, J. 1974. Occupational stress and coronary heart disease: A review an* theoretical integration. J. Health Soc. Behav. 15:12-27.

House, J. S. 1981. Work, Stress, and Social Support. Reading, Mass.: Addison Wesley.

House, J. S. and R. Kahn. 1984. Measuring social support. In Social Suppor and Health, ed. S. Cohen and L. Syme. New York: Academic Press.

House, J., C. Robbins, and H. Metzner. 1982. The association of socia relationships and activities with mortality: Prospective evidence from th Tecumseh Community Health Study. Amer. J. Epid. 116:123-40.

Howard, A. 1966. Plasticity, achievement and adaptation in developin economies. Human Organization 25:265-72.

Howard, A. 1970. Learning To Be Rotuman. New York: Teacher's College Press Columbia Univ.

Howard, A. 1971. Life style, education, and Rotuman character. In Polynesia Readings in a Culture Area, ed. A. Howard. Scranton, Pa: Chandle Publishing Co.

Howard, A. 1974. Ain't No Big Thing. Honolulu: Univ. Press of Hawaii.

Howard, A. 1976. Demographic socialization: Direct and indirect. Anthrop. an* Educ. Quarterly 8:1-5.

Howard, A. 1979. Polynesia and Micronesia in psychiatric perspective Transcult. Psych. Res. Rev. 16:123-45.

Howard, A. 1982. International psychology. Amer. Anthrop. 84:37-57.

Huizinga, J. 1972. Casual blood pressure in populations. In Human Biology o Environmental Change, ed. D. J. M. Vorster. London: Internationa Biological Programme.

Hull, T. 1983. Decision making process in the demography of the family Proceedings of the International Population Conference, 1981 4:IUSSI 117-26.

Hunt, E. E., Jr., N. R. Kidder, and D. H. Schneider. 1954. The depopulation o Yap. Hum. Biol. 26:21-51.

Hypertension Detection and Follow-up Program Cooperative Group (HDFP) 1977. Race, education and prevalence of hypertension. Amer. J. Epid 106:351-61.

Hytten, F. E. and I. Leitch. 1971. The Physiology of Human Pregnancy. Oxford: Blackwell.

Inkeles, A. and O. H. Smith. 1974. Becoming Modern: Individual Change in Six Developing Countries. Cambridge, Mass.: Harvard Univ. Press.

Jackson, L., R. Taylor, S. Faaiuso, S. P. Ainuu, S. Whitehouse, and P. Zimmet. 1981. Hyperuricemia and gout in Western Samoa. J. Chron. Dis. 34:65-75.

James, G. N.d. A preliminary report on the relationship between body build, body composition and blood pressure and indices of the psychosocial environment in a traditional Western Samoan community. Dept. of Anthropology, The Pennsylvania State Univ., University Park.

James, G. D. 1984. Stress response, blood pressure and lifestyle differences among Western Samoan men. Ph.D. diss., The Pennsylvania State Univ., University Park.

James, G. D., D. A. Jenner, G. A. Harrison and P. T. Baker. 1985. Differences in catecholemine excretion rates, blood pressure and lifestyle among young Western Samoan men. Hum. Biol. 57:635-647.

James, G. D., S. T. McGarvey, and P. T. Baker. 1983. The effect of modernization on spouse concordance in American Samoa. Hum. Biol. 55:643-52.

Janes, C. R. 1984. Migration and hypertension: An ethnography of disease risk in an urban Samoan community. Ph.D. diss., Univ. of California, San Francisco and Berkeley.

Janes, C. and I. G. Pawson. In press. Migration and biocultural fitness: Samoans in California. Soc. Sci. Med. 1985.

Jenkins, C. D. 1971. Psychologic and social precursors of coronary disease. New Engl. J. Med. 284:244-55, 307-17.

Jenkins, C. D. 1976. Recent evidence supporting psychologic and social risk factors for coronary disease. Parts 1, 2. New Eng. J. Med. 294:987-94, 1033-38.

Jenner, D. A., M. J. Brown, and F. M. J. Lhoste. 1981. Determination of methyldopa, methylnorasdrenoline, noradrenalin and adrenaline in plasma using high performance liquid chromatography with electrochemical detection. J. Chromat. 224:507-12.

Jenner, D. A., V. Reynolds, and G. A. Harrison. 1980. Catecholemine excretion rates and occupation. Ergonomics 23:237-46.

Jerome, N. W., R. F. Kandel, and G. H. Pelto. 1980. Nutritional Anthropology. Pleasantville, N.Y.: Redgrave.

Johannes, R. E. 1975. Exploitation and degradation of shallow marine food resources in Oceania. In The Impact of Urban Centers in the Pacific, ed. R. W. Force and B. Bishop. Honolulu: Pacific Science Association.

Kagan, A., B. R. Harris, W. Winkelstein, Jr., K. G. Johnson, H. Kato, S. L. Syme, G. C. Rhoads, M. L. Gay, M. Z. Nichaman, H. B. Hamilton, and J. Tillotson. 1974. Epidemiological studies of coronary heart disease and stroke in Japanese men living in Japan, Hawaii, and California: Demographic, physical, dietary, and biochemical characteristics. J. Chron. Dis. 27:345-64.

Kalimo, E., T. Bice, and M. Hovosel. 1970. Cross-cultural analysis of selected emotional questions from the Cornell Medical Index. Brit. J. Prev. Soc. Med. 24:229-40.

Kannel, W. B. 1976. Some lessons in cardiovascular epidemiology from Framingham. Amer. J. Cardiology 37:269-82.

Kannel, W. B., W. P. Castelli, and T. Gordon. 1979. Cholesterol in the prediction of atherosclerotic disease. Ann. Int. Med. 90:85-91.

Kannel, W. B., T. Gordon, and M. J. Schwartz. 1971. Systolic versus diastolic blood pressure and risk of coronary heart disease: The Framingham Study. Amer. J. Cardiology 27:335-46.

Kannel, W. B. and T. Gordon, eds. 1970. The Framingham Study: An Epidemiological Study of Cardiovascular Disease, Section 26: Some Characteristics Related to the Incidence of Cardiovascular Disease and Death, Framingham Study 16-Year Follow-Up. U.S. DHEW, Public Health Service, National Institute of Health, Washington, D.C.

Kannel, W. B. and P. Sorlie. 1979. Some health benefits of physical activity: The Framingham Study. Archives of Internal Med. 139:857-61.

Kanner, A. D., J. C. Coyne, C. Schaefer, and R. S. Lazarus. 1981. Comparison of two modes of stress measurement: Daily hassles and uplifts versus major life events. J. Behav. Med. 14:1-39.

Kaplan, H. B., C. Robbins, and S. S. Martin. 1983. Antecedents of psychological distress in young adults: Self-rejection, deprivation of social support, and life events. J. Health Soc. Behav. 24:230-44.

Karvonen, M. J. 1974. Work and activity classifications. In Fitness, Health and Work Capacity: International Standards, ed. L. A. Larson. New York: Macmillan.

Kasl, S. V. and L. Berkman. 1983. Health consequences of the experience of migration. Annual Rev. Publ. Health 4:69-90.

Kasl, S. V. and S. Cobb. 1970. Blood pressure changes in men undergoing job loss: A preliminary report. Psychosom. Med. 32:19-38.

Katch, F. I. and W. D. McArdle. 1983. Nutrition, Weight Control and Exercise. 2d. ed. Philadelphia: Lea and Febiger.

Katz, S. H., M. Hediger, J. Schall, E. J. Bowers, W. F. Barker, S. Aurand, P. B. Eveleth, A. B. Gruskin, and J. S. Parks. 1980. Blood pressure, growth and maturation from childhood through adolescence. Hypertension 2 (suppl. 1):55-69.

Kear, D. and B. L. Wood. 1962. Structure, landforms and hydrology. In Western Samoa: Land, Life and Agriculture in Tropical Polynesia, ed. J. W. Fox and K. B. Cumberland. Christchurch, New Zealand: Whitcombe and Tombs, Ltd.

Keene, D. T. P. 1978. Houses without walls: Samoan social control. Ph.D. diss., Univ. of Hawaii, Honolulu.

Keesing, F. M. 1934a. Samoa: Islands of conflict. Foreign Policy Reports 9:293-304.

Keesing, F. M. 1934b. Modern Samoa: Its government and changing life. Stanford, Calif.: Stanford Univ. Press.

Keesing, F. M. and M. Keesing. 1956. Elite Communication in Samoa. Palo Alto, Calif.: Stanford Univ. Press.

Kelley, J. 1977. A social anthropology of education: The case of Chiapas. Anthrop. and Educ. Quarterly 8:210-20.

Kessler, G. and H. Lederer. 1965. Flourometric measurement of triglycerides. In Automation in Analytic Chemistry, ed. L.T. Skeggs. New York: Mediad Inc.

Kessler, R. C. and J. A. McRae. 1981. Trends in the relationship between sex and psychological distress: 1957-1976. Amer. Soc. Rev. 46:443-52.

Kessler, R. C., R. H. Price, and C. B. Wortman. 1985. Social factors in psychopathology: Stress, social support, and coping processes. Annual Rev. Psychol. 36:531-72.

Keys, A. 1975. Coronary heart disease: The global picture. Atherosclerosis 22:149-92.

Keys, A. 1980a. Overnutrition, obesity, coronary heart disease and mortality. Nutr. Rev. 38:297-307.

Keys, A., ed. 1980b. Seven Countries: A Multivariate Analysis of Death and Coronary Heart Disease in Ten Years. Cambridge, Mass.: Harvard Univ. Press.

Keys, A., H. L. Taylor, H. Blackburn, J. Brozek, J. T. Anderson, and E. Simonson. 1963. Coronary heart disease among Minnesota business and professional men followed fifteen years. Circulation 28:381-95.

Keys, A., ed. 1970. Coronary heart disease in seven countries. Circulation 41 (suppl. 1):1-210.

Kimball, S. T. 1974. Culture and the Educative Process: An Anthropological Perspective. New York: Teachers College Press.

Kimball, S. T. 1982. Community and hominid emergence. Anthrop. and Educ. Quarterly 13:125-31.

King, C. M. 1975. Nutritional status of preschool children in Western Samoa. Dept. of Health, Apia, Western Samoa.

Kinloch, P. J. 1980. Samoan health practices in Wellington. Occasional Paper No. 12, Dept. of Health, New Zealand.

Kirch, P. V. 1979. Subsistence and ecology. In The Prehistory of Polynesia, ed. J. Jennings. Cambridge, Mass.: Harvard Univ. Press.

Kirk, R. L. 1976. Genetic differentiation in Australia and the Western Pacific and its bearing on the origin of the first Americans. Prepared in advance for participants in the Burg Wartenstein Symposium No. 72, Wenner-Gren Foundation for Anthropological Research, New York.

Kiste, R. C. 1974. The Bikinians: A Study in Forced Migration. Menlo Park, Calif.: Cummings Publishing.

Kitagawa, E. M. 1966. Theoretical considerations in the selection of a mortality index, and some empirical comparisons. Hum. Biol. 38:293-308.

Kitagawa, E. M. and P. M. Hauser. 1973. Differential Mortality in the United States: A Study in Socioeconomic Epidemiology. Cambridge, Mass.: Harvard Univ. Press.

Klapstein, S. 1978. Obesity and Related Seriological Variables in a Migrar Samoan Population. Master's thesis, Univ. of Hawaii, Honolulu.

Kleinbaum, D. G., L. L. Kupper, J. C. Cassel, and H. A. Tyroler. 197 Multivariate analysis of coronary heart disease in Evans County, Georgi; Archives of Internal Med. 128:943-48.

Knodel, J., H. Havonen, and A. Pramualratana. 1984. Fertility transition i Thailand. Population & Development Review 10:297-328.

Knuiman, J. T., R. J. J. Hermus, and J. G. A. J. Hautvast. 1980. Serum tot; and high-density lipoprotein (HDL) cholesterol concentrations in rural an urban boys from 16 countries. Atherosclerosis 36:529-37.

Kotchek, L. 1977. Ethnic visibility and adaptive strategies: Samoans in th Seattle area. J. of Ethnic Studies 4:29-38.

Kotchek, L. 1978. Samoans in Seattle. In New Neighbors: Islanders i Adaptation, ed. C. Macpherson, B. Shore, and R. Franco. Santa Cru; Center for South Pacific Studies, Univ. of California, Santa Cruz.

Kraemer, A. 1902. Die Samoan-inseln. Stuttgart: E. Schweizerbart., vol 1, 2, 3.

Krauss, R. M. 1982. Regulation of high density lipoprotein levels. Med. Clin. N Amer. 66:403-30.

Kurihara, M., T. Matsuyama, and M. Segi. 1970. Diabetes mellitus mortality i Japan compared with other countries. In Diabetes Mellitus in Asia, 197(ed. S. Tsuji and M. Wada. Princeton, N.J.: Excerpta Medica.

LaBarthe, D., D. Reed, J. Brody, and R. Stallones. 1973. Health effects c modernization in Palau. Amer. J. Epid. 98:161-76.

Lee, J., R. Lauer, and W. Clarke. 1984. The distribution of blood pressure i: young persons. In NHLBI Workshop in Juvenile Hypertension, ed. J Loggie, M. Horan, and A. Gruskin. Proceedings from a Symposium, Ma 1983, Bethesda, Md.: Biomedical Information Corp.

Lee, R. B. and I. Devore. 1976. Kalahari Hunter-Gatherers. Cambridge, Mass. Harvard Univ. Press.

Lemert, E. M. 1964. Forms and pathology of drinking in three Polynesia; societies. Amer. Anthrop. 66:361-74.

Levin, M. J. and G. R. Hayes. 1983. How many Samoans? Asian & Pacifi Census Forum 10(4).

Levin, M. J. and P. N. D. Pirie. 1974. Report on the 1974 Census of Americar Samoa. Honolulu: East-West Population Institute.

Levin, M. J. and P. N. D. Pirie. 1976. Report on the 1976 Census of America: Samoa. Honolulu: East-West Population Institute.

Levin, M. J. and P. A. Wright. 1978. Report on the 1974 Census of America; Samoa. Part 2: Analysis. Honolulu: East-West Population Institute.

Levine, R. 1975. Parental goals: A cross-cultural view. In The Family a; Educator, ed. H. J. Leichter. New York: Teacher's College Press.

Levy, M. J., Jr. 1966. Modernization and the Structure of Societies: A Settin; for the Study of International Affairs. Princeton, N.J.: Princeton Univ Press.

Levy, R. 1978. Tahitian gentleness and redundant controls. In Learning Non Aggression: The Experience of Non-Literate Societies, ed. A. Montagu Oxford: Oxford Univ. Press.

Lew, E. A. and L. Garfinkel. 1979. Variations in mortality by weight among 750,000 men and women. J. Chron. Dis. 32:563-76.

Lewthwaite, G. R. 1962. Land, life and agriculture to mid-century. *In* Western Samoa: Land, Life and Agriculture in Tropical Polynesia, ed. J. W. Fox and K. B. Cumberland. Christchurch, New Zealand: Whitcombe & Tombs, Ltd.

Lewthwaite, G. R., C. Mainzer, and P. J. Holland. 1973. From Polynesia to California: Samoan Migration and its sequel. J. Pacific History 8:133-57.

Lichton, I., L. Bullard, and B. Sherrell. 1983. A conspectus of research on nutritional status in Hawaii and Western Samoa — 1960-1980 — with references to diseases in which diet has been implicated. Wld. Rev. Nutr. Diet 41:40-70.

Lieberman, L. S. and P. T. Baker. 1976. The Samoan Migrant Project. Preliminary report from the Human Biology Program, Dept. of Anthropology, The Pennsylvania State Univ., University Park.

Literacy and Language Program, Northwest Regional Education Laboratory. 1984. Study of Unemployment, Poverty, and Training Needs of American Samoans. Prepared for the Employment and Training Administration, U.S. Dept. of Labor.

Lockwood, B. A. 1969. Produce marketing in a Polynesian society: Apia, Western Samoa. *In* Pacific Market Places, ed. H. C. Brookfield. Canberra: Australian National Univ. Press.

Lockwood, B. A. 1971. Samoan Village Economy. Melbourne: Oxford Univ. Press.

Lohman, T. G. 1981. Skinfolds and body density and their relation to fatness: A review. Hum. Biol. 53:181-225.

Lopes-Virella, M., F. P. Stone, S. Ellis, and J. A. Colwell. 1977. Cholesterol determination in high density lipoproteins separated by three different methods. Clin. Chem. 23:882-84.

Lowenstein, F. W. 1961. Blood pressure in relation to age and sex in the tropics and subtropics: A review of the literature and an investigation in two tribes of Brazil Indians. Lancet 1:389-92.

Lowenstein, F. W. 1964. Epidemiological investigations in relation to diet in groups who show little atherosclerosis and are almost free of coronary ischaemic heart disease. Nutritio et Dieta 6:40-51.

Lukaski, H. C. 1977. Some Observations on the Coronary Risk Status of Male Samoan Migrants on Oahu, Hawaii. Master's thesis, The Pennsylvania State Univ., University Park.

McArthur, N. A. 1956. The Populations of the Pacific Islands. Part 3: American Samoa. Part 4: Western Samoa and the Tokelau Islands. Canberra: Australian National Univ. Press.

McArthur, N. A. 1967. Island Populations of the Pacific. Canberra: Australian National Univ. Press.

McCarron D. A., C. D. Morris, H. J. Henry, and J. L. Stanton. 1984. Blood pressure and nutrient intake in the United States. Science 224:1392-98.

McCuddin, C. R. 1974. Samoan medicinal plants and their usage. Dept. of Medical Services, Govt. of American Samoa. Mimeo.

McGarvey, S. T. 1980. Modernization and cardiovascular disease amon Samoans. Ph.D. diss., The Pennsylvania State Univ., University Park.

McGarvey, S. T. 1983. Pre-migration differences in stature, weight, skinfold and blood pressure between migrant and non-migrant Samoan adults Amer. J. Phys. Anthrop. 60:223-24. (Abstr.).

McGarvey, S. T. 1984a. Subcutaneous fat distribution and blood pressure i Samoans. Amer. J. Phys. Anthrop. 63:192. (Abstr.).

McGarvey, S. T. 1984b. Psychosocial Factors and Blood Pressure in Modernizing Samoan Population. Master's thesis, Yale Univ., New Haven Ct.

McGarvey, S. T. and P. T. Baker. 1979. The effects of modernization an migration on Samoan blood pressure. Hum. Biol. 51:461-75.

McGarvey, S. T., D. Schendel, and P. T. Baker. 1980. Modernization effects o familial aggregation of Samoan blood pressure. Med. Anthrop. 4:321-38.

McGarvey, S. T. and M. Weishaar. 1984. Modernization, emotions and bloo pressure in Samoan women. Paper presented at the Society of Behaviora Medicine meetings, May 1984, Philadelphia.

McKeown, T. 1976. The Modern Rise of Population. New York: Academic Press.

McNicoll, G. 1980. Institutional determinants of fertility change. Populatio and Development Review 6:441-.

MacArthur, R. H. 1972. Geographical Ecology. New York: Harper & Row.

Macpherson, C. and L. Macpherson. 1985. Suicide in Western Samoa: A sociological perspective. In Culture, Youth and Suicide in the Pacific Papers from an East-West Center Conference, ed. F. X. Hezel, D. H Rubenstein, and G. M. White. Honolulu: East-West Center.

Macpherson, C., B. Shore, and R. Franco. 1978. New Neighbors: Islanders i Adaptation. Santa Cruz: Center for South Pacific Studies, Univ. o California, Santa Cruz.

Mack, R. W. and F. E. Johnston. 1976. The relationship between growth in infancy and growth in adolescence: Report of a longitudinal study among urban black adolescents. Hum. Biol. 48:693-711.

Ma'ia'i, F. 1957. Study of the developing pattern of education and the factors influencing that development in New Zealand Pacific dependencies. Master's thesis, Victoria Univ., Wellington, New Zealand.

Malcolm, S. 1954. Diet and nutrition in American Samoa. Technical Report No. 63, South Pacific Commission, Noumea.

Mangenot, G. 1963. The effect of man on the plant world. In Man's Place in the Island Ecosystem, ed. F. R. Fosberg. Honolulu: Bishop Museum Press.

Mann, G. V. 1974. The influence of obesity on health. Parts 1,2. New Eng. J. Med. 291:178-85, 226-32.

Markoff, R. A. and J. R. Bond. 1980. The Samoans. In People and Cultures of Hawaii: A Psychocultural Profile, ed. J. F. McDermott, Jr., T. Wen-Shing, and T. W. Maretzki. Honolulu: John A. Burns School of Medicine and the Univ. Press of Hawaii.

Marks, H. H. 1966. Vital statistics. In Chronic Diseases and Public Health, ed. A. Lilienfeld and A. Gifford. Baltimore, Md.: The Johns Hopkins Univ. Press.

Marmot, M. G. and S. L. Syme. 1976. Acculturation and coronary heart disease in Japanese-Americans. Amer. J. Epid. 104:225-47.

Martz. J. M. 1982. Adult blood pressure on Tutuila, American Samoa: An analysis by distance to the Pago Pago harbor area. Master's thesis, The Pennsylvania State Univ., University Park.

Martz, J. M., J. M. Hanna, and S. A. Howard. 1984. Stress in daily life: Evidence from Samoa. Amer. J. Phys. Anthrop. 63:191-92. (Abstr.).

Masland, L. and G. Masland. 1975. The Samoan ETV project: Some cross-cultural implications of educational television. Parts 1, 2. Educational Broadcasting March/April 1975, May/June 1975.

Mason, J. W. 1975. A historical view of the stress field. Part 1. J. Hum. Stress 1:6-12.

Mason, J., E. Bueschner, M. Belfer, M. Artenstein, and E. Maugey. 1979. A prospective study of corticosteroid and catecholamine levels in relation to viral illness. J. Hum. Stress 5:18-27.

Matsuda, M., K. P. Perko, and R. G. Johnston. 1972. Physiological activity and illness history. J. Psychosom. Res. 16:129-36.

Maude, H. E. 1981. Slavers in Paradise: The Peruvian Labour Trade in Polynesia 1862-1864. Stanford, Calif.: Stanford Univ. Press.

Maxwell, R. J. 1969. Samoan Temperament. Ph.D. diss., Cornell Univ., Ithaca, N.Y.

Maxwell, R. J. 1970. The changing status of elders in a Polynesian society. Aging and Human Development 1:137-46.

Mead, M. 1930. Social Organization of Manu'a. Bernice P. Bishop Museum Bulletin No. 76, Honolulu.

Mead, M. 1968. Coming of Age in Samoa: A Psychological Study of Primitive Youth for Western Civilization. New York: Dell Publishing Co.

Mead, M. 1969. Social Organization of Manu'a. 2d. ed. Honolulu: Bernice P. Bishop Museum.

Mechanic, D. 1974. Discussion of research progress on the relationship between stressful life events and episodes of physical illness. In Stressful Life Events: Their Nature and Effects, ed. B. S. Dohrenwend and B. P. Dohrenwend. New York: Wiley.

Meredith, H. V. 1970. Body weight at birth of viable human infants: A comparative worldwide treatise. Hum. Biol. 42:217-64.

Miller, N. E., O. H. Forde, D. S. Thells, and O. D. Mjos. 1977. The Tromso Heart Study. Lancet 1:965-67.

Monge, M. C. 1948. Acclimitization in the Andes. Baltimore, Md.: The Johns Hopkins Univ. Press.

Morgen, S. and B. Caan, eds. 1979. Applied Nutrition in Clinical Medicine. Philadelphia: W.B. Saunders.

Morgenstern, H. 1980. The changing association between social status and coronary heart disease in a rural population. Soc. Sci. Med. 14A:191-201.

Morton, N. E. 1955. The inheritance of human birth weight. Annals Hum. Genetics 20:125-34.

Munro, L. A. 1976. Samoans in Hawaii: Problems with education. Univ. of Hawaii, Honolulu.

Murray, M. J., A. B. Murray, N. J. Murray, and M. B. Murray. 1978. Serum cholesterol, triglycerides and heart disease of nomadic and sedentary tribesmen consuming isoenergetic diets of high and low fat content. Brit. J. Nutr. 39:159-63.

Nag, M. and N. Kak. 1984. Demographic transition in a Punjab village. Population and Development Rev. 10:661-78.

Nagle, F. J., B. Balke, and J. P. Naughton. 1965. Gradational step-tests for assessing work capacity. J. Appl. Physiol. 20:745-48.

Nardi, B. 1983. Breastfeeding and women's work: The case of Western Samoa. Paper presented at the Association for Social Anthropology in Oceania Conference, Spring 1983.

National Center for Health Statistics. 1980. Serum cholesterol of persons 4-74 years of age by socioeconomic characteristics: U.S. 1971-74. Vital and Health Statistics, Series 11, No. 217, DHEW Publ. No. (PHS) 80-1667.

National Center for Health Statistics. 1983. Dietary intake and cardiovascular risk factors, Part 2: Serum urate, serum cholesterol and correlates: U.S. 1971-75. Vital and Health Statistics, Series 11, No. 227, DHHS Publ. No. (PHS) 83-1677.

Neave, M. 1969. The nutrition of Polynesian children. Trop. Geogr. Med. 21:311-22.

Neel, J. V. 1962. Diabetes mellitus: A thrifty genotype rendered detrimental by progress. Amer. J. Hum. Genetics 14:353-62.

Neel, J. V. 1982. The thrifty genotype revisited. In The Genetics of Diabetes Mellitus, ed. J. Kobberling and R. Tattersall. London: Academic Press.

Neff, W. S. 1985. Work and Human Behavior. 3d. ed. New York: Aldine.

Nestel, P. and P. Zimmet. 1981. HDL levels in Pacific Islanders. Atherosclerosis 40:257-62.

Niswander, K. and E. C. Jackson. 1974. Physical characteristics of the gravida and their association with birth weight and perinatal death. Amer. J. Obstet. Gynec. 119:306-13.

Oakey, B. 1980. American Samoan Families in Transition. American Samoan Families in Transition Project, San Francisco, Calif.

Ochse, J. J., M. J. Soule, M. J. Dykeman, and G. Wehlburg. 1961. Tropical and Subtropical Agriculture. New York: Macmillan.

Ogbu, J. 1978. Minority Education and Caste: The American System in Cross-Cultural Perspective. New York: Academic Press.

Oliver, D. L. 1961. The Pacific Islanders. Garden City, N.Y.: Doubleday.

Oliver, D. 1985. Reducing suicide in Western Samoa. In Culture, Youth and Suicide in the Pacific: Papers from an East-West Center Conference, ed. F. X. Hezel, D. H. Rubinstein, and G. M. White. Honolulu: East-West Center.

Oliver, W. J., E. Cohen, and J. V. Neel. 1975. Blood pressure, sodium intake and sodium related hormones in the Yanomamo Indians, a "no-salt" culture. Circulation 52:146-51.

Omran, A. R. 1971. The epidemiological transition. A theory of population change. Millbank Memorial Fund Quarterly 49:509-38.

Onwueme, I. C. 1978. The Tropical Tuber Crops. New York: Wiley.

Orans, M. 1978. Heirarchy and happiness in a Western Samoan Community.

Prepared for the Burg Wartenstein Symposium No. 80, Social Inequality — Comparative and Developmental Approaches, Aug. 25 - Sept. 3 1978.

Orans, M. 1981. Hierarchy and happiness in a Western Samoan community. *In* Social Inequality: Comparative and Developmental Approaches, ed. G. Berreman. New York: Academic Press.

Orans, M. 1985. Hierarchy and experience: An experiment with survey research. Dept. of Anthropology, Univ. of California, Riverside. Photocopy.

Ostfeld, A. and D. D'Atri. 1977. Rapid sociocultural change and high blood pressure. Adv. Psychosom. Med. 9:20-.

Ostfeld, A. and R. Shekelle. 1967. Psychological variables and blood pressure. *In* The Epidemiology of Hypertension, ed. J. Stamler, R. Stamler, and T. Pullman. New York: Grune & Stratton.

Page, L. B. 1979. Hypertension and atherosclerosis in primitive and acculturating societies. *In* Dialogues in Hypertension. Vol 1. Hypertension Update, ed. J. C. Hunt. N.J.: Health Learning Systems.

Page, L. B., A. Damon, and R. Moellering. 1974. Antecedents of cardiovascular disease in six Solomon Island societies. Circulation 49:1132-46.

Page, L. B. and J. Friedlaender. N.d. Blood pressure changes in the survey populations. A chapter to appear in a volume edited by J. Friedlaender and J. Rhodes on the Solomon Islands Harvard Research Project.

Page, L. B., J. S. Friedlaender, and R. Moellering. 1977. Culture, human biology and disease in the Solomon Islands. *In* Population Structure and Human Variation, ed. G. A. Harrison. New York: Cambridge Univ. Press.

Palmore, E. 1970. Physical, mental and social factors in predicting longevity. *In* Normal Aging: Reports from the Duke Longitudinal Study, 1955-1969, ed. E. Palmore. Durham, N.C.: Duke Univ. Press.

Palmore, E. 1971. The relative importance of social factors in predicting longevity. *In* Prediction of Lifespan: Recent Findings, ed. E. Palmore and F. C. Jeffers. Lexington, Mass.: Heath Lexington Books.

Park, C. B. 1972. Population Statistics of American Samoa: A Report to the Government of American Samoa. Honolulu: East-West Center.

Park, C. B. 1979. The Population of American Samoa. Country Monograph Series No. 7.1, Economic and Social Commission for Asia and the Pacific, Bangkok and South Pacific Commission, Noumea.

Park, C. B. 1980. The Population of American Samoa. Honolulu: Univ. of Hawaii School of Public Health and East-West Population Institute.

Parsons, C. J. 1979. Samoan blood genetics: A comparison of two migrant and one sedente population. Amer. J. Phys. Anthrop. 50:470. (Abstr.).

Parsons, C. J. 1982. Samoan Migrants and Nonmigrants: A Morphological and Genetic Comparison. Ph.D. diss., The Pennsylvania State Univ., University Park.

Patkai, P. 1971. The diurnal rhythm of adrenaline secretion in subjects with different working habits. Acta Physiol. Scand. 81:30-34.

Patrick, R. C., I. Prior, J. C. Smith, and A. H. Smith. 1983. Relationship between blood pressure and modernity among Ponapeans. Int. J. Epid. 12:36-44.

Pawson, I. G. and C. R. Janes. 1981. Massive obesity in a migrant Samoan population. Amer. J. Publ. Health 71:508-13.

Pawson, I. G. and C. R. Janes. 1982. Biocultural risks in longevity: Cancers in California. Soc. Sci. Med. 16:183-90.

Pawson, I. G. and C. R. Janes N.d. The California Samoan population: Structure, health risks and outcomes.

Pearlin, L. I., M. A. Lieberman, E. G. Meneghan, and J. T. Mullen. 1981. The stress process. J. Health Soc. Behav. 22:337-56.

Pearlin, L. I. and C. Schooler. 1978. The structure of coping. J. Health Soc. Behav. 19:2-21.

Pelletier, D. L. 1984. Diet, Activity and Cardiovascular Disease Risk Factors in Western Samoan Men. Ph.D. diss., The Pennsylvania State Univ., University Park.

Pelletier, D. L. 1985. The relationship of energy intake and expenditure to body fatness in Western Samoan men. Submitted to Ecol. Food Nutr.

Pelletier, D. L. and J. R. Bindon. 1986. Patterns of growth in weight and length among American Samoan infants and its relationship to type of feeding. Ecol. Food. Nutr. 18:145-47.

Pereira, J. 1985. Bibliography of Samoa. 2 vols. Apia, Western Samoa: Univ. of the South Pacific, Western Samoa Centre.

Pirie, P. 1970. Samoa: Two approaches to population and resource problems. In Geography and a Crowding World, ed. W. Zelinsky, L. Kosinski, and R. M. Prothero. New York: Oxford Univ. Press.

Pitt, D. 1970. Tradition and Economic Progress in Samoa: A Case Study of the Role of Traditional Social Institutions in Economic Development. Oxford: Clarendon Press.

Pitt, D. 1977. Samoan immigrants and population policies in New Zealand. Paper presented at the 8th Summer Seminar in Population, June 1977, East-West Population Institute, Honolulu.

Pitt, D. and C. Macpherson. 1974. Emerging Pluralism: The Samoan Community in New Zealand. Auckland: Longman Paul, Ltd.

Pollock, N. J. 1974. Breadfruit or rice: Dietary choice on a Micronesian atoll. Ecol. Food Nutr. 3:107-15.

Preston, S. H. 1976. Mortality Patterns in National Populations. New York: Academic Press.

Preston, S. H., N. Keyfitz, and R. Schoen. 1972. Causes of Death Life Tables for National Populations. New York: Seminar Press.

Prior, I. A. M. 1970. Population studies in New Zealand and the South Pacific. In WHO Report on Cardiovascular Epidemiology in the Pacific, ed. Z. Fejfar. Geneva: World Health Organization.

Prior, I. A. M. 1971. The price of civilization. Nutr. Today 6:2-11.

Prior, I. A. M. 1973. Cardiovascular epidemiology in Polynesians in the Pacific. Singapore Med. J. 14:223-.

Prior, I. A. M. 1974. Cardiovascular epidemiology in New Zealand and the Pacific. New Zealand Med. J. 80:245-52.

Prior, I. A. M. 1977a. Nutritional problems in Pacific Islanders. The 1976

Muriel Bell Memorial Lecture. Wellington: Nutrition Society of New Zealand.

Prior, I. A. M. 1977b. Migration and physical fitness. *In* Advances in Psychosomatic Medicine. Epidemiologic Studies in Psychosomatic Medicine, ed. S. V. Kasl and F. Reichsman. Basel, Switzerland: Karger.

Prior, I. A. M. 1981. The Tokelau Island Migrant Study: A progress report, 1979. *In* Migration, Adaptation and Health in the Pacific, ed. C. Fleming and I. A. M. Prior. Epidemiology Unit Wellington Hospital, Wellington, New Zealand.

Prior, I. A. M. and F. Davidson. 1966a. The epidemiology of diabetes in Polynesians and Europeans in New Zealand and the Pacific. New Zealand Med. J. 65:375-83.

Prior, I. A. M., H. P. B. Harvey, M. N. Neave, and F. Davidson. 1966b. The health of two groups of Cook Island Maoris. New Zealand Dept. of Health Special Report Series No. 26. Wellington: Govt. Printer.

Prior, I. A. M., B. Rose, and F. Davidson. 1966c. Metabolic maladies in New Zealand Maoris. Brit. Med. J. 1:1065.

Prior, I. A. M., F. Davidson, C. Salmond, and Z. Czochanska. 1981. Cholesterol, coconuts and diet on Polynesian atolls: A natural experiment: The Pukapuka and Tokelau Island studies. Amer. J. Clin. Nutr. 34:1552-61.

Prior, I. A. M., J. Grimley-Evans, H. Harvey, F. Davidson, and M. Lindsey. 1968. Sodium intake and blood pressure in two Polynesian populations. New Eng. J. Med. 279:515-20.

Prior, I. A. M., A. Hooper, J. W. Huntsman, J. M. Stanhope, and C. E. Salmond. 1977. The Tokelau Migrant Study. *In* Population Structure and Human Variation, ed. G. A. Harrison. New York: Cambridge Univ. Press.

Prior, I. A. M., J. M. Stanhope, J. G. Evans, and C. E. Salmond. 1974. The Tokelau Island Migrant Study. Int. J. Epid. 92:94-112.

Prior, I. A. M. and C. Tasman-Jones. 1981. New Zealand Maori and Pacific populations. *In* Western Diseases: Their Emergence and Prevention, ed. H. C. Trowell and D. P. Burkitt. Cambridge, Mass.: Harvard Univ. Press.

Rabkin, J. G. and E. L. Struening. 1976. Life events, stress and illness. Science 194:1013-20.

Rahe, R. H. 1969. Multi-cultural correlations of life change scaling: America, Japan, Denmark and Sweden. J. Psychosom. Res. 13:191-95.

Ralston, C. 1976. The beach community. *In* Pacific Islands Portraits, ed. J. W. Davidson and D. Scarr. Canberra: Australian National Univ. Press.

Reed, D., D. Labarthe, and R. Stallones. 1970. Health effects of westernization and migration among Chamorros. Amer. J. Epid. 92:96-112.

Reid, C. F. 1941. Education in the Territories and Outlying Possessions of the U.S. New York: Teachers College Columbia Univ. Bureau of Publications.

Reisin, E., R. Abel, M. Modan, D. Silverberg, H. Eliahou, and B. Modan. 1978. Effect of weight loss without salt restriction on the reduction of blood pressure in overweight hypertensive patients. New Eng. J. Med. 298:1-6.

Reynolds, V., D. Jenner, C. Palmer, and G. Harrison. 1981. Catecholamine excretion rates in relation to life-styles in the male population of Otmoor, Oxfordshire. Ann. Hum. Biol. 8:197-209.

Rhoads, E. C. 1984. Reevaluation of the aging and modernization theory: The Samoan evidence. The Gerontologist 24:243-50.

Richie, T. R. 1927. Diet of Samoans. Med. J. Austral. Suppl. 6:160.

Riegel, K. F. 1971. The prediction of death and longevity in longitudinal research. In Prediction of Lifespan: Recent Findings, ed. E. Palmore and F. C. Jeffers, 139-52. Lexington, Mass.: Heath Lexington Books.

Ritchie, J. and J. Ritchie. 1979. Growing Up in Polynesia. Sydney: George Allen & Unwin.

Rivers, J. P. W. and P. R. Payne. 1982. The comparison of energy supply and energy need: A critique of energy requirements. In Energy and Effort, ed. G. A. Harrison. London: Taylor & Francis.

Roberts, J. and K. Maurer. 1977. Blood pressure levels of persons 6-74 years of age in the United States, 1971-74. Vital and Health Statistics Series 11, No. 203, DHEW Publ. no. (HRA) 78-1648.

Rolff, K. 1978a. The Samoan community in Oxnard, California. In New Neighbors: Islanders in Adaptation, ed. C. Macpherson, B. Shore, and R. Franco. Santa Cruz: Center for South Pacific Studies, Univ. of California, Santa Cruz.

Rolff, K. 1978b. Fa'a Samoa: Tradition in transition. Ph.D. diss., Univ. of California, Santa Barbara.

Rolland-Cachera, M. F., M. Deheeger, F. Bellisle, M. Sempe, M. Guilloud-Bataille, and E. Patois. 1984. Adiposity rebound in children: A simple indicator for predicting obesity. Amer. J. Clin. Nutr. 39:129-35.

Rosa, F. W. 1974. Birth weight in Fiji and Western Samoa: A Pacific prescription for pregnancy? Amer. J. Obstet. Gynec. 119:1121-24.

Rose, G., P. J. S. Hamilton, H. Keen, D. D. Reid, P. McCartney, and R. J. Jarrett. 1977. Myocardial ischaemia, risk factors and death from coronary heart disease. Lancet 1(8003):105-9.

Rothwell, N. J. and M. Stock. 1979. A role for brown adipose tissue in diet-induced thermogenesis. Nature 281:31-35.

Runeborg, R. E. 1980. Western Samoa and American Samoa: History, Culture and Communication. East-West Communication Institute Pre-print Paper Series. Honolulu: East-West Center.

Ryan, B. F. 1969. Social and Cultural Change. New York: The Ronald Press Co.

SAS Institute, Inc. 1983. The SAS Supplemental User's Guide, 1983. Cary, N.C.

Sahlins, M. D. 1972. Stone Age Economics. New York: Aldine.

Scheder, J. C. 1981. Diabetes and stress among Mexican-American migrants. Ph.D. diss., Univ. of Wisconsin, Madison.

Scheder, J. C. 1983. Migration, stress, and changing health patterns among Samoans and Mexican-Americans. Paper presented at the 1983 Annual Meeting of the Society for Applied Anthropology, San Diego, Calif.

Schendel, D. E. 1980. Age changes in blood pressure among three populations of Samoan children. Master's thesis, The Pennsylvania State Univ., University Park.

Schendel, D. E., D. L. Pelletier, and P. T. Baker. 1983. Variation in blood pressure and body composition among rural Samoan adults. Amer. J. Phys. Anthrop. 60:251. (Abstr.).

Schramm, W., L. M. Nelson, and M. T. Betham. 1981. Bold Experiment: The Story of Educational Television in American Samoa. Stanford, Calif.: Stanford Univ. Press.

Schwarz, W., D. C. Trost, S. L. Reiland, B. M. Rifkind, and G. Heiss. 1982. Correlates of low density lipoprotein cholesterol: Associations with physical, chemical, dietary and behavioral characteristics. Arteriosclerosis 2:513-22.

Scotch, N. A. 1963. Sociocultural factors in the epidemiology of Zulu hypertension. Amer. J. Publ. Health 52:1205-13.

Scotch, N., B. Gampel, J. H. Abramson, and S. Slome. 1961. Blood pressure measurements of urban Zulu adults. Amer. Heart J. 61:173-77.

Scott, R. and A. Howard. 1970. Models of stress. In Social Stress, ed. S. Levin and N. A. Scotch. Chicago: Aldine.

Scribner, S. and M. Cole. 1973. Cognitive consequences of formal and informal education. Science 182:553-59.

Scrimshaw, N. S., A. Balsam, and G. Arroyave. 1957. Serum cholesterol levels in school children from three socioeconomic groups. Amer. J. Clin. Nutr. 5:629-33.

Scudder, T. 1982. No Place To Go: The Effects of Compulsory Relocation on Navajos. Philadelphia: Philadelphia Institute for the Study of Human Issues.

Seligmann, M. E. P. 1975. Helplessness. San Francisco: Freeman.

Seltzer, C. C. 1966. Some re-evaluations of the Build and Blood Pressure Study, 1959 as related to ponderal index, somatotype and mortality. New Eng. J. Med. 274:254-59.

Setchell, W. A. 1924. Ethnobotany of the Samoans. Washington, D.C.: Carnegie Institute.

Sexton, J. D. and C. M. Woods. 1977. Development and modernization among highland Maya: A comparative analysis of ten Guatemalan towns. Hum. Organization 36:156-72.

Shankman, P. 1976. Migration and Underdevelopment: The Case of Western Samoa. Boulder, Colo.: Westview Press.

Shaper, A, G. 1972. Cardiovascular disease in the tropics 3. Blood pressure and hypertension. Brit. Med . J. 3:805-07.

Shephard, R. J. 1978. Human Physiological Work Capacity. Cambridge: Cambridge Univ. Press.

Shephard, R. J. 1980. Population aspects of human working capacity. Ann. Hum. Biol. 7:1-28.

Shephard, R. J., C. H. Weese, and J. E. Meriman. 1971. Prediction of maximal oxygen intake from anthropometric data. Arbeitsphysiol. 29:119-30.

Shore, B. 1977. A Samoan theory of action: Social control and social order in a Polynesian paradox. Ph.D. diss., Univ. of Chicago.

Shore, B. 1982. Sala'ilua: A Samoan Mystery. New York: Columbia Univ. Press.

Shore, B. 1983. A response by Bradd Shore. Pacific Studies 7:145-56.

Shore, B. and M. Platt. 1984. Communication barriers to Samoans training an employment in the U.S. Washington, D.C. Paper prepared for the U.S Dept. of Labor, Employment, and Training Administration.

Shryock, H. S. and J. S. Siegel. 1975. The Methods and Materials (Demography. 3d printing, rev. Bureau of the Census. Washington, D.C G.P.O.

Siegal, B. 1979. Television: Its changing life in Samoa. Los Angeles Times, Jun 14, 1979. Part 1, pp. 22-23.

Simpson, J. W., R. W. Lawless, and A. C. Mitchell. 1975. Responsibility of th obstetrician to the fetus. 2. Influence of prepregnant weight and pregnanc weight gain on birthweight. Obstet. and Gynec. 45:481-87.

Sinnet, P. F. and H. M. Whyte. 1973. Epidemiological studies in a tota highland population, Tukisenta, New Guinea. Cardiovascular disease an relevant clinical, electrocardiographic, radiological and biochemica findings. J. Chron. Dis. 26:265-90.

Smith, D. H. and A. Inkeles. 1966. The OM scale: A comparative socio psychological measure of individual modernity. Sociometry 29:353-77.

Smith, T. 1967. Sociocultural incongruity and change: A review of empirica findings. In Social Stress and Cardiovascular Disease, ed. S. Syme and L Reeder. Milbank Memorial Fund Quarterly 45:23-39.

Sorlie, P., T. Gordon, and W. B. Kannel. 1980. Body build and mortality: Th Framingham Study. JAMA 243:1828-31.

Spicer, E. H. 1971. Persistent cultural systems: A comparative study of identit; systems that can adapt to contrasting environments. Science 174:795-800.

Spiegelman, M. and H. H. Marks. 1966. Empirical testing of standards for th age adjustment of death rates by the direct method. Hum. Biol. 38:280-92

Spindler, G. D. 1973. The transmission of culture. In Culture in Process, 2d. ed. ed. A. R. Beals, G. D. Spindler, and L. Spindler. New York: Holt, Rinehar & Winston.

Spindler, G. 1974a. Schooling in Schonhausen: A study of cultural transmissio and instrumental adaptation in an urbanizing German village. I Education and Cultural Process, ed. G. Spindler. New York: Holt, Rinehar & Winston.

Spindler, G. 1974b. From omnibus to linkages: Cultural transmission models Anthrop. and Educ. Quarterly 5:1-7.

Spindler, L. L. 1977. Culture Change and Modernization. Prospect Heights, Ill. Waveland Press.

Spoehr, A. 1960. Port town and hinterland in the Pacific Islands. Amer. Anthrop. 62:586-592.

Spoehr, A., ed. 1963. Pacific Port Towns and Cities: A Symposium. Honolulu: Bishop Museum Press.

Spurr, G. B. 1983. Nutritional status and physical work capacity. Yrbk. Phys. Anthrop. 26:1-35.

Stair, J. B. 1897. Old Samoa, or Flotsam and Jetsam from the Pacific Ocean. London: Religious Tract Society.

Stallones, L., W. Mueller, and B. Christensen. 1982. Blood pressure, fatness and fat patterning among USA adolescents from two ethnic groups. Hypertension 4:483-86.

Stamler, J. 1983. Nutrition-related risk factors for the atherosclerotic diseases — present status. In Progress in Biochemical Pharmacology Vol. 19, ed. R. J. Hegyeli.

Stanhope, J. M. and I. A. M. Prior. 1976. The Tokelau Island Migrant Study: Prevalence of various conditions before migration. Int. J. Epid. 5:256-66.

Stanhope, J. M. and V. M. Sampson. 1977. High density lipoprotein cholesterol and other serum lipids in a New Zealand biracial adolescent sample. Lancet 1(8019): 968-70.

Stanhope, J. M., U. M. Sampson, and I. A. M. Prior. 1981. The Tokelau Island Migrant Study: Serum lipid concentrations in two environments. J. Chron. Dis. 36:45-55.

Stanton, M. 1978. Mormons, matais, and modernization: Stress and change among Samoans in Laie, Hawaii. In New Neighbors: Islanders in Adaptation, ed. C. Macpherson, B. Shore, and R. Franco. Santa Cruz: Center for South Pacific Studies, Univ. of California, Santa Cruz.

State of Hawaii. 1985. Hawaii's felons: A statistical report on Hawaii's prison population. Honolulu: State Intake Service Center.

Steegman, A. T., ed. 1983. Boreal Forest Adaptations. The Northern Algonkians. New York: Plenum Press.

Stini, W. A. 1971. Evolutionary implications of changing nutritional patterns in human populations. Amer. Anthrop. 73:1019-30.

Stunkard, A. 1975. From explanation to action in psychosomatic medicine: the case of obesity. Psychosom. Med. 37:195-236.

Summers, K. M., G. A. Harrison, D. Hume, and C. Palmer. 1983. Urinary hormone levels: A population study of associations between steroid and catecholamine excretion rates. Ann. Hum. Biol. 10:99-110.

Sutter, F. K. 1980. Communal versus individual socialization at home and in school in rural and urban Western Samoa. Ph.D. diss., Univ. of Hawaii, Honolulu.

Swedlund, A. 1980. Historical demography: Applications in anthropological genetics. In Current Developments in Anthropological Genetics, ed. J. Mielke and M. Crawford. New York: Plenum Press.

Swedlund, A. 1984. Historical studies of human mobility. In Migration and Mobility: Biosocial Aspects of Human Movement, ed. A. J. Boyce. London: Taylor & Francis.

Syme, S. L. 1979. Psychosocial determinants of hypertension. In Hypertension: Determinants, Complications and Intervention, ed. G. Onesti and C. Klimt. New York: Grune & Stratton.

Syme, S. L. and L. F. Berkman. 1976. Social class, susceptibility, and sickness. Amer. J. Epid. 104:1-8.

Syme, S. L., N. Borhani, and R. Buechley. 1965. Cultural mobility and coronary heart disease in an urban area. Amer. J. Epid. 82:334-46.

Syme, S.L., T. Oakes, G. Friedman, B. Feldman, A. Siegelaub, and M. Collen. 1974. Social class and racial differences in blood pressure. Amer. J. Publ. Health 64:619-20.

Syme, S. L. and C. Torfs. 1978. Epidemiological research in hypertension: A critical appraisal. J. Hum. Stress 4:43-48.

Tanner, J. M. 1962. Growth at Adolescence. 2d ed. Oxford: Blackwell Scientific Publications.

Task Force Report on Blood Pressure Control in Children. 1977. Pediatrics 58 (suppl.):797-820.

Taylor, R. and K. Thoma. 1983. Nauruan Mortality 1976-1981 and a Review of Previous Mortality Data. South Pacific Commission, Noumea.

Teitlebaum, M. S. 1975. Relevance of demographic transition theory for developing countries. Science 188:420.

Thelle, D. S., O. H. Forde, K. Try, and E. H. Lehmann. 1976. The Tromso Heart Study: Methods and main results of the cross-sectional study. Acta Med. Scand. 200:107-18.

Thoits, P. A. 1983. Dimensions of life events that influence psychological distress: An evaluation and synthesis of the literature. In Psychosocial Stress: Trends in Theory and Research, ed. H. B. Kaplan. New York: Academic Press.

Thomas, D. P. 1978. Cooperation and competition among children in the Pacific islands and New Zealand: The school as an agent of social change. J. of Research and Development in Educ. 12:88-96.

Thomas. P. 1981. Western Samoa: Social consequences of government acquisition. In Land, People and Government: Public Lands Policy in the South Pacific, ed. P. Larmour, R. Crocombe, and A. Taungenga. Suva, Fiji: Univ. of the South Pacific.

Thomas, R. B. 1975. The ecology of work. In Physiological Anthropology, ed. A. Damon. New York: Oxford Univ. Press.

Thomas, R. M. 1980. The rise and decline of an educational technology: Television in American Samoa. Educ. Comm. and Technol. 28:155-67.

Thompson, J. A. 1921. The geology of Western Samoa. New Zealand J. of Science and Technology 4:49-66.

Tiffany, S. W. 1975. Giving and receiving: Participation in chiefly redistribution activities in Samoa. Ethnology 14:267-86.

Tiffany, S. W. and W. W. Tiffany. 1978. Optation, cognatic descent, and redistribution in Samoa. Ethnology 17:367-90.

Tiffany, W. W. 1979. High court influences on land tenure patterns in American Samoa. Oceania 49:258-69.

Tilly, C. and C. H. Brown. 1967. On uprooting, kinship, and the auspices of migration. Intl. J. Comp. Sociology 8:139-64.

Tipps, D. C. 1973. Modernization theory and the comparative study of societies: A critical perspective. Comp. Stud. Soc. Hist. 15:199-226.

Tobias, P. V. 1972. Growth and stature in southern African populations. In Human Biology of Environmental Change, ed. D. J. M. Vorster. London: International Biological Programme.

Tofa, L. 1980. Trends and patterns of female migration in Western Samoa: 1971-1976. Paper presented at 11th Summer Seminar in Population, East-West Center, Honolulu.

Torrence, E. P. 1962. Cultural discontinuities and the development of originality of thinking. Exceptional Children 29:2-13.

Trowell, H. C. 1981. Hypertension, obesity, diabetes mellitus and coronary heart disease. *In* Western Diseases: Their Emergence and Prevention, ed. H. C. Trowell and D. P. Burkitt. Cambridge, Mass.: Harvard Univ. Press.

Trowell, H. C. and D. P. Burkitt. 1975. Refined Carbohydrate Foods and Disease: Some Implications of Dietary Fibers. New York: Academic Press.

Trowell, H. C. and D. P. Burkitt. 1981. Western Diseases: Their Emergence and Prevention. Cambridge, Mass.: Harvard Univ. Press.

Trulson, M., R. Clancy, and W. Jessop. 1964. Comparisons of siblings in Boston and Ireland. J. Amer. Diet. Assoc. 45:225-29.

Truswell, A., B. Kennelly, J. Hansen, and R. Lee. 1972. Blood pressure of Kung Bushmen in northern Botswana. Amer. Heart J. 84:5-12.

Turner, G. 1884. Samoa a Hundred Years Ago and Long Before. London: Macmillan.

Tyroler, H. A. 1977. The Detroit project studies of blood pressure. A prologue and review of related studies and epidemiological issues. J. Chron. Dis. 30:613-24.

Tyroler, H. A., S. Heyden, and C. G. Hames. 1975. Weight and hypertension: Evans County studies of blacks and whites. *In* Epidemiology and Control of Hypertension, ed. O. Paul. Miami: Symposia Specialists.

Underwood, B., A. H. Van Arsdell, E. Blumenstiel, and N. S. Scrimshaw. 1981. Implications of available information on breast-feeding worldwide. *In* Infant and Child Feeding, ed. J. T. Bond, L. J. Filer, Jr., G. A. Leveille, A. M. Thomson, and W. B. Weil, Jr. New York: Academic Press.

U.S. Bureau of the Census. 1953. 1950 Census of Population. Vol. 2: Characterstics of the Population, Part 54, Territories and Possessions: American Samoa. Washington, D.C.: G.P.O.

U.S. Bureau of the Census. 1963. 1960 Census of Population. Vol 1: Characteristics of the Population, Part 56, Outlying Areas: American Samoa. Washington, D.C.: G.P.O.

U.S. Bureau of the Census. 1973. 1970 Census of the Population. Vol 1: Characteristics of the Population, Part 56, Outlying Areas: American Samoa. Washington, D.C. : G.P.O.

U.S. Bureau of the Census. 1981. Preliminary Report on the 1980 Census of Population. Washington, D.C.

U.S. Bureau of the Census. 1982. 1980 Census of the Population. Vol I. Characteristics of the Population, Number of Inhabitants: American Samoa. Washington, D.C.: G.P.O.

U.S. Bureau of the Census. 1983a. 1980 Census of the Population. General Social and Economic Characteristics: United States Summary (PC80-1-C1). Washington, D.C.: G.P.O.

U.S. Bureau of the Census. 1983b. 1980 Census of the Population. General Social and Economic Characteristics: California (PC80-1-C6). Washington, D.C.: G.P.O.

U.S. Bureau of the Census. 1983c. 1980 Census of the Population. General Social and Economic Characteristics: Hawaii (PC80-1-C-13). Washington, D.C.: G.P.O.

U.S. Bureau of the Census. 1983d. 1980 Census of the Population. Vol. 1: Characteristics of the Population, General Population Characteristics, Part 56, American Samoa. Washington, D.C.: G.P.O.

U.S. Dept. of Agriculture. 1982. Agricultural Handbook No. 8: The Composition of Foods, Raw, Processed, Prepared. U.S.D.A. Human Nutrition Information Service, Washington, D.C.

U.S. Dept. of Health and Human Services. 1982a. Instruction Manual Part 2a: Instructions for Classifying the Underlying Cause of Death, 1983. National Center for Health Statistics, Hyattsville, Md.

U.S. Dept. of Health and Human Services. 1982b. Instruction Manual Part 2c: ICD-9 ACME Decision Tables for Classifying the Underlying Cause of Death, 1983. National Center for Health Statistics, Hyattsville, Maryland.

U.S. Dept. of Health, Education and Welfare. 1978. Facts of Life and Death. DHEW Publ. No. (PHS) 79-1222.

U.S. Dept. of Health, Education and Welfare. 1980. An Assessment of H. E. W. Services Delivery to American Samoans. San Francisco.

Venkatachalam, P. S., T. P. Susheela, and P. Rau. 1967. Effect of nutritional supplementation during infancy on growth of infants. J. Trop. Pediatr. 13:70-76.

Voors, A., L. Webber, R. Frerichs, and G. Berenson. 1977. Body height and body mass as determinants of basal blood pressure in children — Bogalusa Heart Study. Amer. J. Epid. 106:101-08.

Vorster, D. J. M. 1977. Adaptation to urbanization in South Africa. In Population Structure and Human Variation, ed. G. A. Harrison. New York: Cambridge Univ. Press.

Vranic, M., S. Horvath, and J. Wahren. 1979. Exercise and diabetes: An overview. Diabetes 28:107-10.

Walker, A. R. P. and U. B. Arvidsson. 1954. Fat intake, serum cholesterol concentration, and atherosclerosis in the South African Bantu. Part 1: Low fat intake and the age trend of serum cholesterol concentration in the South African Bantu. J. Clin. Invest. 33:1358-65.

Walker, S. H. and D. B. Duncan. 1967. Estimation of the probability of an event as a function of several independent variables. Biometrika: 54(1 and 2): 167-79.

Wallace, A. F. C. 1961. Schools in revolutionary and conservative societies. In Anthropology and Education. The Martin G. Brumbaugh Lectures, 5th. series, ed. F. C. Gruber. Philadelphia: Univ. of Pennsylvania Press.

Wander, H. 1971. Trends and characteristics of population growth in Western Samoa. United Nations Program of Technical Cooperation Report No. TAO/WESA/3.

Ward, R. G. 1959. The banana industry in Western Samoa. Economic Geography 35:123-37.

Ward, R. G. and A. Proctor. 1979. South Pacific Agriculture. Chpt. 17. Western Samoa.

Ward, R. H. 1983. Genetic and sociocultural components of high blood pressure. Amer. J. Phys. Anthrop. 62:91-105.

Ward, R. H., P. Chin, and I. A. M. Prior. 1979. Genetic epidemiology of blood pressure in a migrating isolate: Prospectus. *In* Genetic Analysis of Common Diseases and Applications to Predictive Factors in Coronary Disease, ed. C. F. Sing and M. Skolnick. New York: Alan Liss.

Ward, R. H. and I. A. M. Prior. 1980. Genetic and sociocultural factors in the response of blood pressure to migration of the Tokelau population. Med. Anthrop. 4:339-66.

Watters, R. F. 1958a. Cultivation in old Samoa. Economic Geography 34:338-54.

Watters, R. F. 1958b. Settlement in old Samoa. The New Zealand Geographer 14:1-18.

Watters, R. F. 1958c. Culture and environment in old Samoa. *In* Western Pacific: Studies of Man and the Environment in the Western Pacific, ed. Dept. of Geography. Wellington: Victoria Univ. of Wellington and the New Zealand Geographical Society.

Way, A. B. 1976. Morbidity and postneonatal mortality. *In* Man in the Andes, ed. P. T. Baker and M. A. Little. Stroudsburg, Pa: Dowden, Hutchinson & Ross.

Weiner, J. S. 1977. Nutritional Ecology. *In* Human Biology, ed. G. A. Harrison, J. Weiner, N. Barnicot and J. Tanner. Oxford: Oxford Univ. Press.

Weiner, J. S. and J. A. Lourie. 1969. Human Biology. Philadelphia: F.A. Davis Co.

Weishaar, M. 1976. The fertility of Samoan migrants in Hawaii. Master's thesis, The Pennsylvania State Univ., University Park.

West, K. M. 1974. Diabetes in American Indians and other native populations of the world. Diabetes 23:841-55.

West, K. M. 1978. Epidemiology of Diabetes and its Vascular Lesions. New York: Elsevier.

Western Samoa Dept. of Statistics. 1976. Census of Population and Housing. Apia, Western Samoa.

Western Samoa Dept. of Statistics. 1978. Annual Statistical Abstract 1977. Apia, Western Samoa.

Western Samoa Dept. of Statistics. 1980. Annual Statistical Abstract 1980. Apia, Western Samoa.

Western Samoa Dept. of Statistics. 1981. Report of the Census of Population and Housing. Apia, Western Samoa.

Western Samoa Ministry of Education. 1981. Annual Report for 1980. Apia, Western Samoa.

Wiehl, D. G. and W. T. Tompkins. 1954. Size of babies of obese mothers receiving nutrient supplements. Millbank Memorial Fund Quarterly 32:125-40.

Wilcox, K. 1982. Ethnography as a methodology and its application to the study of schooling: A review. *In* Doing the Ethnography of Schooling, ed. G. Spindler. New York: Holt Rinehart & Winston.

Wilkins, R. M. 1965. Nutritional Survey in Urban and Rural Villages in Western Samoa. South Pacific Health Service, Suva, Fiji.

Williams, C. N. 1975. The Agronomy of the Major Tropical Crops. Oxford: Oxford Univ. Press.

Williams, J. F. 1939. Dental service in Western Samoa. New Zealand Dental J. 35:115-33.

Williams, T. R. 1983. Socialization. Englewood Cliffs, N. J.: Prentice-Hall.

Wood, C. S. and L. P. Gans. 1981. Hematological status of reproductive women in Western Samoa: An analysis of biometric data. Hum. Biol. 53:269-79.

Wood, C. S. and L. P. Gans. 1984. Some hematological findings in children of Western Samoa. J. Trop. Pediatr. 30:104-10.

World Health Organization. 1949. Demographic Yearbook 1949. United Nations, New York.

World Health Organization. 1951. Demographic Yearbook 1951. United Nations, New York.

World Health Organization. 1952. Demographic Yearbook 1952. United Nations, New York.

World Health Organization. 1977. The Manual of the International Statistical Classification of Diseases, Injuries and Causes of Death. 9th Rev. Geneva.

World Health Organization. 1982. Demographic Yearbook 1980. United Nations, New York.

World Health Organization. 1983. Demographic Yearbook 1981. United Nations, New York.

World Health Organization Expert Committee on Arterial Hypertension and Ischaemic Heart Disease. 1962. World Health Organization Technical Report Series 231. Geneva.

Wright, A. C. S. 1962. The soils. In Western Samoan: Land, Life and Agriculture in Tropical Polynesia, ed. J. W. Fox and K. B. Cumberland. Christchurch, New Zealand: Whitcombe & Tombs, Ltd.

Wybenga, D. R., V. J. Pileggi, P. H. Dristine and J. DiGiorgio. 1970. Direct manual determination of serum total cholesterol with a single stable reagent. Clin. Chem. 16:980-84.

Wyndham, C. H., N. B. Strydom, and W. P. Leary. 1966. Studies of the maximum capacity of men for physical effort. Part 2. The maximum oxygen intakes of young active Caucasians. Arbeitsphysiol. 22:296-303.

Young, F. A. 1972. Stability and change in Samoa. Ph.D. diss., Univ. of Oregon, Eugene.

Young, N. F. 1974. Searching for the Promised Land: Samoans and Filipinos in Hawaii. Honolulu: Univ. of Hawaii Press.

Zak, B., N. Moss, A. J. Boyle, and A. Alatkis. 1954. Reactions of certain unsaturated steroids with acid iron reagent. Anal. Chem. 26:776-77.

Zapata, B. and E. Marticorena. 1968. Presion arterial sistemica en el individuo senil de altura. Archivos del Instituto de Biologia Andina 2:220-28.

Zimmet, P. 1978. Diabetes in Pacific populations — a price for westernization. Proceedings of the Sixth Asia and Oceania Congress on Endocrinology, Singapore.

Zimmet, P. 1979a. Epidemiology of diabetes and its macrovascular manifestations in Pacific populations: The medical effects of social progress. Diabetes Care 2:144-53.

Zimmet, P. 1979b. The epidemiology of diabetes in Micronesia and Polynesia. *In* Epidemiology of Diabetes in Developing Countries, ed. M. M. S. Ahuja. New Delhi: Interprint.

Zimmet, P. 1980. Cardiovascular and metabolic diseases in Western Samoa. World Health Organization, Western Pacific Regional Office, Manila, the Philippines.

Zimmet, P. 1981. Diabetes: The paradigm of affluence in response to modernization. *In* Migration, Adaptation, and Health in the Pacific, ed. C. Fleming and I. A. M. Prior. Epidemiology Unit Wellington Hospital, Wellington, New Zealand.

Zimmet, P., S. Faaiuso, J. Ainuu, S. Whitehouse, B. Milne, and W. DeBoer. 1981. The prevalence of diabetes in the rural and urban Polynesian population of Western Samoa. Diabetes 30:45-51.

Zimmet, P., P. Taft, A. Gionea, W. Guthrie, and K. Thoma. 1977. The high prevalence of diabetes mellitus on a Central Pacific Island. Diabetologia 13:111-15.

Zimmet, P., R. Taylor, L. Jackson, S. L. Whitehouse, S. Faaivaso, and J. Ainuu. 1980. Blood Pressure Studies in rural and urban Western Samoa. Med. J. Austral. 2:202-05.

Zodgekar, A. V. 1979. Mortality. *In* The Population of New Zealand: Interdisciplinary Perspectives, ed. R. J. W. Neville and C. J. O'Neill. Auckland, N.Z.: Longman Paul.

Index

Hawaii, 1974–1978, 110–11
risk factors, 114–16
linkages with, 431–32
summary interpretation, 423, 430–32
Catecholamines. *See* Urinary hormone excretion
Child care. *See* Growth, infants; Infant feeding practices; Socialization process
Child growth. *See* Growth, children and adolescents
Christianity
education system, effects on, 153–54
introduction into Samoan Islands, 27, 153
political influence in traditional culture, 43
Cigarette smoking
blood lipids, rel with, 344
patterns, 186–87
age effects on, 187
Climate. *See* Samoan archipelago, physical environment
Coping behavior. *See also* Aggression; Coping strategies; Stress
definition, 398
of migrants
crisis management, 410
economic strategies, 405–8
employment strategies, 405–8
health care strategies, 408–10
housing strategies, 404–5
social relationships, reliance on, 404–10
Coping strategies, 394–418. *See also* Coping behavior; Stress
anger management, 402–3, 410–15
cognitive styles, 399
definition, 396
social support, 400–402, 414–15
health implications, 415–16
Cornell Medical Index response scores. *See also* Blood pressure, adults; Cornell Medical Index studies
age effects on, 178–81
areal variation, 178–83
interpretation problems, 178–81, 184–85
modernization effects on, 181
mortality, rel with, 119–20
psychosocial section
areal variation, 181–83
blood pressure, rel with, 184
education effects on, 181
education and occupation interaction, effects of, 184
section scores, intercorrelation of, 177–78
sex differences, 178–81
Cornell Medical Index studies. *See also* Cornell Medical Index response scores; Questionnaires, use of

questionnaire
administration, 176–77
analyses, 177, 383–84
applications, 116–17, 174, 176, 181, 379
description, 174, 381
sample description, 177
Coronary-prone behavior, 190. *See also* Stress
characteristics of, 397
urinary hormone excretion, rel with, 220
Cortisol. *See also* Urinary hormone excretion
as a stress measure, 205, 220

Death rates, all causes. *See also* Death rates, cause-specific; Death rates, history of; Mortality patterns
American Samoa, 1950–1981, 95, 96
California, 1978–1982, 99–100
Hawaii, 1974–1978, 110–11
Western Samoa, 1982 and 1983, 98
Death rates, cause-specific. *See also* Death rates, all causes; Death rates, history of; Mortality patterns
American Samoa, 1950–1981
comparison with non-Samoans, 101, 103, 109
degenerative diseases, 105–10
infectious diseases, 100–102
trauma, 102–5, 411
Hawaii, 1974–1978
cardiovascular disease, 110–11
California, 1978–1982
comparison with non-Samoans, 111–13
degenerative diseases, 111–13
Western Samoa, 1981–1983
suicide, 411–12
Death rates, history of. *See also* Death rates, all causes; Death rates, cause-specific; Mortality patterns
nineteenth century, Samoan Islands, 65
twentieth century, American Samoa, 75
twentieth century, Western Samoa, 73
crude death rate, 73
infant mortality, 73
Demographic decision-making process, 64–65, 76–77
'aiga, role of, 64–65
description of, site specific, 77–92
American Samoans, 83–89
California Samoans, 91
Hawaii Samoans, 89–91
Western Samoa, remote villagers, 79–80, 82–83
Western Samoa, suburban villagers, 81–83